Biodeterioration of Stone Surfaces

Biodeterioration of Stone Surfaces

Lichens and Biofilms as Weathering Agents of Rocks and Cultural Heritage

Edited by

Larry L. St.Clair

and

Mark R.D. Seaward

SPRINGER SCIENCE+BUSINESS MEDIA, B.V.

A C.I.P. Catalogue record for this book is available from the Library of Congress.

ISBN 978-90-481-6724-1   ISBN 978-1-4020-2845-8 (eBook)
DOI 10.1007/978-1-4020-2845-8

Cover picture:
Limestone sculpture, University of Caracas, by F. Narvaez showing lichen and cyanobacterial disfiguration about 40 years after its installation. (Photo: Marisa Tabasso)

Printed on acid-free paper

All Rights Reserved
© 2004 Springer Science+Business Media Dordrecht
Originally published by Kluwer Academic Publishers in 2004
Softcover reprint of the hardcover 1st edition 2004
No part of this work may be reproduced, stored in a retrieval system, or transmitted
in any form or by any means, electronic, mechanical, photocopying, microfilming,
recording or otherwise, without written permission from the Publisher, with the exception
of any material supplied specifically for the purpose of being entered
and executed on a computer system, for exclusive use by the purchaser of the work.

Dedication

To Rieta and Vanessa

Contents

Dedication	v
Contributing Authors	ix
Preface	xiii
Acknowledgments	xv
Caption for Image on Book Cover	xvii

Chapter 1   Biodeterioration of Rock Substrata by Lichens: Progress and Problems
LARRY L. ST. CLAIR AND MARK R.D. SEAWARD   1

Chapter 2   Lichens as Subversive Agents of Biodeterioration
MARK R.D. SEAWARD   9

Chapter 3   Limestone Stabilization Studies at a Maya Site in Belize
WILLIAM S. GINELL AND RAKESH KUMAR   19

Chapter 4   Lichens and the Biodeterioration of Stonework: The Italian Experience
ROSANNA PIERVITTORI   45

Chapter 5   Deteriorative Effects of Lichens on Granite Monuments
BENITA SILVA AND B. PRIETO   69

Chapter 6   Microbial Biofilms on Carbonate Rocks from a Quarry and
            Monuments in Novelda (Alicante, Spain)
            CARMEN ASCASO, M.A. GARCIÁ DEL CURA AND ASUNCION DE LOS RÍOS   79

Chapter 7   Lichens on Wyoming Sandstone: Do They Cause Damage?
            GIACOMO CHIARI AND ROBERTO COSSIO   99

Chapter 8   Lichen Encroachment onto Rock Art in Eastern Wyoming:
            Conservation Problems and Prospects for Treatment
            CONSTANCE S. SILVER AND RICHARD WOLBERS   115

Chapter 9   Lichen Biodeterioration at Inscription Rock, El Morro National
            Monument, Ramah, New Mexico, USA
            KATHRYN B. KNIGHT, LARRY L. ST. CLAIR AND JOHN S. GARDNER   129

Chapter 10 Lichens of Different Mortars at Archaeological Sites in Southern
            Spain: An Overview
            X. ARIÑO AND C. SAIZ-JIMENEZ   165

Chapter 11 Observations on Lichens Growing on Artifacts in the Indian
            Subcontinent
            S. SAXENA, D.K UPRETI, AJAY SINGH AND K.P. SINGH   181

Chapter 12 Biodeterioration of Prehistoric Rock Art and Issues in Site
            Preservation
            ALICE M. TRATEBAS   195

Chapter 13 Raman Spectroscopy of Rock Biodeterioration by the Lichen Lecidea
            tessellata Flörke in a Desert Environment, Utah, USA
            HOWELL G.M. EDWARDS, SUSANA E. JORGE VILLAR, MARK R.D. SEAWARD
            AND LARRY L. ST. CLAIR   229

Chapter 14 Lichens and Monuments: An Analytical Bibliography
            ROSANNA PIERVITTORI, ORNELLA SALVADORI AND MARK R.D. SEAWARD   241

Index   283

Contributing Authors

Xavier Arino
Instituto de Recursos Naturales y Agrobiologia, CSIC, Apartado 1052, 41080 Sevilla, Spain, Email: xavier.arino@menta.net

Carmen Ascaso
Departamento de Biología Ambiental y Servicio de Microscopía Electrónica, Centro de Ciencias Mediambientales CSIC, 28006 Madrid, Spain, Email: ascaso@ccma.csic.es

Giacomo Chiari
Getty Conservation Institute 1200 Getty Center Drive, Los Angeles, California 90049, USA, Email: gchiari@getty.edu

Roberto Cossio
Dipartimento di Scienze della Terra, Via Valperga Caluso 35 - I10125 Torino, Italy, Email: roberto.cossio@ unito.it

Asuncion de los Rios
Departamento de Biología Ambiental y Servicio de Microscopía Electrónica, Centro de Ciencias Mediambientales CSIC, 28006 Madrid, Spain, Email: cmcsic@pinar1.csic.es

M.A. Garcia del Cura
Instituto de Geología Económica CSIC-UCM and Laboratorio de Petrología Aplicada, Unidad Asociada CSIC –UA, Alicante, Spain

Howell G. M. Edwards
Department of Chemical and Forensic Sciences, University of Bradford, Bradford BD7 1DP, UK, Email: h.g.m.edwards@bradford.ac.uk

John S. Gardner
Department of Integrative Biology, Brigham Young University, Provo, Utah 84602, USA,
Email: John_Gardner@byu.edu

William S. Ginell
Getty Conservation Institute, 1200 Getty Center Drive, Los Angeles, California 90049, USA
Email: wginell@getty.edu

Kathryn B. Knight
Department of Integrative Biology, Brigham Young University, Provo, Utah 84602, USA,
Email: katybknight@hotmail.com

Rakesh Kumar
Specialty Coatings Company, 2238 Williams Glen Blvd. Zionsville, IN 46077

Rosanna Piervittori
Dipartimento di Biologia Vegetale, Università di Torino viale Mattioli 25, I-10125 Torino, Italy, Email: Rosanna.piervittori@unito.it

Beatriz Prieto
Departamento Edafología y Química Agrícola, Facultad Farmacia, Univ. Santiago de Compostela, 15782-Santiago de Compostela, Spain, Email: edprieto@uscmail.usc.es

Cesareo Saiz-Jimenez
Instituto de Recursos Naturales y Agrobiologia, CSIC, Apartado 1052, 41080 Sevilla, Spain

Ornella Salvadori
Soprintendenza Speciale per il Polo Museuale Veneziano-Laboratorio Scientifico, Cannaregio 3553, I-30131 Venice, Italy

S. Saxena
Lichenology Laboratory, Plant Biodiversity and Conservation Biology Division National Botanical Research Institute, Lucknow-1, India

Mark R.D. Seaward
Department of Environmental Science, University of Bradford, Bradford BD7 1DP, UK,
Email: m.r.d.seaward@bradford.ac.uk

Benita Silva
Departamento Edafología y Química Agrícola, Facultad Farmacia, Univ. Santiago de Compostela, 15782-Santiago de Compostela, Spain, Email: edbsilva@usc.es

Constance S. Silver
Preservar, Inc. 310 Riverside Dr. New York, NY 10025, USA, Email: C.S.Silver@worldnet.att.net

Ajay Singh
Lichenology Laboratory, Plant Biodiversity and Conservation Biology Division National Botanical Research Institute, Lucknow-1, India

Krishna P. Singh
Botanical Survey of India, Central Circle, 10 Chetham Lines, Allahabad, India

Larry L. St. Clair
Department of Integrative Biology, Brigham Young University, Provo, Utah 84602 USA, Email: larry_stclair@byu.edu

Alice M. Tratebas
Bureau of Land Management, 1101 Washington Blvd., Newcastle, WY 82701, USA, Email: Alice_Tratebas@blm.gov

Dalip K. Upreti
Lichenology Laboratory, Plant Biodiversity and Conservation Biology Division National Botanical Research Institute, Lucknow-1, India, Email: upretidk@rediffmail.com

Susana E. Jorge Villar
Department of Chemical and Forensic Sciences, University of Bradford, Bradford BD7 1DP, UK (On leave from: Area de Geodinamica, Facultad de Humanidades y Educacion, University of Burgos, Calle Villadiego S/N, 09001, Burgos, Spain)

Richard Wolbers
Winterthur Museum Winterthur, DE 19735, Email: Wolbers@earthlink.net

Preface

This is a timely volume in view of the considerable interest currently shown in the preservation of our cultural heritage, and the extensive and growing literature on the subject. Unfortunately, the latter is to be found in a wide variety of published sources, some aimed at a very specific readership and therefore not all that accessible to those who need this resource. The present volume draws together a spectrum of biodeterioration work from across the world to provide an overview of the materials examined and the methodologies employed to elucidate the nature of the problems, as well as an extensive and current bibliographical resource on lichen biodeterioration.

Biodeterioration of historical and culturally important stone substrata is a complex problem to be addressed. Easy, risk-free solutions are simply not available to be dealt with by other than a wide range of expertise. Successful resolution of this issue will inevitably require a multidisciplinary effort, where biologists work in close cooperation with ecologists, geologists, geochemists, crystallographers, cultural property conservators, archaeologists, anthropologists, and historians in order to recommend the most effective management scheme. The advantage of this approach is obvious: multidisciplinary management teams with good leadership can ask more appropriate questions while developing much more thoughtful and informed decisions.

The current volume is the first treatment of the subject of biodeterioration that includes a careful consideration of the role of the above mentioned disciplines. This combination of disciplines makes this book valuable not only as a solid scientific treatise, but equally important as a serious resource for evaluating both impact processes and preservation options related to biodeterioration of culturally significant rock substrata. In conclusion, it is hoped that this volume will provide not only background information, but also practical advice on the detection, measurement and control of lichens and biofilms on a wide range of cultural heritage, that will prove of value to historians, archaeologists and conservators, as well as specialist biologists.

Acknowledgments

The editors would like to express their sincere gratitude to all those who participated in the Biodeterioration Symposium held in Albuquerque, New Mexico in August 2001, and to acknowledge the dedication and patience of all of the contributors to this volume. We also wish to express sincere appreciation to Marissa Tabasso for permission to reproduce her photograph on the book cover. Without a generous grant from the Samuel H. Kress Foundation of New York City the symposium and this book would not have been possible. Most importantly we wish to acknowledge Kathryn B. Knight for her very considerable help in processing the edited manuscripts.

Caption for Image on Book Cover

Limestone sculpture, University of Caracas, by F. Narvaez showing lichen and cyanobacterial disfiguration about 40 years after its installation. (Photo: Marisa Tabasso)

Chapter 1

BIODETERIORATION OF ROCK SUBSTRATA BY LICHENS: PROGRESS AND PROBLEMS

LARRY L. ST. CLAIR[1] and MARK R.D.SEAWARD[2]
[1]*Department of Integrative Biology,Brigham Young University, Provo Utah 84602, USA;*
[2]*Department of Environmental Science,University of Bradford, Bradford BD7 1DP, UK*

1. INTRODUCTION

This is a timely volume in view of the considerable interest currently shown in the preservation of our cultural heritage and the extensive and growing literature on the subject. Unfortunately, the latter is to be found in a wide variety of published sources, some aimed at a very specific readership. The present volume draws together a spectrum of biodeterioration work from across the world to provide an overview of the materials examined and the methodologies employed to elucidate the nature of the problems, as well as an extensive and current bibliographical resource on lichen biodeterioration.

Generally, rock surfaces are not regarded as being particularly conducive to the growth and development of living things. Occasionally, grasses or forbs or even, more rarely, a small shrub or stunted tree growing from a crack in a large boulder or rock wall may be encountered, but by most people, rock is perceived as dry, sterile, impenetrable, and generally uninviting. However, to the experienced eye rock surfaces are often teeming with life – lichens, bryophytes, a host of small, invertebrate animals, as well as a vast array of microscopic organisms including bacteria, cyanobacteria, algae and non-lichenized fungi. The longevity and structural stability of most rocks superficially suggest that rock surface inhabitants are benign; however, slowly and steadily all rock dwelling organisms contribute to the relentless decomposition of rock surfaces – augmented by the natural physical forces associated with changing seasons, weather patterns, and in some localized settings the caustic effects of air pollution.

Rock-dwelling communities vary in complexity and composition depending on the specific structural and chemical features of the rock. Even man-made or man-influenced stone supports to some degree a living community – and herein are found the real issues and concerns related to biodeterioration of rock substrata. In a natural setting biodecomposition of rock is accepted as normal and even desirable; however, in the human environment biodeterioration of monuments, buildings, artwork, statues and grave markers is regarded as a serious problem. Even in natural settings, culturally significant prehistoric and historic rock art is subject to the same processes of biodeterioration.

2. A LICHEN PRIMER

Lichens are symbiotic systems consisting of a fungus (usually an ascomycete) and a eukaryotic alga and/or a cyanobacterium. The fungal symbiont, commonly referred to as the mycobiont, accommodates and facilitates the alga and/or cyanobacterium, commonly called the photobiont, by producing a suitable and somewhat rigorously controlled greenhouse-like habitat. More specifically, for most lichen associations the mycobiont organizes and concentrates the photobiont into a distinct layer within the fungal superstructure, regulates the quality and quantity of light reaching the photobiont, delivers water and mineral nutrients to the photobiont, promotes gas exchange, chemically controls herbivores, and effectively neutralizes some groups of potentially toxic airborne elements that the lichen may accumulate from atmospheric outwash. In return the photobiont fixes carbon through photosynthesis, a significant portion of which is transferred to the mycobiont. The mycobiont in turn uses carbon skeletons from the photobiont to meet its own basic energy needs and construct a truly unique and complex structure, which accommodates and enhances the operation of the photobiont. In some cases, a mycobiont may exclusively employ a cyanobacterium as its carbon-fixing partner, while in others it may use a combination of a eukaryotic alga and a cyanobacterium. Many lichen cyanobionts are also able to fix atmospheric nitrogen and in these cases organic nitrogen is also transferred to the mycobiont.

Lichens are eminently successful and enjoy a worldwide distribution. They occur in every conceivable habitat and in more than a few cases play a central role in the operation of some natural systems, growing on a variety of substrata, including most natural substrata as well as a host of human-manipulated or manufactured substrata. Common natural lichen substrata include all categories of rocks and bark, decorticated wood, decomposing wood, living and dead branches of trees and shrubs, evergreen leaves, cactus spines, exoskeletons of insects, bleached bones, and both mineral and organically enriched soils. Common man-made substrata supporting lichens include asphalt, rubber, plastic, glass, stonework, concrete, plaster, ceramic and terracotta tiles, bricks, processed wood products, cloth, canvas and various types of metals (Brightman and Seaward 1977).

3. LICHENS IN URBAN AREAS

Lichen communities in urban areas are frequently lacking in diversity and complexity. This condition appears to be related to three factors: 1) local accumulation of high concentrations of toxic airborne contaminants produced by various urban-related activities and processes, 2) removal and/or replacement of natural substrata, and 3) alteration of natural hydrologic cycles. In most cases these three factors act synergistically to yield an environment that generally discourages the growth and development of lichen communities. The impact of urban development, especially air pollution, often extends into adjacent natural lichen communities. Prior to the onset of the industrial age, stone and wooden buildings, gates, fences and tombstones, as well as glass, bricks and tiles, in urban areas supported quite diverse lichen communities. This condition is still typical of older, more rural human population centers, which have experienced little or no industrial development. As industrial activity developed in and around human population centers, lichen communities on natural substrata declined, but man-made substrata supported their own, albeit limited, lichen assemblages. However, in recent years many urban centers have experienced a recovery of lichen populations due to a widespread switch to cleaner fuels and greater efforts to control, or at least reduce, emissions from some of the more damaging air pollution sources (Seaward 1997).

4. PEDOGENESIS

The term pedogenesis refers to the process of soil formation. The decomposition of rock is a natural phenomenon. Soils are derived in large measure from the breakdown of parent material. This process is inherently slow, but nevertheless relentless, usually operating on a geologic time scale and involving a complex and interactive combination of physical, chemical, and biological activities.

Some aspects of the process are strictly abiotic, such as wetting/drying, heating/cooling, and freezing/thawing, whilst others are clearly biological in nature, such as the encroachment of roots, rhizines, and hyphae into cracks and fissures in rock surfaces, the bio-mediated chemical erosion of cementing agents, and the bio-transformation of the molecular structure of rock substrata. Both biological and abiotic factors operate together, the effects of one accentuating the effects of the other, relentlessly breaking down rocks, large and small, to form the mineral component of soil. The nature and rate of pedogenesis is dictated by several factors: 1) the chemical and physical attributes of the rock, including elemental composition, molecular structure, cementing agents, density, porosity, pH and nature of origin, 2) the composition and degree of development of the epilithic and endolithic biological community, and 3) local and regional climatic patterns.

The dynamics of rock decomposition apply to both natural rocks and man-made 'stone' substrata. With the onset of the industrial era, human-induced changes to the environment, especially changes in air and water quality, have altered the dynamics and rates of rock degradation. Specifically, increasing levels of air pollutants accompanied by acidification of precipitation have dramatically altered or in some cases eliminated some biological communities in the urban setting. In some cases, there has also been an impact on adjacent natural communities. However, in the urban context, whilst environmental degradation has reduced the impact of the biological components on rock decomposition, it has dramatically enhanced the impact of some abiotic factors.

5. LICHENS AS BIODETERIORATORS OF ROCK SUBSTRATA

Lichens play an important role in the process of pedogenesis. Historically, their contribution to the breakdown of rock substrata was perhaps somewhat overrated; however, today it is often seriously underestimated. Frequently, lichens are the first living things to occupy newly exposed rock surfaces. Invasion of rock substrata by lichens is dictated by several factors including 1) proximity of native lichen communities on similar substrata, 2) reproductive strategies (sexual and/or asexual) of local lichens, and 3) dispersal capacity of local lichens.

Many saxicolous lichen communities undergo regular patterns of successional change; for example, one assemblage of species may occupy a given rock surface for several years, steadily altering the substratum in ways that eventually better accommodate a new combination of species. Thus, over time, the changing lichen community relentlessly changes the rock surface.

Lichens contribute to mechanical weathering of rocks in four ways: 1) penetration of mycobiont hyphae (up to 15-20 mm) and rhizines into naturally occurring crevices and cracks in rock surfaces; 2) expansion and contraction of lichen thalli with daily and seasonal changes in ambient temperature and humidity (Nash 1996); 3) swelling action of organic salts produced by lichens; and 4) fracturing and incorporation of mineral fragments by lichen thalli (Chen *et al.* 2000).

Water is essential for many of the chemical reactions associated with breakdown of rock substrata. Because lichens are able to absorb water in either the liquid or vapor phase, chemical and physical weathering processes are expedited on lichen covered rock surfaces. The mixing of respiratory CO_2 with water in lichen tissues results in the formation of carbonic acid, which also enhances the solubility of rock

1. Biodeterioration of Rock Substrata By Lichens

surfaces by lowering the pH of the substratum microenvironment adjacent to lichen thalli (Chen *et al.* 2000).

Lichens commonly produce secondary chemicals, including various weak organic acids, which actively chelate substrate cations, and thus modify the chemical and physical structure of mineral substrata (Jones 1988); for example, oxalic acid, produced in significant amounts by many lichen species, forms chemical complexes with substrate cations (Chen *et al.* 2000). Specifically, oxalic acid reacts with rock substrata containing calcium carbonate to form the insoluble compound calcium oxalate, which accumulates on the surface and within lichen thalli, or at the lichen-rock interface (Seaward and Edwards 1997). Residues of calcium oxalate often remain on such substrata, leaving significant and often unsightly white deposits, after the death of the lichen. This phenomenon can be particularly serious in the case of delicate and intricate rock monuments, as calcium oxalate deposits often obscure the detail and historical significance of such structures (Seaward and Edwards 1995). Oxalic acid has also been shown to solubilize magnesium silicates (Jones 1988).

Several lichen species typically associated with recovering urban lichen communities appear to be unusually aggressive in exploiting, altering and, in some cases, destroying important historical and cultural structures. Furthermore, some species are able to significantly degrade rock surfaces over relatively short periods of time. Of particular concern is the fact that many of these unusually aggressive rock-degrading species appear to thrive under conditions typical of modern urban environments.

Various methods for controlling or eliminating lichen growth on rocks have been investigated. Typically, control measures range from mechanical removal to application of various types of biocides. Since many lichen species are able to regenerate from thallus fragments, mechanical removal of lichens may only temporarily reduce lichen coverage while causing significant physical damage to rock surfaces. Application of biocides has yielded mixed results, including: 1) generally poor treatment response, 2) changes in community dynamics, with surviving species aggressively exploiting open space, 3) persistent dead thalli, which decompose slowly, especially in arid habitats, and 4) good results on lichen removal but damage to substrate surfaces, with changes ranging from discoloration to structural issues requiring use of chemical consolidants. The problem is further complicated by the fact that in some situations lichen thalli may play a central role in consolidating and protecting the substrate surface, with removal of the lichen actually reducing the structural integrity of the substratum.

6. OTHER IMPORTANT CONSIDERATIONS

Recently, biodeterioration of prehistoric and historic images and notations on natural rock formations has been investigated. These situations often present unusually complex problems usually involving some combination of: 1) laws and ordinances enacted with the intent of protecting and maintaining the resource, 2) altered local resource use patterns, which sometimes result in environmental conditions that enhance development of biological communities on or near critical cultural resources, 3) consideration of threatened and endangered and endemic lichen species, and 4) religious and cultural concerns and expectations of aboriginal populations, which in some cases may run counter to government-mandated management programs.

7. DEVELOPING A MULTIDISCIPLINARY APPROACH TO BIODETERIORATION ISSUES

Biodeterioration of historical and culturally significant stone substrates is a complex problem. Easy, risk free solutions are simply not available. The dynamics of the problem are too complicated to address effectively with only one kind of expertise. Successful resolution of this issue will inevitably require a multidisciplinary effort, where biologists join with ecologists, geologists, geochemists, crystallographers, cultural conservators, archaeologists, anthropologists and historians to recommend the most effective management scheme. The advantage of this approach is obvious: multidisciplinary management teams with good leadership can ask more appropriate questions while developing much more thoughtful and informed management decisions.

With this approach in mind, a symposium entitled 'Lichen biodeterioration: progress and problems' was convened in August 2001 as part of the Conference 'Plants and People', co-hosted by the American Bryological and Lichenological Society and the Botanical Society of America, held in Albuquerque, New Mexico. This symposium brought together experts from various fields in order to initiate a meaningful dialogue concerning the progress and problems in this subject area, particularly in respect of culturally significant rock and monument substrata. The chapters of this book embrace not only many of those presentations made at the symposium, but also papers from several invited contributors in order to provide a wide spectrum of knowledge, particularly that related to the cultural property of more remote areas, often in developing countries, which have hitherto been understudied.

Following an introduction (Chapter 1) which provides background information on biodeterioration processes, together with a structural framework of the topics covered and the general rationale for this book, three case studies are detailed to

1. Biodeterioration of Rock Substrata By Lichens

illustrate the impact of lichen species on culturally important substrata (Chapter 2); these show how significant biodeterioration of stonework can take place over a relatively short time-scale and also demonstrate the conservation issues when considering aggressive *versus* aesthetic damage.

The next contribution is concerned with tests undertaken in order to investigate the stability of limestone substrata at an important Mayan site and to determine the effectiveness of selected biocides (Chapter 3). Further work on Spanish substrata demonstrates the differences between lichen and free-living microorganism colonization of quarried carbonate rock and of city monuments constructed of similar material (Chapter 6).

Rock art and petroglyph studies are extensively covered in four contributions. A review of the conservation issues relating to biodeterioration of ancient rock at selected sites in the western United States includes a promising non-caustic treatment for lichen control (Chapter 8). A highly detailed investigation of petroglyphs of a national monument in New Mexico establishes the role of lichens (Chapter 9), but a further study of petroglyphs on sandstone questions their role at a Wyoming site (Chapter 7). Archaeological, cross-cultural and resource management issues related to the biodeterioration of petrogyphs are analysed in detail (Chapter 12).

Relatively little is known of the impact of biodeteriorative processes on monuments in developing countries, much of the very considerable disfiguration observed in tropical and subtropical areas being due not only to lichens but also to cyanobacteria which create dark biofilms. It is hoped that the review of lichens growing on man-made substrata in India (Chapter 11) will stimulate others to investigate this under-researched subject.

An FT-Raman spectroscopic study of the biodeterioration process on sandstone in an arid Utah environment clearly demonstrates the impact of one particular species on native rock uninfluenced by human manipulation or by human-impacted environments (Chapter 13).

The extensive lichen biodeterioration work undertaken in Italy is portrayed by means of a detailed analysis of published sources, suitably annotated (Chapter 4), and the volume is appropriately rounded-off by a major bibliographic survey of 658 worldwide titles, supported by an analytical index categorizing their content into eight subject areas (Chapter 14).

The recently published work on *Cultural Heritage and Aerobiology* (Mandrioli *et al.* 2003), covering methods and measurement techniques for biodeterioration monitoring, complements in part this symposial volume.

The current volume is the first treatment of the subject of biodeterioration that includes a careful consideration of the role of related disciplines, such as geology, crystallography, cultural conservation, archaeology, and resource management. This combination of disciplines makes this book valuable not only as a solid scientific treatise, but equally important as a serious resource for evaluating both impact processes and preservation options related to biodeterioration of culturally significant rock substrata. In conclusion, it is hoped that this volume will provide not only background information, but also practical advice on the detection, measurement and control of lichens and biofilms, on a wide range of cultural heritage, that will prove of value to historians, archaeologists and conservators, as well as specialist biologists.

REFERENCES

Brightman, F.H. and Seaward, M.R.D. (1997) Lichens of man-made substrates. In: *Lichen Ecology* (M.R.D. Seaward ed.): 253-293. Academic Press, London.

Chen, J., Blume, H. and Bayer, L. (2000) Weathering of rocks induced by lichen colonization – a review. *Catena* 39: 121-146.

Jones, D. (1988) Lichens and pedogenesis. In: *Handbook of Lichenology, Volume 3* (M.Galun, ed.): 109-124. CRC Press, Boca Raton.

Mandrioli, P., Caneva, G. and Sabbioni, C., eds. (2003) *Cultural Heritage and Aerobiology*. Kluwer Academic, Dordrecht.

Nash, T.H., ed. (1996) *Lichen Biology*. Cambridge University Press, Cambridge.

Seaward, M.R.D. (1997). Urban deserts bloom: a lichen renaissance. *Bibliotheca Lichenologica* 67: 297-309.

Seaward, M.R.D. and Edwards, H.G.M. (1995) Lichen-substratum interface studies, with particular reference to Raman microscopic analysis. I. Deterioration of works of art by *Dirina massiliensis* forma *sorediata*. *Cryptogamic Botany* 5: 282-287.

Seaward, M.R.D. and Edwards, H.G.M. (1995) Biological origin of major chemical disturbances on ecclesiastical architecture studied by Fourier transform Raman spectroscopy. *Journal of Raman Spectroscopy* 28: 691-696.

Chapter 2

LICHENS AS SUBVERSIVE AGENTS OF BIODETERIORATION

M. R. D. SEAWARD
Department of Environmental Science, University of Bradford, Bradford BD7 1DP, UK

Abstract: Lichens play major roles in shaping the natural world, both physically and biologically. Their function as biological agents in soil development used to be considered only in a geological context, but recent research has shown that lichens are capable of biodeteriorating stone substrata within a relatively short time-scale. Chemical alteration of the substratum is brought about by the disruptive action of many species, particularly those capable of producing oxalate(s) at the thallus - substratum interface. Raman microscopic analysis has proved invaluable in the interpretation and characterization of the physical and chemical nature of this interface. The oxalate contributes significantly to the bulk and composition of the thallus itself and persists as an obvious encrustation after the lichen's death. On ancient monuments, these disfiguring oxalate residues have been variously interpreted in the past as inorganic residues resulting from earlier physico-chemical renovative treatments of their surfaces, from atmospheric pollution and/or from climatic weathering processes. Human-influenced environments appear to be conducive to many lichen species exhibiting this aggressive behavior and the damaging effects on monuments subjected to this type of biodeterioration should give grave concern to archaeologists and conservators of our cultural heritage.

1. INTRODUCTION

Modern electron microscopy and chemical techniques have made it possible to recognize that many lichens contribute to the deterioration of a wide range of materials, particularly rocks and stonework, as a result of physical and/or chemical processes (Seaward, 1997a). In the past, attention was drawn to the possible effect of dissolved carbon dioxide, derived from lichen respiration, attacking the substratum to produce pits and channels for easier penetration of hyphae, with attendant loosening of mineral particles and their incorporation into lichen tissue.

Such effects, although important on a geological time-scale, have until recently considered to be minimal in terms of the life of stone buildings and monuments.

Many lichen species create microclimatic effects at the thallus/substratum interface, particularly in terms of water retention, which undoubtedly lead to mechanical damage to stonework on a short time-scale of ten, or even fewer, years. Various crustose and squamulose lichens are implicated, their aggressive behavior no doubt promoted by particular man-made environmental conditions. Furthermore, forces generated by climatic wetting and drying of lichen thalli cause them to expand and contract in conjunction with the chemical breakdown of substrata by lichen acids.

Lichen acids have a relatively low solubility, but they are effective chelators, forming metal complexes with silicates, etc., derived from the substratum. X-ray powder diffraction and transmission electron microscopy have clearly demonstrated the presence of characteristic alteration products at the interface between rocks and various lichens. Experiments involving pure lichen acids or lichen fragments incubated with different types of rock have confirmed these observations.

2. SUBSTRATA

Oxalic acid secreted by the mycobiont is extremely soluble in water and acts as a chelator of metal ions, and oxalates formed at the thallus/substratum interface are closely related to the chemical composition of the rock (Purvis 1996); thus, species growing on serpentinite, mainly composed of magnesium silicate, form magnesium oxalate dihydrate at the interface (Wilson *et al.* 1981). Other alteration products have been shown to be incorporated into the thallus and/or precipitated at the lichen - substratum interface, such as manganese oxalate (Wilson and Jones 1984) and copper oxalate (Purvis 1984) on manganese-rich and copper-rich rocks respectively.

However, the commonest oxalate found in lichens is calcium, the thallial content ranging from 1 to 50% according to the species and its underlying substratum (Syers *et al.* 1967, Edwards *et al.* 1994); it should be noted that this compound exists in two hydrated forms, monoclinic monohydrate (whewellite) and tetragonal dihydrate (weddellite). The monohydrate form is the major bio-deterioration product and the equilibrium (ratio between monohydrate and dihydrate forms) is dependent on various environmental factors (Edwards *et al.* 1992). Many lichens known to contain calcium oxalate undoubtedly cause extensive corrosion of a range of rock substrata.

It has been shown that lichens are capable of producing calcium oxalate on a wide variety of substrata (Edwards *et al.* 1997). *Dirina massiliensis* forma *sorediata* has the capacity to produce encrustations of various thicknesses, with different concentrations of calcium oxalate and a variable chemistry derived from the following range of substrata: gypsum/calcite underlying a fresco, stucco of a church wall, a Roman brick wall, a lead/glass interface of a church window, mortar and acidic stone. However, it was also noticeable that the effectiveness of this lichen to degrade its particular substratum was determined by the various environmental conditions obtaining (see below) in the different habitats and geographical settings studied.

The ability of lichens to produce significant levels of calcium oxalate even when growing on non-calcareous substrata is particularly interesting. FT-Raman spectroscopic studies of lichens on granitic monuments in Spain, for example, have shown that not all species are capable of producing calcium oxalate, and those that do have significant levels derive some of the calcium for its production from atmospheric sources or from leachates of neighboring substrata (Prieto *et al.* 2000).

3. ENVIRONMENTAL CONDITIONS

The production of calcium oxalate dihydrate by thalli (see above), which is a measure of a lichen's capacity to biodeteriorate its substratum, is related to microclimatic conditions such as temperature and humidity of the air as well as the chemical and physical nature of the substratum. It would appear, for example, that warmer and drier sites are more conducive to the production of calcium oxalate dihydrate by *Diploicia canescens* (Edwards *et al.* 1995) and by *Ochrolechia parella* (Prieto *et al.* 1999); this is in accordance with the findings of Wadsten and Moberg (1985) who observed that lichens growing on humid sites produced calcium oxalate monohydrate, whereas those from drier sites produced a mixture of the monohydrate and dihydrate.

There is strong evidence that recent environmental changes have been conducive to increasingly detrimental invasion by certain aggressive lichens, as in the case of the establishment of nitrophilous species due to environmental hypertrophication (Seaward and Coppins 2004). Such evidence could help to explain why it is that monuments, undamaged for many centuries, appear in recent years to be vulnerable to lichen attack, in addition to the known problems resulting from air pollution.

Whilst acknowledging the recent effects of air pollution (especially acid rain) on monuments, it must also be recognized that *Dirina massiliensis* forma *sorediata* is a relative newcomer: its dramatic spread in Europe, and more particularly England, in recent years has been facilitated by new environmental regimes, including qualitative changes in air pollution, which have allowed it to dominate substrata in the wake of the rapid disappearance of other more pollution-sensitive species. *D. massiliensis* forma *sorediata* is by no means the only organism implicated in short-term deterioration processes: other lichens, and indeed other microorganisms, capable of adapting to man-made environmental disturbances can be equally destructive when ecosytem equilibrium is disrupted.

Such observations constitute the necessary first phase of any program aimed at quantifying the actual role played by lichens in the deterioration of archaeological materials. Stone- and art-work in exposed and partially-enclosed situations, giving rise to a variety of environmental conditions, need to be examined in detail in order to elucidate the relationships between particular lichen species and the physical and chemical nature of their substrata in order to determine the relative importance of those species in biodeterioration processes obtaining in specific circumstances. It is necessary to determine, for example, which species are disfiguring but intrinsically harmless, and which cause actual physical damage.

4. CASE STUDIES

4.1 Museo Nationale Romano, Rome, Italy

A grassed area surrounded by cloistered museum buildings of the Museo Nationale Romano in Rome contains numerous archaeologically important monuments and artifacts constructed from a variety of materials. In interpreting the deterioration of these materials, the ambient urban climate and associated atmospheric pollutants should be taken into consideration, particularly since the latter dramatically affects the lichen flora. Several of the more toxi-tolerant species (poleophiles) with an aggressive behavior were actively colonizing stone monuments and other archaeological artifacts.

The most dramatic and alarming case of lichen attack was observed in the case of several large terracotta Roman pots scattered about the grassed area (Seaward 1988). The rims and shoulders of these pots were lightly colonized by relatively innocuous species such as *Lecanora dispersa* and *Candelariella vitellina*, but here and there, thalli of *Lecanora muralis*, mainly 4-7 cm in diameter, probably representing 6 to 15 years' growth, were causing demonstrable damage (Fig. 2-1). A section through one of the thalli clearly reveals the results of such damage: a central blister, created by the crowding of apothecia, pulled away a fragment of the substratum, two or more mm in thickness, over an area of almost 12 cm^2. *L. muralis* appears to be a highly successful lichen in urban environments into which it has spread dramatically in recent years, due in part to lack of competition from other species; a change in the nature of air pollution in Rome in that period may be a contributory factor to this aggressive behavior.

2. Lichens as Subversive Agents of Biodeterioration

Figure 2-1. Blister created by *Lecanora muralis* on the shoulder of a terracotta pot

4.2 Palazzo Farnese, Caprarola, Italy

The Palazzo Farnese, a beautiful mansion on a hillside at Caprarola in central Italy, built by Vignola in 1547-1549, features a circular courtyard surrounded by cloisters on ground and first floor levels, the inner walls of which bear frescoes by Zuccari painted in the 1560s.

Examination of the water-based paintwork, which had relatively recently shown alarming signs of biodeterioration, revealed that a single lichen species, *Dirina massiliensis* forma *sorediata*, was responsible for the disfigurement. The latter gave great cause for concern, the attack being very pronounced in many places, and clearly demonstrating the predilection of this lichen for the brown and yellow pigments, rather than the red pigment which contained one or more metals antagonistic to its growth.

The distribution pattern of this lichen was not only dictated by the color of the paintwork: when first viewed by the author in 1986 at several meters distant from the frescoes, it was obvious that some recent cleaning activity had distributed lichen propagules from an inoculum, or dispersed them from existing thalli, to create distinctive areas of lichen invasion across the frescoes (Edwards and Seaward 1993).

Using FT-Raman spectroscopy, it was shown that *D. massiliensis* forma *sorediata* on the above frescoes (Fig. 2-2) can produce calcium oxalate encrustations at the thallus - substratum interface up to 1.8 mm in thickness in less than 12 years (c. 0.15 mm per annum). From such measurements it was deduced that a 60% obliteration of the fresco by this lichen, which was not uncommon, generated more than 1 kg of calcium oxalate over a similar time period; furthermore, with the incorporation of calcite and gypsum into the thallus encrustation, it is likely that more than four times this amount of the underlying substratum has been chemically and physically disturbed (Seaward and Edwards 1995, 1997).

Figure 2-2. Renaissance fresco attacked by *Dirina massiliensis* forma *sorediata*, its removal by light brushing from the central area also taking away the underlying paintwork

4.3 Fiskerton Church, Lincolnshire, England

The short-term biodeteriorative capacity of *Dirina massiliensis* forma *sorediata* is not specific to the above-mentioned frescoes; for example, detailed studies of this lichen on exterior stonework of English churches, as demonstrated by a 14th century church at Fiskerton in Lincolnshire, England, have shown its similarly destructive nature (Seaward and Edwards 1997).

In the past, encrustations generated by this, and no doubt other lichen species, have been misinterpreted as the remaining traces of a whitish coating or rendering applied as a decorative or protective surface in a 19th or 20th century restoration program. It is now clear that this is not so, since these 'renderings' consist almost entirely of calcium oxalate (Fig. 2-3) and more often than not evidence remains of the thalli producing them (Fig. 2-4). These encrustations are usually more than 0.5 mm in thickness and cover considerable areas of many church walls throughout England.

Figure 2-3. Calcium oxalate coating of church wall

2. Lichens as Subversive Agents of Biodeterioration

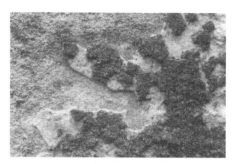

Figure 2-4. Close up of calcium oxalate encrustation with in situ thalli of *Dirina massiliensis* forma *sorediata* which have generated it

5. STONEWORK CONSERVATION

The presence of lichens on stonework is variously interpreted by the lay public and by specialists in different disciplines, whose attitudes are inevitably colored by differing aesthetic and practical considerations.

The lichenologist regards the appearance of a lichen mosaic as a natural feature of ancient monuments, finding the diversity of species present aesthetically pleasing, besides being both taxonomically and ecologically interesting. There is a direct correlation between the composition of the flora and the passage of time, the different lichen communities established on buildings and monuments reflecting the various materials employed in their construction and often correlating to the chronology of successive building phases, therefore assisting in archaeological interpretation. Furthermore, since lichens are exceedingly sensitive to environmental change, the diversity of the flora can be a reliable indication of the level of air pollution which in itself is one of the most serious factors in the deterioration of ancient monuments (see above).

It is ironic that in a bland, homogenous urban environment, where a lichen mosaic would be a welcome relief to the eye, the higher levels of air pollution prevent its establishment, only allowing the existence of a monotonous flora composed of a few algae and lichen crusts. However, it is pleasing to note that the undoubted improvement in air quality of many cities in recent years is reflected in the continued recovery of their lichen floras (Seaward 1997b).

On the other hand, fine art specialists concerned with the conservation of ancient monuments view the encroachment of lichens from a very different standpoint: inscriptions and fine details may be obscured, and, depending on the nature of the substratum, and in some cases the ambient conditions (see above), serious physical damage is often caused through lichen-induced biodeterioration. The lichen floras vary considerably according to the spatial differences in the chemical properties of stone surfaces, the micro-environmental conditions and the overall influence of air pollution.

It is self-evident that base-line work is a prerequisite to laboratory research designed to establish the nature of the interface between problematic lichens and their substrata, and field trials intended to test the relative effectiveness of differing techniques and treatments for the removal and discouragement of lichens from stonework. Any treatment should be selected with care, since although immediately effective, the long-term effects are highly likely to be deleterious.

Mechanical methods involving scraping and brushing, usually followed by washing, are tedious, damaging and often ineffective. Absorbed water may adversely affect the monument, particularly under fluctuating temperature regimes; although penetration can be minimized by the use of water repellents, entrapped water and rising damp can nevertheless prove highly destructive.

A wide range of biocides have been tried, many of which have since been rejected due to side-effects such as crystallization of soluble salts which have penetrated the stonework, staining and discoloration of monuments where the chemicals used have interacted with particular metals present in the substratum, and the promotion of secondary biological growths, which may be even more unsightly than the primary growths. Furthermore, regular treatments are likely to be necessary which are expensive both in terms of the chemicals used and the labor employed for the mechanical removal of only partially detached and brittle lichen growths which remain. The biocides employed may also be harmful to the operators and, not surprisingly, dangerous to wildlife. Some success has been achieved using organo-metallic compounds, quaternary ammonium compounds and borates, but the latter have proved problematic when used in air-polluted environments where, of course, many of the monuments it is desired to conserve are to be located.

The subtle coloration of a varied lichen mosaic can be retained for its aesthetic appeal provided it does not produce disfigurement or unduly obscure detail. It also has to be acknowledged that the lichen flora itself may be intrinsically interesting, in some instances a strong case being made to conserve a monument on lichenological as well as historical grounds. Furthermore, it has been shown that lichens, in certain situations, can afford a protective barrier, shielding the stonework from external weathering agents (Viles and Pentecost 1994, Mottershead and Lucas 2000).

In the light of the above, any decision to remove lichens from stonework must not be undertaken over-hastily or without very careful consideration of the wider implications of long-term effects. Unfortunately, it has to be acknowledged that the problem is under-researched and much of the work published to date is of a largely empirical nature, which has yet to be adequately substantiated by long-term experimentation. It remains for future generations to judge the relative effectiveness of the various conservation techniques currently employed.

6. POSTSCRIPT

The growing interest in the role of lichens in the biodeterioration of natural rocks and man-made stonework has seen a corresponding proliferation of the literature on the subject: for detailed reviews see Syers and Iskandar (1973), Jones and Wilson (1985) and Jones (1988), and for recent comprehensive surveys of the literature see Nimis *et al.* (1992) and Chapter 14 of this volume which incorporates titles from the on-going series in *The Lichenologist* by Piervittori *et al.* (1994, 1996, 1998, 2004).

REFERENCES

Edwards, H.G.M., Edwards, K.A.E., Farwell, D.W., Lewis, I.R. and Seaward, M.R.D. (1994) An approach to stone and fresco lichen biodeterioration through Fourier Transform Raman microscopic investigation of thallus - substratum encrustations. *Journal of Raman Spectroscopy* 25: 99-103.

Edwards, H.G.M., Farwell, D.W., Jenkins, R. and Seaward, M.R.D. (1992) Vibrational Raman spectroscopic studies of calcium oxalate monohydrate and dihydrate in lichen encrustations on Renaissance frescoes. *Journal of Raman Spectroscopy* 23: 185-189.

Edwards, H.G.M., Farwell, D.W. and Seaward, M.R.D. (1997) FT-Raman spectroscopy of *Dirina massiliensis* forma *sorediata* encrustations growing on diverse substrata. *The Lichenologist* 29: 83-90.

Edwards, H.G.M., Russell, N.C., Seaward, M.R.D. and Slarke, D. (1995) Lichen biodeterioration under different microclimates: and FT-Raman spectroscopic study. *Spectrochimica Acta* 51A: 2091 2100.

Edwards, H.G.M. and Seaward, M.R.D. (1993) Raman microscopy of lichen – substratum interfaces. *Journal of the Hattori Botanical Laboratory* 74: 303-316.

Jones, D. (1988) Lichens and pedogenesis. In: *Handbook of Lichenology, Vol.III* (M.Galun ed): 109-124. CRC Press, Boca Raton.

Jones, D. and Wilson, M.J. (1985) Chemical activity of lichens on mineral surfaces - a review. *International Biodeterioration* 21: 99-104.

Mottershead, D. and Lucas, G. (2000) The role of lichens in inhibiting erosion of a soluble rock. *The Lichenologist* 32: 601-609.

Nimis, P.L., Pinna, D. and Salvadori, O. (1992) *Licheni e Conservazione dei Monumenti.* Cooperativa Libraria Universitaria Ediatrice Bologna, Bologna.

Piervittori, R., Salvadori, O. and Laccisaglia, A. (1994) Literature on lichens and biodeterioration of stonework. I. *The Lichenologist* 26: 171-192.

Piervittori, R., Salvadori, O. and Laccisaglia, A. (1996) Literature on lichens and biodeterioration of stonework. II. *The Lichenologist* 28: 471-483.

Piervittori, R., Salvadori, O. and Isocrono, D. (1998) Literature on lichens and biodeterioration of stonework. III. *The Lichenologist* 30: 263-277.

Piervittori, R., Salvadori, O. and Isocrono, D. (2004) Literature on lichens and biodeterioration of stonework. IV. *The Lichenologist* 36: 145-157.

Prieto, B., Edwards, H.G.M. and Seaward, M.R.D. (2000) A Fourier Transform-Raman spectroscopic study of lichen strategies on granite monuments. *Giomicrobiology Journal* 17: 55-60.

Prieto, B., Seaward, M.R.D., Edwards, H.G.M., Rivas, T. and Silva, B. (1999) Biodeterioration of granite monuments by *Ochrolechia parella* (L.) Mass.: an FT-Raman spectroscopic study. *Biospectroscopy* 5: 53-59.

Purvis, O.W. (1984) The occurrence of copper oxalate in lichens growing on copper sulphide- bearing rocks in Scandinavia. *The Lichenologist* 16: 197-204.

Purvis, O.W. (1996) Interactions of lichens with metals. *Science Progress* 79: 283-309.

Seaward, M.R.D. (1988). Lichen damage to ancient monuments: a case study. *The Lichenologist* 20: 291-294.

Seaward, M.R.D. (1997a) Major impacts made by lichens in biodeterioration processes. *International Biodeterioration and Biodegradation* 40: 269-273.

Seaward, M.R.D. (1997b) Urban deserts bloom: a lichen renaissance. In: *New Species and Novel Aspects in Ecology and Physiology of Lichens* (L. Kappen, ed.). *Bibliotheca Lichenologica* 67: 297-309.

Seaward, M.R.D. and Coppins, B.J. (2004) Lichens and hypertrophication. In: *Festschrift Hannes Hertel: Contributions to Lichenology* (P. Döbbeler and G. Rambold, eds). *Bibliotheca Lichenologica* 88: 561-572.

Seaward, M.R.D. and Edwards, H.G.M. (1995) Lichen-substratum interface studies, with particular reference to Raman microscopic analysis. 1. Deterioration of works of art by *Dirina massiliensis* forma *sorediata*. *Cryptogamic Botany* 5: 282-287.

Seaward, M.R.D. and Edwards, H.G.M. (1997) Biological origin of major chemical disturbances on ecclesiastical architecture studied by Fourier Tranform Raman spectroscopy. *Journal of Raman Spectroscopy* 28: 691-696.

Syers, J.K., Birnie, A.C. and Mitchell, B.D. (1967) The calcium oxalate content of some lichens growing on limestone. *The Lichenologist* 3: 409-414.

Syers, J.K. and Iskandar, I.K. (1973) Pedogenetic significance of lichens. In: *The Lichens* (V. Ahmadjian and M.E. Hale, eds): 225-248. Academic Press, New York.

Viles, H.A. and Pentecost, A. (1994) Problems in assessing the weathering action of lichens, with an example of endoliths on sandstone. In *Rock Weathering and Landform Evolution* (D.A. Robinson and R.G.G. Williams, eds): 99-116. Wiley, Chichester.

Wadsten, T. and Moberg, R. (1985) Calcium oxalate hydrates on the surface of lichens. *The Lichenologist* 17: 239-245.

Wilson, M.J. and Jones, D. (1984) The occurrence and significance of manganese oxalate in *Pertusaria corallina*. *Pedobiologica* 26: 373-379.

Wilson, M.J., Jones, D. and McHardy, W.J. (1981) The weathering of serpentinite by *Lecanora atra*. *The Lichenologist* 13: 167-176.

Chapter 3

LIMESTONE STABILIZATION STUDIES AT A MAYA SITE IN BELIZE

WILLIAM S. GINELL[1] and RAKESH KUMAR[2]
[1]*Getty Conservation Institute 1200 Getty Center Drive, Los Angeles California 90049, USA;*
[2]*Specialty Coatings Company, 2238 Williams Glen Blvd. Zionsville, IN 46077*

Abstract: Stone used in the construction of the 8th-11th century Maya structures at Xunantunich in Belize is a low strength, porous limestone that is nearly pure calcium carbonate. Degradation of archaeologically excavated stone structures in the humid, tropical environment of Belize is caused mainly by wind and water erosion and the wide cyclic variations of humidity and temperature. However, damage to the limestone is accelerated to varying extents by the chemical and mechanical effects of lichens, mosses, algae, fungi, and bacteria that are endemic to the region. To evaluate the effectiveness of possible stabilization treatments, tests were conducted in which stone-penetrating consolidant solutions were applied to limestone samples, which were then exposed to both sunny and shaded environments over a period of four years. The results of these treatments were evaluated visually and by particle and water erosion resistance measurements on the aged samples. The effectiveness of several biocides in controlling the establishment and growth of microflora on the exposed samples and on in situ, ancient stonewalls was also studied. Some tests were conducted to determine if the organic polymer consolidants would support, or even accelerate, the growth of microflora on the stone and how the concurrent use of biocides would affect the results. Several consolidant solutions were found that could penetrate and stabilize the usually moist limestone and, in combination with biocides, would minimize the growth of the local microflora.

Keywords: limestone stabilization; Maya structures; lichens; biodegradation; limestone consolidation; biocides

1. INTRODUCTION

The ancient Maya city of Xunantunich was a small administrative and ceremonial center in western Belize near the Guatemala border (Foster 1992). It sits on a hilltop in the Belize River Valley of the Cayo District, near the Mopan River and the towns of San Ignacio, San Jose Succotz, and Benque Viejo del Carmen. The site, which was occupied by the Maya during the Later and Terminal Classic periods from about 700 to 1100 AD, was first described by Thomas Gann in 1895. Xunantunich consists of a fairly compact center of large architectural structures grouped around several plaza areas that are oriented North-South (Fig. 3-1). Surrounding the central area of this city are numerous outlying settlements that have remained largely uninvestigated until recently.

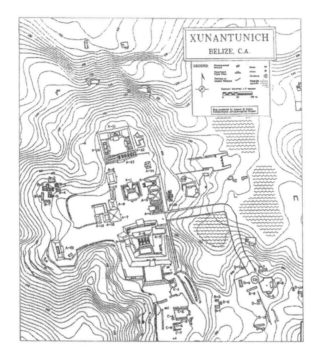

Figure 3-1. Map of Xunantunich, Maya site in Belize

The site is dominated by a monumental central building, "El Castillo" (Structure A-6), a 41m (135 foot) high pyramid, which remains one of the tallest buildings in Belize today (Fig. 3-2). It is likely that El Castillo once displayed a three-dimensional stucco frieze on all four sides of the pyramidal structure. The restored remains of the once spectacular frieze, which depicted ancient Maya cosmological symbols, are visible today on the east side of the building (Fig. 3-3). Extensive archaeological excavations began in 1938 and on exposure to the torrential, wind-swept rains and high relative humidity and temperatures, excavated limestone structures began to deteriorate and active biological growth covered exposed limestone surfaces (Figs. 3-4a,b,c,and d).

3. Limestone Stabilization Studies at a Maya Site in Belize

Figure 3-2. Structure A-6: El Castillo

Figure 3-3. El Castillo – east face frieze

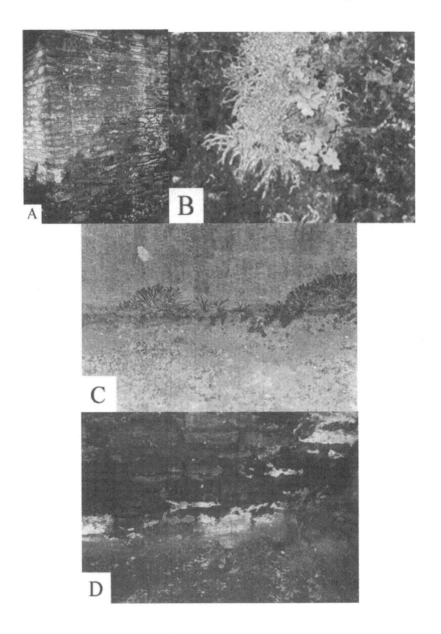

Figure 3-4. Microflora on excavated limestone at Xunantunich

In 1991, the Xunantunich Archaeological Project (XAP) was initiated. The work carried out by XAP personnel represented the first comprehensive study of the ancient city and the surrounding settlements. This research, excavation and conservation project was under the joint direction of Richard Leventhal, Director of The Institute of Archaeology, University of California, Los Angeles, and the Belize

3. Limestone Stabilization Studies at a Maya Site in Belize

Department of Archaeology. As part of the project, the west face of El Castillo was excavated and mechanically stabilized. During this work, the relatively well-preserved remaining section of the stucco frieze was uncovered. This section was conserved and was subsequently reburied to minimize environmental degradation. A replica of this frieze was made and is now visible directly in front of the location of the buried original frieze (Fig. 3-5). This impressive building and the remaining structures at the site of Xunantunich continue to attract visitors from around the world.

Figure 3-5. El Castillo – west face frieze

In 1992, the Getty Conservation Institute began a collaborative effort with the archaeologists and government authorities in Belize to study some of the general problems associated with conservation of archaeological sites in humid, tropical areas. The objectives of this study were to: (a) investigate the nature of the construction materials used by the Maya in Belize; (b) determine the factors that affect the materials degradation processes; (c) determine if the application of consolidants could reduce the rate of environmental degradation of the local limestone; (d) study the effectiveness of several biocides in controlling microfloral growth on consolidated and unconsolidated masonry surfaces; and (e) determine if exposure to the differing environmental conditions in both jungle shade and in full sun could affect the performance of consolidants and biocides.

Typically, the walls of major Xunantunich structures consisted of outer and inner masonry layers made up of tooled limestone blocks and separated by rubble masonry, which was composed of irregular limestone blocks laid in lime mortar. Often, wall surfaces were covered by lime-based plaster renders or stuccos that were decorated, carved, modeled, or were left unadorned. Building floors were plastered with a finely graded lime plaster that was tooled to produce a dense, low permeability, water resistant surface.

Much of the limestone used in the construction of the Maya buildings at Xunantunich is highly porous, mechanically weak, easily crushed (Fig. 3-6a,b), and where it is exposed to the environment, is in an advanced state of surface deterioration. The principal reasons for this condition are: generally, the poor quality of the original limestone construction materials; the solubility of limestone in CO_2-containing rainwater (Crankovic 1985); the use of Portland cement in new repair mortar and stucco; active microfloral growth; wide-ranging cyclic changes in the humidity and temperature in the region; and exposure of the stone structures to the erosive effects of wind, torrential rainfall, and flowing water. Water, absorbed by limestone from rain, dissolves the calcium carbonate to form a solution of soluble calcium bicarbonate, which diffuses to the stone surface during the dry period, evaporates, and forms a crust. Eventually, the crust peels away leaving a fresh stone surface and this repetitive process results in the gradual deterioration of surface features.

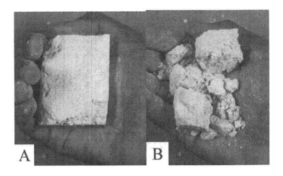

Figure 3-6. Limestone block is easily crushed

Both limestone and lime-containing masonry, which have been exposed to humid, tropical environments for any length of time, are subject to biodeterioration from the lichens, mosses, algae, fungi, bacteria, and macroscopic flora that are endemic to the region. Physical intrusion by lichen hyphae and rhizinae and chemical attack by fungi-generated chelating organic acids all contribute to the degradation of plaster, stucco, mortar and limestone. The organic acids are capable of dissolving both limestone and the intergrain cementing materials between particles and the shear forces created by temperature changes and cyclic wetting and drying result in loss of debonded mineral fragments. In addition, much of the masonry and stucco immediately below the surface is damp throughout the year and this enhances the growth of microflora and reduces resistance to mechanical degradation. Although it has been stated that the existence of a microfloral layer on stone provides a stabilizing, humid environment and limits the loss of stone surface, degradation below the surface may still be in progress and, eventually, the biomass and surface layers will be detached and lost.

One widely reported means of stabilizing building materials of this type has been through consolidation procedures that involve impregnation of the porous stone with appropriate consolidants. However, conservators have accepted only a few materials as being effective for the consolidation and strengthening of dry, calcareous masonry, and damp materials present additional complications. Moreover, it is not known if consolidation materials found to be acceptable in other areas where tests have been carried out will be able to resist the cyclical environmental changes of the hot, humid, tropical jungle. Further, it is not known whether or not the selected consolidants (generally organic polymers) will provide nutrients that can enhance the growth of microflora. Consolidation of moist limestone, mortar, or plaster requires a consolidant system that can penetrate the masonry, cure, and bond in the presence of moisture or, at times, liquid water. In addition, the performance of candidate consolidants under these conditions and over long time periods also is not generally known. Microflora, such as lichens, algae, fungi, or mosses, can penetrate, chemically attack, and weaken the masonry (Martin and Johnson 1992). Identification of biocides that are effective in tropical environments and that will be long-lasting, inexpensive, and relatively non-toxic to humans is an area in which definitive information is meager (Kumar and Kumar 1999).

The effectiveness of consolidating materials is best determined by carrying out diagnostic tests on construction materials following natural aging tests or, if this is not possible, after artificial aging under conditions that produce the deteriorating effects of the natural environment (Robertson 1982). It is important to know which performance criteria the consolidated materials are required to meet and how they can be evaluated. Since the artificial test conditions are intended to reproduce the environmental conditions responsible for deterioration, measurement of changes in materials properties after aging should serve to assess the relative effectiveness of the consolidants when they are used in the field. However, synergistic effects are likely to occur in complex environments and one is never sure whether the real environment has been simulated adequately. The results of the long-term aging in the natural environment provide more reliable information regarding the suitability of a given consolidant, but the acquisition of these data requires long exposure times and many samples. However, if agreement between the data from natural and artificial aging tests is obtained, the results can serve to validate the chosen laboratory procedures and environments.

The work described in this paper presents results of studies to determine: (a) if certain water-tolerant consolidants that are resistant to tropical environments could penetrate and cure within moist limestone; (b) if a group of biocides that was found effective at both Copan and Tikal archeological sites (Hale 1980; Richardson 1976, 1988) would be effective also against the microflora found at Xunantunich; (c) if these and other biocides would be capable of preventing microfloral infestation of virgin limestone; and (d) if the consolidants used would provide nutrients that enhance the growth of microflora on treated stone. More complete details of this study are given in Ginell and Kumar (2002).

2. MATERIALS AND PROCEDURES

Much of the limestone in Belize is of marine origin ranging in age from Upper Pennsylvanian to Recent. Limestone and the other masonry materials derived from limestone that were used in Maya buildings, such as stucco, mortar, and plaster, are composed of the mineral calcite. Xunantunich stone is a fine-grained limestone containing foraminifera and some shell material. Limestone recovered from recent excavations or obtained from the local Cayo quarries exhibits a wide range of homogeneity and often occurs as layers of hard, soft, and porous regions. Plant roots and hyphae can easily penetrate and disrupt the fragile stone.

Laboratory and fields tests of consolidant and biocide effectiveness were carried out on stone samples, on *in situ* new limestone blocks, and old excavated walls. Test samples were cut from excavated rubble limestone and from newly quarried stone. The samples were treated by spraying with consolidant solutions, with and without admixed biocides, or just with biocide solutions or suspensions. After treatment, samples were exposed on racks to two types of environments: a sunny open area and the deep, shaded rainforest. Meteorological conditions at both sites were monitored and recorded by solar panel-activated stations (Maekawa 2000). Sensors monitored relative humidity, air and sample temperatures, wind velocity and direction, rainfall, and solar irradiance. Laboratory measurements determined changes in limestone porosity and erosion resistance following consolidation, the ability of the consolidant to cure in moist stone, and the consolidant penetration depth into the stone (Kumar and Ginell 1995).

Initially, three types of consolidants were selected for testing: water-borne, water-miscible, and non-aqueous systems. The former two groups would be suitable for use on limestone that was either dry or moist near the stone surface or within the consolidant penetration depth of the surface. Consolidant curing in water-borne systems requires the evaporation of water and, where applicable, the coalescence of emulsion or dispersion-phase, polymer particles (Phillips 1992, personal communication). At high relative humidities, water-borne consolidants are stable but curing will not occur. For water-miscible systems, curing via polymerization or cross-linking of oligomers (Phillips 1982, 1995,) can occur in solution and polymer precipitation occurs when the molecular weight of the polymer exceeds a minimum value and its solubility decreases. Hence, curing can take place at high relative humidities. The last group, comprising non-aqueous consolidants, was included in the event that the consolidation of moist stone was not feasible and that only dry limestone could be treated successfully. The consolidants used in the tests are listed in Table 3-1 and the biocides are shown in Table 3-2.

3. Limestone Stabilization Studies at a Maya Site in Belize

Table 3-1. Consolidants used in tests

Consolidant Category	Consolidant Type	Consolidant Trade Name	Solvent	Concentration Used	Supplier
Water-Borne	Acrylic/Epoxy*	Carboset 514H+ ERL 4221	Isopropanol-Water	3, 11%	Goodrich, Union Carbide
		Neocryl BT-520+ ERL 4221	Isopropanol-Water	3, 11%	Imperial Chemical Industries, Union Carbide
	Acrylic Emulsion	Rhoplex AC-630	Water	3%	Rohm & Haas
		Rhoplex AC-33 (Primal AC-33)	Water	3%	Rohm & Haas
		El Rey Superior 200 (ERS 200)	Water	3%	El Rey
	Polyurethane Dispersion	Bayhydrol 121	Water	3, 5, 11, 12%	Mobay
	Lithium Silicate (Inorganic)	Sinak 125	Water	As received (12%)	Sinak
Water-Miscible	Epoxy	Eponex 1510 (resin) Jeffamine T403 (hardener)	Isopropanol	3%	Shell Texaco
		Eponex 1510 (resin) Jeffamine T403 (hardener) Jeffamine 399 (accelerator)	Isopropanol	3, 11, 12%	Shell Texaco Texaco
Water-Immiscible. Non-aqueous	Polysilicate	Silbond 40	Mineral Spirits	As received (40%)	Silbond
	Silane	Conservare-H (Stone Strengthener-H)	Methyl Ethyl Ketone	As received (75%) 12%	ProSoCo
	Acrylic Solution	Paraloid B-67 (Acryloid B-67)	Toluene	3%	Rohm & Haas
		Paraloid B-72 (Acryloid B-72)	Toluene	3,5,12%	Rohm & Haas
	Epoxy	Epon 828 (resin) Epicure 3274 (hardener)	Toluene	12%	Shell Shell
		Epon 828 (resin) HY 956 (hardener)	Toluene	3, 12%	Shell Ciba-Giegy

Table 3-2. Biocides Tested

Biocide	Year	Concentration
Polybor (complex cyclic borates)	1993	4% in water
	1994	5% in water
	1995	5% in water
	1996	5% in water
Cetyldimethyl benzyl ammonium chloride (CDMBQ)	1993	1% in water-IPA*
	1993	5% in water-IPA
	1995	1% in water-IPA
Cetyltrimethyl ammonium chloride (CTMQ)	1993	1% in water
	1993	5% in water
	1994	2% in water
	1995	2% in water
	1996	2% in water
Bis(tri N-butyltin) oxide (TBTO) + CDMBQ	1993	0.5% TBTO + 1% CDMBQ
	1994	0.5% TBTO + 2% CDMBQ
Bis(tri N-butyltin) oxide (TBTO) + CTMQ	1993	0.5% TBTO + 1% CTMQ
	1994	0.5% TBTO + 2% CTMQ
Copper (I) thiocyanate CuSCN	1994	2% in water
Copper (I) chloride CuCl	1994	1% in water
	1995	1% in water
	1996	0.7% in water
Copper (II) fluoride $CuF_2 \cdot H_2O$	1994	1% in water
	1995	1% in water
	1996	1% in water
	1996	0.1% in water
Copper (II) hydroxyphosphate $Cu(OH)PO_4$	1995	1% in water
Silver Nitrate $AgNO_3$	1996	20 ppm in water

* IPA = Isopropanol

In situ tests were performed to determine the effectiveness of the selected biocides for controlling damaging microflora on previously excavated limestone walls. Also, new limestone blocks, which were installed during conservation of damaged Maya structures, were treated *in situ* with biocide solutions to evaluate the rate at which microfloral colonization would occur on freshly exposed stone.

No attempt was made in these *in situ* tests to quantify the biocidal action toward individual species or to determine precisely how long the *ad hoc* treatment would be effective. Only the overall visual appearance of the treated areas, as compared with

3. Limestone Stabilization Studies at a Maya Site in Belize

adjacent, similarly appearing untreated stone or control stone samples, was determined. Only a preliminary effort was made in these studies to identify microflora initially present and no effort was made to distinguish between those present originally and those that developed with time. However, a preliminary analysis was performed by R. Mitchell (1995, personal communication) to identify microorganisms found on some consolidated stone samples returned to GCI after exposure on racks in both sunny and shaded locations. Table 3-3 lists the types of organisms that had become established on consolidated stone in 1 to 2 years. Significant differences were noted between organisms identified on stone exposed at the two locations. Samples of microorganisms, which were obtained from a number of *in situ* walls of various structures at Xunantunich, were identified by SEM. Identification of these species is given in Table 3-4 and photographs of some of the microflora found are included in Figures 3-7a and b.

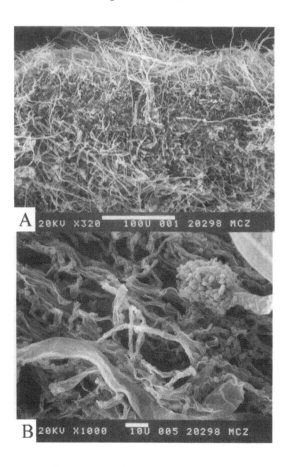

Figure 3-7. a) Penicillium; b) Aspergillus

Table 3-3. Microorganisms Identified on Stone Samples

Exposure Years	Sample #	Consolidant	Location Sun	Location Shade	Description	Organism Type
2	112	A/E 514	√		Yellow colonies with red spores/ White colonies w/ ext. hyphae	*Aureobasidium* sp.
2	116	121	√		Tan mucoid col./brown colonies with white hyphae	*Streptomyces* sp.
1	147	SS-H	√		Pink mucoid colonies/white rounded colonies	Bacteria (rods)
1	153	Silbond	√		No growth	
1	157	Control	√		No growth	
2	224	B-72	√		Large white colonies with ext. green hyphae	Actinomycete
1	331	1510	√		Yellow-brown colonies w/ hyphae	*Streptomyces* sp.
1	333	828	√		Orange and tan colonies w/ hyphae	*Fusarium?* *Aureobasidium?*
1	343	Sinak	√		Pink and tan colonies	Bacteria (rods)
2	526	1510	√		Yellow-brown colonies w/ hyphae	Actinomycete

3. Limestone Stabilization Studies at a Maya Site in Belize

Table 3-3.1. (Continued) Microorganisms Identified on Stone Samples

Exposure Years	Sample #	Consolidant	Location Sun	Location Shade	Description	Organism Type
2	526	1510	√		Yellow-brown colonies w/ hyphae	Actinomycete
2	131	121		√	Green-black colonies w/ white hyphae	Non-sporulating fungus
1	161	1510		√	Yellow w/ green/brown colonies	*Aureobasidium* sp.
1	163	828		√	Small tan mucoid colonies	Yeast-like fungus
1	165	SS-H		√	Tan streaks and large colonies	Yeast-like fungus
1	171	Silbond		√	Orange, green and purple hyphae	*Aureobasidium* sp.
2	241	B-72		√	Small tan round colonies	Bacteria (cocci)
2	241	B-72		√	Small tan round colonies	Bacteria (cocci)
1	361	Sinak		√	Yellow colonies with ext. hyphae	Non-sporulating fungus
1	363	Control		√	Yellow and tan colonies	Bacteria (rods & cocci)
2	529	A/E 514		√	Dark green colonies/resting spores	Actinomycetes
2	543	1510		√	Small white col./Yellow col w/ hyphae	Bacteria (rods)

Table 3-4. Microorganisms Found on Xunantunich Structures

Sample #	Structure Location*	Color	Predominant Microorganisms
1	Area B Wall AA	Black	*Aspergillus* (flavus/glaucus) *Penicillium* sp. *Fusarium* sp.
2	Area B Wall AA	Black	*Aspergillus* (flavus/glaucus) *Fusarium* sp. *Candida albicans*
3	Area B Wall AA	Black	*Aspergillus* (flavus/glaucus)
4	Area B Wall AA	Black-brown	*Cladosporium* sp.
5	Area B Wall AA	Black	*Aspergillus* (flavus/glaucus)
6	Area B Wall AA	Black	*Fusarium* sp.
7	A-6 summit south face	Brown	*Aspergillus* (flavus/glaucus)
8	A-6 north face	Mixed color	Yeast
9	A-6 north face inside room	Black, greasy	*Aspergillus* (flavus/glaucus)
10	A-16	Orange	*Aspergillus* (flavus/glaucus)
11	A-1 south face	Black	*Aspergillus* (flavus/glaucus)
12	A-1 north face	Black	*Penicillium* sp.
13	Area B near Wall AA	Yellow-green	Yeast
14	Near Bodega	Green-black	*Aspergillus clavatus* *Cladosporium* sp.

* See Map Figure 3.1

To determine if some of the polymers being considered as consolidants were susceptible to attack by the fungi isolated from the limestone samples, tests were performed in which polymer films were exposed to fungal cultures (Mitchell and Gu, 1997). The polymers tested included: Paraloid B-67; Bayhydrol 121; Eponex 1510, Jeffamine T403, Accelerator 399; and Epon 828/Epicure 3274. After exposure for one week, growth of fungi was observed on all polymer samples, suggesting that these organic polymers provided useable nutrient sources for the fungi. Parallel tests using a biocide-impregnated cellulose acetate film demonstrated that fungal colonization could be inhibited by this procedure.

3. Limestone Stabilization Studies at a Maya Site in Belize

The penetration of two polymer films, Bayhydrol 121 and Eponex 1510, by the microorganisms *Aspergillus versicolor* and *Chaetomium* was studied by electrochemical impedance spectroscopy (Mitchell *et al.* 1996). It was found that water began to penetrate the polyurethane film in 60 days and the epoxy film in 75 days, indicating that degradation of the film by the microorganisms had occurred.

Variously treated stone samples were prepared and set out each year over a four-year period. Observations were made of the relative erosion resistance provided by the consolidants and the color and intensity of the biological growth. These parameters were correlated with the exposure time by calculation of a Figure of Merit (FOM) for each sample. Numerical values were assigned for: slight or no erosion loss (E); for medium, slight, and no surface color development (A); and for exposure time. The FOM was then given by: A+E-T. A high FOM was indicative of superior performance.

Figure 3-8. Limestone samples treated with CTMQ, CDMBQ, TBTO + CTMQ, TBTO + CDMBQ, or Polybor and aged in the shade for one year. No visible microbiological colonization

3. RESULTS AND DISCUSSIONS

Field-tests that were initiated in 1993 on unconsolidated, soft limestone samples involved spraying the samples with a biocide solution and exposing them on the racks along with the consolidated samples. Figure 3-8 shows the samples after exposure in the shade for one year. The four light-colored samples on the upper shelf were treated with CTMQ, CDMBQ, TBTO + CTMQ, TBTO + CDMBQ, and the one on the lower shelf, with Polybor. The remaining samples on the rack were consolidated.

Figure 3-9a shows the two TBTO-treated samples after exposure for two years (top row). Most of the untreated, but consolidated, samples were covered by green, black, or yellow fungi and some were eroded but the two biocide-containing samples were not eroded and remained clear. After four years, the TBTO-treated samples continued to resist erosion and remained essentially free of fungal deposits (Fig. 3-9b). In this and in many of the other tests, the quaternary ammonium compounds, CTMQ and CDMBQ, performed effectively as biocides in the shade or sun for one to two years. However, the TBTO-containing samples remained clear of fungi for at least four years in either the sun or shade.

Figure 3-9. a) Limestone samples treated with TBTO + CDMBQ or TBTO + CTMQ and exposed in the shade for two years; b) samples after four year exposure

When considering the coordinated application of consolidants and biocides to effect simultaneous erosion resistance and microorganism control, a choice of one of three applications procedures is possible: the biocide can be applied to the stone before, after, or mixed with the consolidant assuming, in the latter case, that the solvent carriers for the consolidant and the biocide are compatible. Accordingly, to determine which of these modes was most effective, three parallel test matrices were set up in 1995 and in 1996. Consolidants and biocides were used that had been shown earlier to exhibit the most effective actions in either the sunny or shaded locations.

In Test Matrix #1, the stone samples were treated first with the consolidants Eponex 1510, Bayhydrol 121, and acrylic 514H/epoxy, and were allowed to cure for several days before being sprayed with the biocide. The treatment sequence was reversed in Test Matrix #2, in that the biocide impregnation was performed first and after drying, the consolidant was applied. For Test Matrix #3, the two completely water-soluble biocides, Polybor and CTMQ, were each mixed with each of the water-compatible consolidants before impregnation of the stone samples. The control samples included as part of Matrix #1 were treated only with the biocides being tested.

3. Limestone Stabilization Studies at a Maya Site in Belize

After exposure for 1-2 years in both sun and shade, it was found that, on average, the FOM values for Matrix #2 in the sun were slightly better than those for Matrix #1, but in the shade, just the reverse was the case (Table 3-5). Apparently, the surface deposits of biocides were more easily dispersed than those that had penetrated the stone and were protected to some extent by the consolidant. The procedure and specific biocides used in Matrix #3 turned out to be less effective than either of the other two. It was interesting to note that most of the FOM values for these tests indicated that the treatments were either highly effective or offered little protection to the stone and no intermediate values were found.

Table 3-5. FOM Values for Biocide-Consolidant Matrices 1, 2 and 3

Biocide	Consolidant	Sun Matrix # 1	2	3	Shade Matrix # 1	2	3
Polybor	Eponex 1510	*	7	*	*	6	8
	A/E 514	*	*	*	*	*	*
	Bayhydrol 121	7	7	7	*	*	*
CTMQ	1510	*	*	*	*	7	*
	514	*	*	*	*	*	8
	121	7	7	*	*	*	*
Copper (I) Chloride	1510	7	8	-	8	*	--
	514	*	9	-	7	*	--
	121	8	9	-	7	*	-
Copper (II) Fluoride	1510	9	*	-	9	7	--
	514	9	8	-	8	6	--
	121	9	9	-	*	*	--
Copper (II) Hydroxy Phosphate	1510	*	*	-	7	*	--
	514	*	7	-	*	*	--
	121	*	*	-	*	*	--

*FOM values were less than 5

Water soluble copper (II) compounds have long been known to be potent biocides, however their residence time on limestone surfaces in regions of high rainfall should be short and frequent re-treatment would be necessary to maintain continuous effectiveness. Accordingly, copper compounds selected for these tests were either suspensions of slightly soluble copper compounds [copper(II) fluoride and copper(II) hydroxy phosphate] or compounds that disproportionate [copper(I) chloride and copper(I) thiocyanate] to form finely divided metallic copper (Remy 1961) or are air-oxidized to form complex, slightly soluble copper(II) hydroxy chlorides (Scott 2002). These particles lodge in the limestone pores and dissolve or gradually oxidize and release biocidal copper(II) ions over time. Copper (II) fluoride and copper (I) chloride were particularly effective. Although staining of stone in contact with massive copper has been observed at other sites before, the low, but effective, concentrations of the compounds tested here, did not result in noticeable discolorations of the limestone samples. Similarly, silver nitrate, another effective biocide, can cause staining but when used at a concentration of 20 ppm, no color change effects were observed in the tests.

Figure 3-10. Structure A–1: south view before excavation

Consolidant testing continued for four years and the results indicated that the cycloaliphatic epoxy resin, Eponex 1510 with Jeffamine 403T hardener, was most effective in minimizing erosion loss of soft limestone. This epoxy, in an isopropanol-water solution, penetrated and cured in moist stone without discoloration. Both the polyurethane dispersion Bayhydrol 121 in water and the standard bisphenol-A epoxy resin, Epon 828 in toluene, performed well on dry stone. The acrylic emulsion solutions did not penetrate the stone but formed surface films that were nonprotective.

Biocides were applied to several excavated walls to determine both the effectiveness of the treatment with regard to the eradication of existing microflora and the duration of the biocidal action. Tests were also performed on newly installed limestone blocks to evaluate the use of biocides for retarding the establishment of new microfloral colonies on the stone surfaces.

One of the activities of the UCLA Institute of Archaeology at Xunantunich was the excavation and stabilization of structure A-1 (Fig. 3-10). Stabilization involved the partial reconstruction of the lower north and south walls of the structure using newly quarried limestone blocks set in lime mortar. New blocks were installed on the north face of A-1 to the left and right of the stairway and are easily recognizable by their lighter color (Fig. 3-11a). An overall view of the north face after one year is shown in Figure 3-11b. It is apparent that the new stone surfaces had darkened considerably. A close-up view of a stone block illustrates the effect of drying on the adherence of the dark fungal layer, Figure 3-11c. Contraction, cupping, and eventually loss of the organic surface layer occurred. The surface layer of stone was attached to the dark layer and was also lost at the same time. Repetition of this process resulted in considerable stone surface loss.

Figure 3-11. Structure A–1: a) north face following installation of new limestone; b) after one year; c) cupping and loss of bio layer and stone surface

Tests were performed to determine if a biocide treatment of newly quarried limestone replacement blocks would reduce the initiation rate of microfloral growth or would permit the establishment of more adherent species. It is also important that benign growth not be eliminated entirely and that the general appearance of new stone should match that of ancient stone after a short exposure time. Accordingly, newly installed limestone blocks on the east side of the south face of A-1 were treated with a series of biocides. Figure 3-12a shows the general area after structural conservation. A close-up of the treated blocks and the biocides used are shown in Figures 3-12b-1 and 3-12b-2, respectively. The appearance of the treated blocks after 6 months, 1 year and 2 years are shown in Figures 3-12c, d and e, respectively. With the exception of stones 6, 7 and 8, and possibly 5, fungal growth covered all new limestone areas in 6 months and only silicate rocks used in surrounding areas seem to be free of microbiological growths. About one year after the biocide treatments, dark fungal growth had covered all of the treated limestone blocks (Fig. 3-12d), with the possible exception of portions of stones 4 to 8 and stone 3, which appeared to be lighter in appearance than the control stone 2 and other untreated stones. The final photograph of the area, taken after 2 years, shows only minor differences between the appearances of treated and untreated limestone blocks, (Fig. 3-12e). At that time, the surfaces of the stones 6, 7 and 8 that had been treated with copper compounds, appeared to be intact with little of the cupping loss shown in

Figure 3-11c. Because the site has not been visited since 1997, it is not known how long this observed biocidal action will continue to be effective.

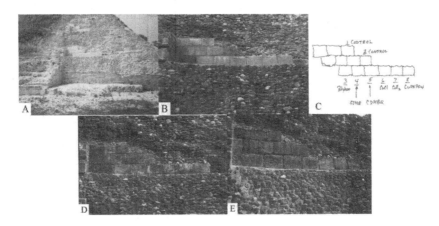

Figure 3-12. Structure A–1: a) south face after new limestone installation; b) after six months; c) map of treated stones d) after one year; e) after two years

In situ excavated walls were used as substrates for a practical test of the ability of biocides to control existing microflora. Various biocide solutions were sprayed on adjacent stone blocks and the regrowth of microflora was followed over the four-year duration of the project. Other adjacent stone blocks served as untreated controls. The wall selected for these tests was located in an area that was partially shaded by high trees and the environment was neither full sun nor deep, rainforest shade, but was intermediate between these extremes. The limestone wall, before treatment, is shown in Figure 3-13a. Biocide solutions that were sprayed on specific individual wall blocks are identified in Figure 3-13c. Degenerative effects on the microflora covering several of the stones were evident within 24 hours.

Photographs of the wall after 13, 20, and 38 months show the progression of effects with time (Fig. 3-13b,d,e). Application of the TBTO-CTMQ solution resulted in the rapid deterioration of existing biological growth that left the stone with a cleaned appearance. No brushing or other mechanical removal of degraded microflora was necessary. The quaternary ammonium compounds, CTMQ and CDMBQ, on the other hand, destroyed much of the microflora but the resulting stone surface exhibited a brownish gray color that was not as starkly different from the original surface as was the TBTO-treated stone. Some regrowth of microflora and broad-leafed plants was also observed.

3. Limestone Stabilization Studies at a Maya Site in Belize

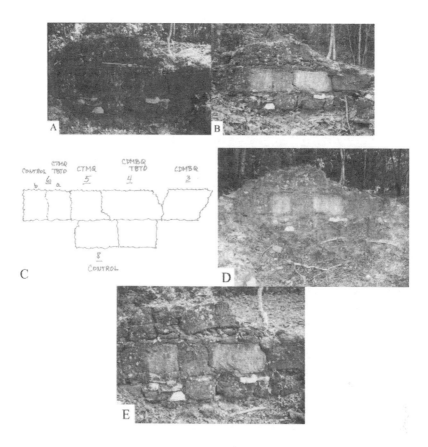

Figure 3-13. Wall A: a) before biocide treatment; b) after 13 months; c) map of biocides used on individual stone blocks; d) after 20 months; e) after 38 months

The superior effectiveness of TBTO in quaternary ammonium salt solutions (CTMQ and CDMBQ) was immediately apparent (stones 4 and 6a) and residual effects continued for at least four years. This may have been due to the relative insolubility of the tin compound, which resulted in a gradual release of the biocide over an extended time period. The two quaternary ammonium compounds alone were somewhat less effective in destroying the surface microflora and, as a result, did not alter the initial limestone's appearance as drastically as did the tin-containing solutions. After 1 to 2 years, the appearance of stones 3 and 5, which had been treated with CDMBQ and CTMQ, respectively, was similar to that of the stones before treatment but it was not determined whether the original microfloral species had returned. After about one year, large-leafed plants started growing back on stone 3 but this did not occur on stone 5 until the third year after treatment. No lichen regrowth was visible.

Polybor acted more slowly than the TBTO-CTMQ or the CTMQ alone. The effects of this biocide were not visible immediately following application but biocidal activity appeared gradually and continued for 2-3 years. Denuding of the stone was not as drastic an effect as was produced by the TBTO treatments.

The copper biocides, Polybor, CTMQ, and CTMQ + TBTO were used to treat another wall in the same area. Some of the stone surfaces, which were partially covered with lichens and fungi, showed distinct changes within 24 hours. After 3 years, no biological regrowth was observed on the TBTO, CuCl, and CuF_2-treated stones but some green fungus layers appeared on the CTMQ, CuSCN, and the Polybor stones. Again, no lichen regrowth was visible.

A room in structure A-5 (adjacent to El Castillo) was selected for an *in situ*, combined biocide-consolidant test (Fig. 3-14a). All walls, except F, were brushed clean and Figure 3-14b shows walls D and E before cleaning and in Figure 3-14c, walls D and E after cleaning. Walls B and D were sprayed with Polybor solution and copper (II) fluoride suspension, respectively. After drying, the biocide-treated walls were consolidated with a 12% solution of Eponex 1510 in isopropanol. Walls A, C, and E were retained as untreated controls. Figure 3-14d shows the condition of the north-facing, Polybor-treated wall B and the east-facing, untreated wall, A, after exposure to the Xunantunich environment for one year. South-facing wall D and east-facing wall E are shown in Figure 3-14e. As expected, the untreated walls, A and E, developed the black, microfloral fungal coating that had been shown to be readily established on clean, porous limestone in sunny areas. Both walls B and D, which had been treated with the biocides and consolidated with the epoxy solution, were hard and firm to the touch and in good condition with no noticeable surface erosion. There was no biological growth on wall B and wall D showed only a slight blackening after one year. Based on the results of the other *in situ* stone and rack tests, the durability of the biocidal action of the two biocides is likely to be 2-4 years. Because the project ended in 1997, no further observations of the A-5 test areas were made to confirm this estimate.

3. Limestone Stabilization Studies at a Maya Site in Belize

Figure 3-14. Structure A–5: a) Plan view of room; b) corner between walls D and E before cleaning; c) walls D and E after cleaning; d) walls A and B after exposure for one year; e) walls D and E after one year

4. CONCLUSIONS

It is somewhat risky to draw unequivocal conclusions from tests in which the control of experimental and environmental parameters was not possible. We frequently came up with conflicting observations that, by virtue of hindsight, we could attribute to some unforeseen, but apparently important parameter variation. This is especially the case when conducting *in situ* field tests where the significance and interrelationships among individual parameters is not known and moreover, are not under our control. As examples; the wind direction and velocity as functions of season and orientation of test samples could influence the probability of infestation by air-borne microflora and hence, the effectiveness of biocides; the cyclic frequency of wet/dry and foggy periods, relative humidity changes, cloud cover, and obscuration of samples by leaves could all affect the residence time of biocides, which is also a function of sample orientation; the environmental parameters may be known accurately at the sensor location but could vary from one local area to another within the site, and so on. In the laboratory aging tests, where parameters can be controlled much more easily, difficulties were still encountered because of stone inhomogenieties. Many replicate observations were necessary to obtain statistically significant data that apply to a particular set of test conditions. Another factor is that a synthetic environment and an arbitrary aging cycle was used in the laboratory tests and to the extent that this was not realistic, the test results may not represent accurately what would happen in the field. These uncertainties and many others need to be considered when evaluating the results of laboratory and *in situ* test data.

So, it is with these factors in mind that the following conclusions are presented:

1. Polymer consolidant films are easily penetrated by microfloral hyphae and rhizinae.
2. Cycloaliphatic epoxy and bisphenol-A epoxy resins, when applied as dilute solutions in water-compatible solvents, can consolidate soft limestone and increase the resistance of historic limestone to particle and water erosion.
3. Cycloaliphatic epoxy resins in isopropanol solution can consolidate moist limestone.
4. A standard bisphenol-A epoxy resin in an organic solvent can consolidate dry limestone.
5. An acrylic/epoxy resin in water solution can consolidate dry limestone.
6. A polyurethane dispersion resin increases erosion resistance of soft limestone.
7. Aqueous acrylic emulsion solutions are not effective as penetrating consolidants for soft limestone.
8. Biocides can be effective when applied before or after consolidation, or when dissolved in compatible aqueous consolidant solutions.
9. SS-H and most likely other silanes are not effective consolidants for soft limestone. SS-H performs as a biocide either because of the presence of an organic, tin-containing catalyst or because of its hydrophobic surface properties.
10. TBTO plus CTMQ in dilute aqueous solution prevents the establishment of microfloral growth on clean limestone for at least 4 years. The quaternary ammonium compounds seem to be effective for 1-2 years.
11. Slightly soluble copper compounds or finely divided copper metal derived from CuCl are effective against existing microflora on limestone and can prevent growth on clean limestone for at least 2 –3 years.
12. A low concentration of aqueous silver nitrate is an effective biocide.
13. Polybor is a useful biocide in the shade but is not effective in the sun or when exposed to rain. It is less effective than copper, silver, or organotin-containing compounds, which are long -lasting under both sunny and shady conditions.

In common with all conservation treatments of historic or culturally significant materials, it is important to stress that preliminary tests should be carried out at unobtrusive locations prior to general application of the procedure. Specifically, the treatment of stone surfaces that are decorated with frescoes or wall paintings requires special care and the application of either consolidants or biocides should not be carried out without extensive research on the possible deleterious consequences of the treatments.

ACKNOWLEDGEMENTS:

The authors would like to express their gratitude and appreciation to the following diverse group of individuals who contributed their skills and knowledge to this effort:

GCI Scientific Program: Frank Preusser, Director; Richard Coffman, Eric Doehne, Gary Mattison, Carlos Navarro, and Michael Schilling.
Consultants: Ralph Mitchell and Ji-Dong Gu, Harvard University; Morgan Phillips, Ron Schmidtling, and Charles Selwitz.
GCI Special Projects Program: Neville Agnew, Director; Martha Demas, Project Leader; Rudy Larios.
UCLA Institute of Archaeology: Richard Leventhal, Director, Xunantunich Archaeological Project, and Staff.
Belize Department of Archaeology: Allan Moore, John Morris, and Harriot Topsey, Commissioners; and Ruben Panados and the Xunantunich Staff.

REFERENCES

Crankovic, B. (1985) *Changes in Porous and Soft Limestone under the Influence of Rainwater.* Durability of Building Materials, No. 2. Elsevier Science Publishers, Amsterdam, pp. 229-242. In Gann, T.W.F. (1928). Maya Cities: A Record of Exploration and Adventure in Middle America. Scribners. New York.

Foster, B., ed. (1992) *Warlords and Maizemen.* Cubola Productions, Benque Viejo del Carmen, Belize.

Ginell, W.S., and Kumar, R. (2002) *Conservation of Maya Limestone at Xunantunich, Belize: Final Report.* Getty Conservation Institute, Los Angeles.

Hale, M.E. Jr. (1980) *Control of Biological Growths on Mayan Archaeological Ruins in Guatemala and Honduras.* In National Geographic Research Reports, 1975 Projects: 305-321. National Geographic Society, Washington D.C.

Kumar, R. and Kumar, A. (1999) *Biodeterioration of Stone in Tropical Environments – An Overview.* Research in Conservation Series, Getty Conservation Institute, Los Angeles.

Kumar, R. and Ginell, W.S. (1995) Evaluation of consolidants for stabilization of weak Maya stone. In: *International Colloquium on Methods of Evaluating Products for the Conservation of Porous Building Materials in Monuments*, ICCROM Rome, 1995.

Maekawa, S. (2000) *Report on the Environmental Monitoring Conducted at Xunantunich, Belize.* Getty Conservation Institute Report, Los Angeles.

Martin, A.K. and Johnson, G.C. (1992) Chemical control of lichen growths established on building materials: a compilation of the published literature. *Biodeterioration Abstracts* 6: 101-117.

Mitchell, R. and, Gu, Ji-Dong. (1997) *Microbial Colonization of GCI Consolidant Polymers and Inhibition of Microbial Growth by Biocides.* Microbial Ecology Laboratory Report, Harvard University, Cambridge, Mass.

Mitchell, R., Gu, Ji-Dong, and Ginell, W.S. (1996) Biofilms and degradation of protective coatings. In: Proceedings of the 8th International Union Microbiology Society Congress Jerusalem. 1996.

Phillips, M. (1982) Acrylic precipitation consolidants. In: *Science & Technology in the Service of Conservation* (edited by Norman S. Brommelle and Garry. Thomson): 56-60. International Institute for Conservation of Historic and Artistic Works, London.

Phillips, M. (1995) Aqueous acrylic/epoxy consolidants, *The Journal of Preservation Technology* 26 (2-3): 68-75.

Remy, H. (1961) *Treatise on Inorganic Chemistry.* Volume 2. Elsevier Publishing, New York.

Richardson, B.A. (1976) Control of moss, lichen and algae on stone. In: *Conservation of Stone 1: Proceedings of the International Symposium, Bologna, June 19-21, 1975* (R. Rossi-Manaresi ed.): 225-231. Centro Per La Conservazione Della Sculpture All'aperto, Bologna, Italy.

Richardson, B.A. (1988) Control of microbial growths on stone and concrete. In *Biodeterioration* 7: 101-106.

Robertson, W.D. (1982) Evaluation of the durability of limestone masonry in historic buildings. In: *Science and Technology in the Service of Conservation Science.* Brommelle, Norman S. and Thomson, Garry. : 51-55. Preprints of the Contributions to the Washington Congress, 3-9 September 1982. The International Institute for Conservation of Historic and Artistic Works. London, England.

Scott, D. A. (2002) *Copper and Bronze in Art: Corrosion, Colorants, Conservation.* Getty Conservation Institute, Los Angeles.

Chapter 4

LICHENS AND THE BIODETERIORATION OF STONEWORK: THE ITALIAN EXPERIENCE

ROSANNA PIERVITTORI
Dipartimento di Biologia Vegetale, Università di Torino viale Mattioli, 25, I-10125 Turin, Italy e-mail: rosanna.piervittori@unito.it

Abstract: Lichens play an important role as biogeophysical and biogeochemical agents in the degradation of stone surfaces, particular problems arising at the lichen-substratum stratum interface on stonework of artistic value. This work presents a detailed, mainly bibliographical, review of research carried out to date on the lichen biodeterioration of Italian archeological and monumental stonework. This survey presents the main methodologies used in this field: systematic and ecological studies, lichen-substratum interface qualitative analyses, chemical and physical mechanisms of deterioration by lichens, calcium oxalate films and methods of prevention and control. Lichens have also been investigated in terms of their dispersion by air: sampling methods for the quantitative assessment of the presence of airborne propagules in areas of artistic and archeological interest are proposed.

Key words: lichens; biodeterioration; stonework; bibliographic review; Italy

1. INTRODUCTION

The ability of lichens to colonize rock surfaces takes on greater significance when the colonization occurs on stonework of historical or architectural value. In this case the process of alteration and disintegration of rocks is of primary interest because it can sometimes create serious problems for their restoration and conservation. Lichens which colonize stonework can, in some cases, cause clearly visible manifestations resulting from both chemical and physical alterations of the lithic substrata.

2. LICHENS AND MONUMENTS IN ITALY

Italy is undoubtedly a country with an extremely rich artistic and monumental heritage, with a very considerable quantity of stonework exposed to the action of abiotic and biotic agents. However, only recently have conservationists and researchers from the Agency for the Supervision of Environmental and Architectural Heritage started to tackle the question of biodeterioration of worked stone in the necessary interdisciplinary way. The conservation of such materials also requires specialist biological expertise, including a knowledge of lichens. The presence of lichens in archeological areas and on monuments is normally conditioned by a number of factors, inclduing exposure, type of substratum, angle of the surfaces and contribution of nitrogenous substances. An adequate knowledge of the biological and ecological features of these organisms is therefore necessary to evaluate the most suitable methods of intervention to prevent, control and, if necessary, eliminate the lichen cover (Nimis *et al.*, 1992; Caneva *et al.*, 1994, 1996). To obtain such results it is becoming increasingly necessary to have adequate training. In this area the 'Società Lichenologica Italiana' (S.L.I.) has played a fundamental role since its foundation in 1987.

Certainly a positive signal has come more recently from the academic world with the institution of a degree course in Cultural Heritage. It is interesting to note the initiative of the University of Turin to introduce a course in lichenology, which deals specifically with aspects of biodeterioration, among the compulsory subjects of a new three-year degree course in Science and Technology for Cultural Heritage (http://www.stbeniculturali.unito.it).

This paper presents a synthesis of studies conducted to date in Italy on the subject of lichens and stonework in archaeological and monumental areas on the basis of the blibliographical data collected by the author, as well as some reviews on the problems dealt with.

3. BIODETERIORATION STUDIES IN ITALY

The problem of the biodeterioration of monuments was recognized as early as the first decades of the 20th century, when a study highlighted the presence of lichens on windows of Italian churches in Aosta, Arezzo, Bologna, Cuneo and Verona. At the same time the first rudimentary advice was given on the subject of restorative conservation to treat what was then considered as a fully-fledged "disease of ancient window panes", namely to protect "the window panes from the invasion of lichens, with careful washing with pure water or better with soapy water or water containing another antiseptic, which cannot react with the glass, and paying special attention to the cleaning and touching up of the leadwork". With a certain foresight the same

4. Lichens and the Biodeterioration of Stonework

scholar, in a previous work, expressed the desirability of an inter-disciplinary approach to the subject of monuments and works of art.

Yet the problems concerning lichens and monuments only aroused special certain interest in the early 1970s. This fact can be correlated to the plight of lichenology in Italy that had negative repercussions on the applied side of the discipline too. Thus in this period when the authorities had to tackle problems of the biodeterioration of monuments correlated to the presence of lichens, foreign experts were called in as consultants; of particular importance was the collaboration of the British lichenologist Professor Mark Seaward with the Rome-based 'Istituto Centrale per il Restauro'.

Lichenological research started to pick up again towards the mid-1970s, but it was only with the founding of the 'Società Lichenologica Italiana', as has been mentioned earlier, that there was a real reawakening of interest, documented by an increasingly intense and diversified production of scientific papers.

From a critical analysis of the research published in the last thirty years it has been possible to identify the main methodologies used in this field (the classification of which corresponds to the titles of the following sections and the layout of the bibliographical lists):

1. systematic and ecological studies [3.1]
2. lichenometry [3.2]
3. lichen-substratum relationships: qualitative analyses, chemical and physical mechanisms of deterioration by lichens [3.3]
4. calcium oxalate films [3.4]
5. methods of prevention and control [3.5].

From a geographical point of view Latium (Central Italy) and Venetia (North-East Italy) are the regions where the highest number of studies have been conducted.

3.1 Systematic and ecological studies

The importance of correctly determining the species, as a preliminary phase to each subsequent conservation intervention, is an accepted and fundamental fact. If the early works provide exclusively floristic analyses, most research work in this field has tended to compare the presence and distribution of lichens with ecological factors, which have been indexed and later quantified in a series of values expressed on an ordinal scale, related to the acidity of the substratum (pH index), eutrophication (nitrophytism index), moisture (hygrophytism index) and light (photophytism index).

Floristic and vegetational studies conducted on stonework, if supported by ecological analysis, allows us to acquire, rapidly and economically, biological and environmental data for a more considered appraisal of the specific action to be taken. This approach has been tried out for the first time in 16 areas of archaeological interest in Latium (central Italy). The study allows the acquisition of a series of data: floristic (284 *taxa* at an infrageneric level were identified), vegetational and ecological (especially regarding the rich variety of local and imported building material used), which are fundamental for the continuation of investigations into lichen-induced biodeterioration of stonework. The validity of the use of ecological indices is confirmed by the conservation interventions on the monumental buildings, in that they allow the acquisition of a considerable set of data on interacting abiotic factors, such as in the case of Orvieto Cathedral (Umbria, central Italy). Since then this methodological approach has been widely used throughout Italy.

However, while applying this approach it has been evident that the use of ecological indices, formulated for the climate conditions of Central Europe, can cause problems when applied to investigations in the Mediterranean area. An in-depth study conducted in Sardinia on megalithic works ("nuraghi") has provided an opportunity to introduce a new hygrophytism index expressed on an ordinal scale.

The pH index has also been revised recently since it commonly refers not only to terricolous and corticicolous lichens but, incorrectly, to saxicolous ones as well. Most lichen coenosis found on the ground and on tree bark are closely linked to the pH of the substratum. The same specificity has also emerged for saxicolous lichens, even if the pH cannot be measured easily on the lithic substratum.

Towards the end of the 1980s the organization in Rome of an international conference on the subject "Lichens and Monuments" offered for the first time the occasion for an interesting debate between Italian and foreign scholars, and those who deal with the question of biodeterioration within the Agency for the Supervision of Environmental and Architectonic Heritage and restoration workshops, etc.

Altieri, A., Mazzone, A., Pietrini, A.M., Ricci, S. and Roccardi, A. (2000) Indagini diagnostiche sul biodeterioramento delle fontane. In: *Piazza di Corte, il recupero dell'immagine berniniana* (M. Natoli, ed.): 34-57. Ministero per i Beni e le Attività Culturali, Palombi Editore, Roma.

Altieri, A., Pietrini, A.M., Ricci, S., Roccardi, A. and Piervittori, R. (2000) The temples of the archaeological area of Paestum (Italy): a case study on biodeterioration. In: Proceedings of 9th International Congress on *Deterioration and Conservation of Stone* (V. Fassina, ed.), (Venice, 19-24 June 2000): 433-451. Elsevier, Amsterdam.

Andreoli, C., Caniglia, G., Salvadori, O. and Scapin, C. (1982) Studi preliminari su due forme licheniche sviluppantesi sulle colonne della Basilica di Santa Maria Assunta a Torcello (VE). *Giornale Botanico Italiano* 116: 166-167.

Baroni, E. (1893) Notizie e osservazioni sui rapporti dei licheni calcicoli col loro sostrato. *Bullettino della Società Botanica Italiana* 1893: 136-140.

Bartoli, A. (1990) I licheni della Peschiera dei Tritoni nell'Orto Botanico di Roma, Villa Corsini. *Giornale Botanico Italiano* 125: 87.

4. Lichens and the Biodeterioration of Stonework

Bartoli, A. (1992) Flora e vegetazione lichenica del Porto di Traiano. In: *Il parco archeologico-naturalistico del porto di Traiano*. Ministero Beni Culturali Ambientali, Soprintendenza Archeologica di Ostia: 167-171. Edizioni Gangemi, Roma.

Bartoli, A. (1997) I licheni del Colosseo. *Allionia* 35: 59-67.

Bartoli, A. (1997) I licheni del Colosseo. In: Abstracts del Convegno annuale della Società Lichenologica Italiana *Licheni e Ambiente* (D. Ottonello, ed.), (Palermo, 9-12 Dicembre 1995). *Notiziario della Società Lichenologica Italiana* 10: 67.

Bartoli, A., Massari, G. and Ravera, S. (1997) The lichens of Munazio Planco's Mausoleum (Gaeta). In: Proceeding of 3rd Symposium *Progress and Problems in Lichenology in the Nineties* (R. Türk and R. Zorer, eds.). *Bibliotheca Lichenologica* 68: 145.

Bartoli, A., Massari, G. and Ravera, S. (1998) The lichens of the Mausoleum of *Manatius Plancus* (Gaeta). *Sauteria* 9: 53-60.

Caneva, G. and Roccardi A. (1991) Harmful flora in the conservation of Roman Monuments. In: Proceeding of the International Conference on *Biodeterioration of Cultural Property* (O.P. Agrawal and S. Dhawan, eds.), (Lucknow-India, February 20-25 1989): 212-218. Macmillan, Delhi.

Caneva, G., Roccardi, A., Marenzi, A. and Napoleone, I. (1985) Proposal for a data base on biodeterioration of stone artworks. In: Proceedings of the 5th International Congress *Deterioration and Conservation of Stone*, vol. 2 (G. Félix, ed.): 587-596. Presses Polytechniques Romandes, Lausanne.

Caneva, G., Roccardi, A., Marenzi, A. and Napoleone, I. (1989) Correlation analysis in the biodeterioration of stone artworks. *International Biodeterioration* 25: 161-167.

Caneva G., Nugari, M.P. and Salvadori, O. (1994) *La Biologia nel Restauro*. Nardini Editore, Firenze. [– also 3.3, 3.4, 3.5]

Caneva, G., Gori, E. and Montefinale, T. (1995) Biodeterioration of monuments in relation to climatic changes in Rome between 19th-20th centuries. *The Science of the Total Environment* 167: 205-214.

Caneva, G., Nugari, M.P. and Salvadori, O. (1991) *Biology in the Conservation of Works of Art*. ICCROM, Roma.

Caniglia, G., Cornale, G. and Salvadori, O. (1993) The lichens on the statues in the orangery of "Villa Pisani" at Stra (Venezia). *Giornale Botanico Italiano* 127: 620.

Cengia-Sambo, M. (1939). Licheni che intaccano i mosaici fiorentini. *Nuovo Giornale Botanico Italiano* 46: 141-145.

Ciarallo, A., Festa, L., Piccioli, C. and Raniello, M. (1985) Microflora action in the decay of stone monuments. In: Proceedings of the 5th International Congress *Deterioration and Conservation of Stone*, vol. 2 (G. Félix, ed.): 607-616. Press Polytechniques Romandes, Lausanne.

Del Monte, M. (1989) I monumenti in pietra e i licheni. *Rassegna dei Beni Culturali* 3: 12-17.

Del Monte, M. (1991) Trajan's column: lichens don't live here any more. *Endeavour* 15: 86-93.

Di Benedetto, L. and Grillo, M. (1995) Contributo alla conoscenza dei biodeteriogeni rilevati nel complessso archeologico del Teatro greco-romano ed Anfiteatro romano di Catania. *Quaderni di Botanica Ambientale Applicata* 6: 61-66.

Di Francesco, C., Grillini, G.C., Pinna, D. and Tucci, A. (1989) Il restauro come occasione di studio e documentazione nel caso del campanile del Duomo di Ferrara. In: Atti del Convegno di Studi, *Il Cantiere della Conoscenza, il Cantiere del Restauro* (G. Biscontin, M. Dal Col and S. Volpin, eds.), (Bressanone 1989): 251-264. Libreria Progetto Editore, Padova.

Gallo, L.M. and Piervittori, R. (1991) La flora lichenica rupicola dei Monti Pelati di Baldissero (Canavese, Piemonte). In: Atti del Convegno su *I Monti Pelati di Baldissero. Importanza Paesistica e Scientifica* (P.M. Giachino, ed.): 25-31. Feletto (Torino).

Gallo, L.M., Piervittori, R. and Montacchini, F. (1989) Rapporti con il substrato litologico degli esemplari del gen. *Rhizocarpon* presente nelle Collezioni crittogamiche dell' *Herbarium Horti Taurinensis* (TO). *Bollettino Museo Regionale di Scienze Naturali di Torino* 7: 129-156.

Gargani, G. (1971-1972) Fattori biologici nel degradamento delle opere d'arte. In: Vita e decadenza delle opere d'arte. In: Atti dell'*Accademia di Scienze di Ferrara* 49: 133-142.

Giacobini, C. and Roccardi, A. (1984) Indagini sui problemi biologici connessi alla conservazione dei manufatti storico-artistici. *Restauri nel Polesine*: 34-35. Electa Editrice, Rovigo.

Giacobini, C. and Seaward, M.R.D. (1991) Licheni e monumenti: studi in Veneto e in Puglia. In: Atti del Convegno *Scienza e Beni Culturali, Le pietre nell'architettura: Struttura e Superfici* (G. Biscontin, ed.): 215-224. Libreria Progetto Editrice, Padova.

Giacobini, C., Nugari, M.P., Micheli, M.P., Mazzone, B. and Seaward, M.R.D. (1986) Lichenology and the conservation of ancient monuments: an interdisciplinary study. In: *Biodeterioration 6* (S. Barry, D.R. Houghton, G.C. Llewellyn and C.E. O'Rear, eds.): 386-392. C.A.B. International Mycological Institute, Slough.

Giacobini, C., Roccardi, A., Bassi, M. and Favali, M.A. (1986) The use of electronic microscopes in research on the biodeterioration of works of art. In: Proceedings of the Symposium *Scientific Methodologies Applied to Works of Art* (P. Parrini, ed.): 71-75. Florence.

Giacobini, C., Pietrini, A.M., Ricci, S. and Roccardi, A. (1987) Problemi di biodeterioramento. *Bollettino d'Arte* 41: 53-64.

Giacobini, C., Pedica, M. and Spinucci, M. (1989) Problems and future projects on the study of biodeterioration: mural and canvas paintings. In: Proceeding of the International Conference *Biodeterioration of Cultural Property* (O.P. Agrawal and S. Dhawan, eds.), (Lucknow-India, 20-25 February 1989): 275-286. Macmillan, Delhi.

Guidobaldi, F., Laurenzi Tabasso, M. and Meucci, C. (1984) Monumenti in marmo di epoca imperiale a Roma: indagine sui residui di trattamenti superficiali. *Bollettino d'Arte* 24: 121-134.

Laccisaglia, A., Gallo, L.M. and Piervittori, R. (1994) Criteri terminologici relativi ad acidità e basicità nello studio dei rapporti licheni-substrato litico. *Notiziario Società Lichenologica Italiana* 7: 33-41.

Lazzarini, L. (1979) I rilievi degli arconi dei portali della Basilica di San Marco a Venezia: ricerche tecnico scientifiche. In: *Die Sculpturen Von San Marco in Venedig*: 58-65. Deutscher Kunstverlag, München.

Lloyd, A.O. (1974) Lichen attack on marble at Torcello - Venice. In: Atti del Convegno *Petrolio e Ambiente*: 221-224. Editore Artioli, Modena.

Marchese, E.P., Di Benedetto, L., Luciani, F., Razzara, S., Grillo., M., Stagno F. and Auricchia, A. (1997) Biodeteriogeni vegetali di monumenti del centro storico della città di Noto. *Archivi di Geobotanica* 3: 71-80.

Mattirolo, O. (1928) I licheni e la malattia delle vetrate antiche. *Rivista Archeologica della Provincia e Antica Diocesi di Como* 94-95: 1-23.

Monte, M. (1991) La lichenologia applicata alla conservazione dei monumenti in pietra esposti all'aperto: problemi e prospettive. In: Atti del Convegno *Scienza e Beni Culturali, Le pietre nell'Architettura: Struttura e Superfici* (G. Biscontin, ed.): 287-298. Libreria Progetto Editore, Padova.

4. Lichens and the Biodeterioration of Stonework

Monte, M. (1991) Lichens on the ruins of a Roman aqueduct. *Biodeterioration and Biodegradation* 8: 394-396.
Monte, M. (1991) Multivariate analysis applied to the conservation of monuments: lichens on the Roman aqueduct Anio Vetus in S. Gregorio. *International Biodeterioration* 28: 133-150.
Monte, M. and Ferrari, R. (1996) Biodeterioration of the temple of Segesta (Sicily). In: Proceedings of the International Congress on *Deterioration and Conservation of Stone* (J. Riederer, ed.): 585-591. Möller, Berlin.
Monte, M. and Tretiach, M. (1992) Licheni sui Nuraghi: un cantiere di ricerca. In: Atti del Convegno *Scienze dei Materiali e Beni Culturali. Esperienze e prospettive nel restauro delle costruzioni nuragiche* (C. Atzeni and U. Sanna, eds.): 73-81.
Moriconi, G., Castellano, M.G. and Collepardi, M. (1994) Dégradation de mortiers de murs en mançonnierie dans les édifices historiques. Un exemple: Le Mole de Vamvitelli à Ancona. *Materials and Structure* 27: 408-414.
Nascimbene, J. (1997) Licheni e conservazione dei monumenti. Un'esperienza in campo didattico. In: Abstracts del Convegno annuale della Società Lichenologica Italiana *Licheni e Ambiente* (D. Ottonello, ed.), (Palermo, 9-12 Dicembre 1995). *Notiziario della Società Lichenologica Italiana* 10: 55-56.
Nimis, P.L. (1999) Opere d'arte e di storia: ecosistemi minacciati. In: *Frontiere della Vita* 4: 531-541. Istituto della Enciclopedia Italiana, Roma.
Nimis, P.L. (2001) Artistic and historical monuments: threatened ecosystems. In: *Frontiers of Life*, Part 2 - *Discovery and Spoilation of the Biosphere*, sect. - *Man and the Environment*: 557-569. Academic Press, San Diego.
Nimis, P.L. and Monte, M. (1988) The lichen vegetation on the cathedral of Orvieto (Central Italy). *Studia Geobotanica* 8: 77-88.
Nimis, P.L. and Monte, M., eds. (1988) Lichens and monuments. *Studia Geobotanica* 8: 1-133. [– also 3.3, 3.4, 3.5]
Nimis, P.L. and Zappa, L. (1988) I licheni endolitici calcicoli su monumenti. *Studia Geobotanica* 8: 125-133.
Nimis, P.L., Monte, M. and Tretiach, M. (1987) Flora e vegetazione lichenica di aree archeologiche del Lazio. *Studia Geobotanica* 7: 3-161.
Nimis, P.L. and Martellos, S. (2001) *Checklist of the lichens of Italy 3.0*. Department of Biology, IN2.0/2, University of Trieste (http://dbiodbs.univ.trieste.it).
Nimis, P.L., Pinna, D. and Salvadori, O. (1992) *Licheni e Conservazione dei Monumenti*. CLUEB, Bologna. [– also 3.3, 3.4, 3.5]
Nimis, P.L., Seaward, M.R.D., Ariño, X. and Barreno, E. (1998) Lichen-induced chromatic changes on monuments: a case-study on the Roman amphitheater of Italica (S. Spain). *Plant Biosystems* 132: 53-61.
Normal-3/80 (1980) *Materiali Lapidei: Campionamento*. C.N.R.-I.C.R, Roma.
Normal-19/85 (1985) *Microflora Autotrofa ed Eterotrofa:Tecniche di Indagine Visiva*. C.N.R.-I.C.R, Roma.
Normal-1/88 (1990) *Alterazioni Macroscopiche dei Materiali: Lessico*. C.N.R.-I.C.R, Roma.
Normal-30/89 (1991) *Metodi di Controllo del Biodeterioramento*. C.N.R.-I.C.R, Roma.
Not, R. and Ottonello, D. (1989) Osservazioni sulla colonizzazione lichenica e briofitica del Tempio della Vittoria (Sicilia settentrionale). *Giornale Botanico Italiano* 123: 161.
Ottonello, D. (1990) Osservazioni preliminari sulla flora lichenica gipsicola. *Giornale Botanico Italiano* 124: 91.
Ottonello, D., Alaimo, R., Calderone, S. and Montana G. (1991) Contributo alla conoscenza del rapporto tra i licheni e i substrati litici. *Giornale Botanico Italiano* 125: 263.

Paleni, A. and Curri, S. (1972) Biological aggression of works of art in Venice. In: *Biodeterioration of Materials 2* (A.H. Walters and H. van der Plas E.H., eds.): 392-400. Applied Science Publishers, London.

Paleni, A. and Curri, S. (1972) L'aggression des algues et des lichens aux pierres et moyens pur la combattre. In: 1er Colloque International sur la *Deterioration des Pierres en Oeuvre*: 157-166. La Rochelle, France.

Paleni, A., Curri, S. and Benassi, R. (1973) Aggressione biologica ad una statua del 700 in marmo di Carrara in ambiente collinare campestre. In: Atti del Convegno *Petrolio e Ambiente, sezione Arte*: 29-50.

Piervittori, R. (1992) I popolamenti lichenici rupicoli calcifughi nel settore occidentale delle Alpi (Piemonte e Valle d'Aosta). *Biogeographia* 16: 91-104.

Piervittori, R. and Favero Longo, S. (2002) Fenomenologia dei licheni degli ambienti serpentinitici. In: Atti del Convegno Nazionale *Le ofioliti: isole sulla terraferma* (A. Saccani, ed.), (Fornovo Taro-Parma, 22-23 Giugno 2001): 73-81. Graphital, Parma.

Piervittori, R. and Sampò S. (1987-1988) Colonizzazione lichenica su manufatti litici: la facciata dell'Abbazia di Vezzolano. Asti (Piemonte). *Allionia* 28: 93-101.

Piervittori, R. and Sampò, S. (1988) Lichen colonization on stoneworks: examples from Piedmont and Aosta Valley. *Studia Geobotanica* 8: 73-75.

Piervittori, R., Sampò, S., Appolonia, L., Gallo, L.M. and Polini, V. (1990) Caractéres écologiques d'éspéces licheniques dans les châteaux de la Vallée d'Aoste (Italie). In: Book of Abstracts, VI OPTIMA Meeting (Delphi, 10-16 September 1989): 124.

Piervittori, R., Laccisaglia, A., Appolonia, L. and Gallo, L.M. (1991) Aspetti floristico-vegetazionali e metodologici relativi ai licheni su materiali lapidei in Valle d'Aosta. *Revue Valdôtaine d'Histoire Naturelle* 45: 53-86.

Pinna, D. and Salvadori, O. (1999) Biological growth on Italian monuments restored with organic or carbonatic compounds. In: Book of Abstracts, International Conference *Microbiology and Conservation (ICMC '99) Of Microbes and Art: The role of microbial communities in the degradation and protection of cultural heritage* (O. Ciferri, P. Tiano and G. Mastromei, eds.): 149-154. C.N.R., Firenze.

Poli-Marchese, E., Razzara, S., Grillo, M. and Galesi, R. (1990) Indagine floristica e restauro conservativo di San Nicola l'Arena di Nicolosi (Etna). *Bollettino Accademia Gioenia di Scienze Naturali Catania* 23: 707-720.

Poli-Marchese, E., Luciani, F., Razzara, S., Grillo, M., Auricchia, A. and Stagno F. (1995) Biodeteriogeni di origine vegetale causa del degrado del complesso monumentale dei Benedettini di Catania. *Giornale Botanico Italiano* 129: 58.

Poli-Marchese, E., Di Benedetto, L., Luciani, F., Grillo, M., Auricchia, A. and Stagno, F. (1996) Indagine sui vegetali causa di degrado dei monumenti della città di Noto (Sicilia orientale). *Giornale Botanico Italiano* 130: 457.

Poli-Marchese, E., Di Benedetto, L., Luciani, F., Razzara, S., Grillo, M., Stagno, F. and Auricchia, A. (1997) Biodeteriogeni vegetali di monumenti del centro storico della città di Noto. *Archivio Geobotanico* 3: 71-80.

Poli-Marchese, E., Luciani, F., Razzara, S., Grillo, M., Auricchia, A., Stagno, F., Giacone, G. and Di Martino, V. (1995) Biodeteriorating plant entities on monument and stoneworks structures historic city centre of Catania "Il Monastero dei benedettini". In: Proceedings of the 1st Congress on *Science and Technology for the Safeguard of Cultural Heritage in the Mediterranean Basin*: 1195-1203. Catania.

Realini, M., Mioni, A., Favali, M. A. and Fossati, F. (1994) Lichen-stone surface interaction under different environmental conditions. *Giornale Botanico Italiano* 128: 363.

4. Lichens and the Biodeterioration of Stonework

Rigoni, C. (1990) Le statue di Villa Cordellina. Problemi e indagini sui licheni. In: Atti della Giornata di Studio *Il Prato della Valle e le opere di pietra calcarea collocate all'aperto* (S. Borsella, V. Fassina and A.M. Spiazzi, eds.): 227-232. Libreria Progetto Editore, Padova.

Roccardi, A. and Bianchetti, P. (1988) The distribution of lichens on some stoneworks in the surroundings of Rome. *Studia Geobotanica* 8: 89-97.

Rossi Manaresi, R., Tucci, A., Grillini, G.C., Pinna, D. and Di Francesco, C. (1989) Indagini multidisciplinari per lo studio di un monumento esemplare: casa Romei a Ferrara. In: Atti del Convegno di Studi *Scienza e Beni Culturali, Il Cantiere della Conoscenza, il Cantiere del Restauro* (G. Biscontin, M. Dal Col and S. Volpin, eds.): 403-416. Libreria Progetto Editore, Padova.

Salvadori, O. and Lazzarini, L. (1991) Lichen deterioration on stones of Aquileian monuments (Italy). *Botanika Chronika* 10: 961-968.

Seaward, M.R.D. (1988) Lichen damage to ancient monuments: a case study. *The Lichenologist* 20: 291-295.

Seaward, M.R.D. and Giacobini, C. (1988) Lichen-induced biodeterioration of Italian monuments, frescoes and other archeological materials. *Studia Geobotanica* 8: 3-11.

Seaward, M.R.D. and Giacobini, C. (1989) Lichens as biodeteriorators of archaeological materials, with particular reference to Italy. In: Proceeding of the International Conference *Biodeterioration of Cultural Property* (O. P. Agrawal and S. Dhawan, eds.), (Lucknow-India, 20-25 February 1989): 195-206. Macmillan, Delhi.

Seaward, M.R.D., Capponi, G. and Giacobini, C. (1989) Biodeterioramento da licheni in Puglia. In: Proceedings of the 1st International Symposium *La conservazione dei monumenti nel bacino del Mediterrano* (F. Zezza, ed.): 243-245. Grafo Edizioni, Bari.

Seaward, M.R.D., Giacobini, C., Giuliani, M.R. and Roccardi, A. (1989) The role of lichens in the biodeterioration of ancient monuments with particular reference to Central Italy. *International Biodeterioration* 25: 49-55. [Reprinted in *International Biodeterioration and Biodegradation* 48, 2001, 202-208.]

Seaward, M.R.D., Giacobini, C. and Roccardi, A. (1990) I licheni a Villa Cordellina. In: Catalogo *Problemi di salvaguardia e restauro, Le sculture del Giardino. Vicende e problemi di conservazione* 327-328. Electa, Milano.

Tiano, P. (1986) Biological deterioration in stone on exposed works of art. In: *Biodeterioration of Constructional Materials* (L.G.H. Morton ed.). *The Biodeterioration Society Occasional Publication* 3: 37-44.

Tiano, P. (1986) Problemi biologici nella conservazione del materiale lapideo esposto. *La Prefabbricazione* 22: 261-272.

Tiano, P. (1987) Problemi biologici nella conservazione delle opere in marmo esposte all'aperto. In: *Restauro del marmo-Opere e problemi* (A. Giusti ed.). Quaderni dell'Opificio delle Pietre Dure e Laboratori di Restauro di Firenze, 47-53. Opus Libri, Firenze.

Tiano, P. (1991) Problemi biologici nella conservazione del patrimonio culturale. *Kermes* 10: 56-73.

Tretiach, M. and Monte, M. (1991) Un nuovo indice di igrofitismo per i licheni epilitici sviluppato sui nuraghi di granito della sardegna nord-occidentale. *Webbia* 46: 183-192.

Tretiach, M., Monte, M. and Nimis, P.L. (1991) A new hygrophytism index for epilithic lichens developed on basaltic nuraghes in NW Sardinia (Italy). *Botanika Chronika* 10: 953-960.

Zitelli, A. and Salvadori, O. (1982) Studio dell'azione biodeteriogena e descrizione del lichene colonizzante le colonne della Basilica di Santa Maria Assunta dell'isola di Torcello

(VE) - Considerazioni e proposte. *Rapporti e Studi, Istituto Veneto di Scienze, Lettere e Arti* 8: 153-169.

3.2 Lichenometry

Lichens are characterized by slow growth and longevity. These two important ecophysiological characteristics, which are the result of interaction between numerous environmental and physiological factors, can supply precious information on the age of the colonized surface. Lichenometry, the use of lichens as a dating instrument, is based on geometric-type methodologies. The lichenometric method is basically applied on Holocene dating and all the papers published are not mentioned here, because they are out of the aim of this review [for comprehensive review see Gallo and Piervittori (1993)]. Nevertheless this technique has been reported here, because it could assume great importance if applied to chronological reconstruction in areas of historical and architectural interest.

Gallo, L.M. and Piervittori, R. (1993) Lichenometry as a method for holocene dating: limits in its applications and realibility. *Il Quaternario* 6: 77-86. [– also 5]

Gallo L.M. and Piervittori R. (1995) Lichenometria: sintesi e limiti attuali della teoria. In: Abstracts del Convegno annuale della Società Lichenologica Italiana *Biologia dei licheni: stato dell'arte sull'avanzamento delle conoscenze Lichenologiche in Italia* (R. Piervittori and M. Gallo, eds.), (Torino, 13-15 Ottobre 1994). *Notiziario della Società Lichenologica Italiana* 8: 87-89.

Politi, M.A. (1989) Datare i monumenti studiando i licheni. *Scienza & Vita Nuova* 6: 136-139.

3.3 Lichen-substratum relationships: qualitative analyses, chemical and physical mechanisms of deterioration by lichens

The colonization of a rock surface by a lichen is controlled by factors which bring about the mechanical deterioration (*physical factors*) of the lithic substratum, and its alteration and decomposition (*chemical factors*). Because of this lichens play an important ecological role as biogeophysical and biogeochemical agents in the weathering of stone surfaces. It has been amply demonstrated that rocks and minerals subjected to lichen metabolism (the production of lichen acids and other secondary metabolites) undergo chemical and morphological changes. On the other hand, the physical (structure and hardness of the mineral components), chemical (lithochemism) and ecological (slope and exposure) features of the lithic substratum can influence the colonization and distribution of lichens.

The effects of colonization on stone surfaces can cause serious conservation problems when this occurs on stoneworks and works of art. In order to arrive a better knowledge of lichen-surface relationships, studies on saxicolous crustose forms are important. Their close contact with the rock surface allows us to evaluate the effects of a single organism on the various minerals which make up the rocks. For a better understanding of the role of lichens in deterioration, especially of the mechanisms they set off, investigations performed in natural environments have also proved very useful. In this category we must include research conducted to evaluate the specificity of silicicolous epilithic forms on serpentine rocks, gneiss in the alpine ambient, on carbonate rocks such as pure limestone and dolomite and on volcanic and mafic rocks in the Mediterranean region.

The great advances made in rock and mineral lichen-induced bioalteration use, often in a complementary way, various thin and ultra-thin section techniques, optical (OM) and electron (SEM/TEM) microscopy, microanalysis (EDXRA), X-ray diffractometry (XRD), infrared (FTIR) and Raman spectroscopy, as well as analytical chemistry investigations.

Observation of thin sections under the polarising microscope is an essential step in evaluating the deterioration of the minerals by lichens. In most cases the preparation of the sections (average thickness 30 µm) follows traditional methods: embedding in glutaraldehyde, dehydration in the increasing series of alcohols and embedding in epoxy resins. Some authors have introduced certain methodological changes regarding the preparation (consolidation of the samples in cyanoacrylate and later cold embedded in araldite), which speeds up the process, and microscopical observation using certain accessories (1/4 compensator and first order red interference color gypsum plate) allows the simultaneous examination of the lichen component and the inorganic one.

Particularly effective for observing endolithic thalli in limestone is the technique which employs the staining of polished sections with Periodic Acid Schiff (PAS) and with methylene blue for later observation in stereo- and reflecting microscopes.

Biogeophysical processes may cause the disintegration of the lithic components and are generally due to the penetration of hyphae, rhizines or other thallus parts into the rock. Species such as *Caloplaca crenularia, Candelariella vitellina* and *Diploschistes actinostomus* have hyphae up to 1.5-2.0 mm in certain carbonate sandstones of central Italy ("pietra serena"), while the hyphae of *Lecidea fuscoatra* can penetrate up to 4-5 mm in volcanic clastites. For some endolithic forms (*Caloplaca ochracea, Verrucaria marmorea, Verrucaria sphincritella*), hyphal penetrations up to 1.5-2.0 mm have been observed in dense limestone.

Of further importance are the effects on lichen morphology linked with the phenomena of hydration and rehydration (revivescence) of gelatinous and mucilaginous substances contained within thalli which may induce peeling and pitting. The studies conducted have shown that for a better understanding of the phenomenon of biodeterioration it is essential to focus on the interactions that exist between lichens and the rock substratum. The biogeochemical processes may determine the decomposition and transformation of the substratum. The production of lichen substances determines morphological alteration, corrosion (for example the endolithic lichens are particularly active in dissolving carbonatic substrata) and the release of cations in the rock minerals. Some researchers have focused their attention on the neogenesis of numerous mineralogical units (ferric oxalates, oxides and hydroxides, aluminum silicates).

Adamo, P. (1996) Ruolo dell'attività dei licheni nell'alterazione di substrati rocciosi e nella neogenesi di entità mineralogiche. In: *Atti XIII Convegno Nazionale della Società Italiana di Chimica Agraria*: 13-26. Patron Editore, Bologna.

Adamo, P. (1997) Bioalterazione di rocce di natura vulcanica, dolomitica e mafica indotta da licheni. In: Abstracts del Convegno annuale della Società Lichenologica Italiana *Licheni e Ambiente* (D. Ottonello, ed.), (Palermo, 9-12 Dicembre 1995). *Notiziario della Società Lichenologica Italiana* 10: 63-64.

Adamo, P. (2000) Lo studio della bioalterazione di substrati minerali indotta da licheni: applicazione di nuove metodologie analitiche. In: Abstracts Convegno annuale della Società Lichenologica Italiana La Lichenologia in Italia. Bilancio di fine secolo (P. Adamo and G. Aprile, eds.), (Napoli, 22-24 Ottobre 1999). *Notiziario della Società Lichenologica Italiana* 13: 62.

Adamo, P. and Violante, P. (1989) Bioalterazione di roccia dolomitica operata da una specie lichenica del genere *Lepraria*. *Agricoltura Mediterranea* 119: 460-464.

Adamo, P. and Violante, P. (1991) Bioalterazione di rocce diabasiche, gabbriche e serpentinose operata da licheni. In: Atti XXX Convegno annuale della Società Italiana di Genetica Agraria, *Fertilità del Terreno e Biomassa Microbica*: 77-81. Congedo Editore, Galatina (Lecce).

Adamo, P. and Violante, P. (1991) Weathering of volcanic rocks from Mt Vesuvius associated with the lichen *Stereocaulon vesuvianum*. *Pedobiologia* 35: 209-217.

Adamo, P. and Violante, P. (1992) Biological weathering of volcanic and serpentine rocks. In: Book of Abstracts, *Mediterranean Clay Meeting (MCM '92), Gruppo Italiano A.I.P.E.A.,* (Lipari-Aeolian Islands (Italy), 27-30 September 1992): 7-8. Lipari.

Adamo, P. and Violante, P. (2000) Weathering of rocks and neogenesis of minerals associated with lichen activity. *Applied Clay Science* 16: 229-256.

Adamo, P., Marchetiello, A. and Violante, P. (1990) I licheni *Stereocaulon vesuvianum* Pers. e *Parmelia conspersa* (Ach.) Ach. agenti della bioalterazione di roccia vulcanica. In: Atti del Convegno Nazionale SICA (Società Italiana di Chimica Agraria), 25-27/9/1990, Bari.

Adamo, P., Marchetiello, A. and Violante, P. (1993) The weathering of mafic rocks by lichens. *The Lichenologist* 25: 285-297.

Adamo, P., Colombo, C. and Violante, P. (1995) Occurrence of poorly ordered Fe-rich phases at the interface between the lichen *Stereocaulon vesuvianum* and volcanic rock from Mt. Vesuvius. In: *Euroclay '95, Clays and Clay Materials Sciences*, Book of Abstracts (A. Elsen, P. Grobet, M. Keung, H. Leeman, R. Schoonheydt and H. Toufar, eds.): 341-342. Leuven, Belgium.

Adamo, P., Colombo, C. and Violante P. (1997) Iron oxides and hydroxides in the weathering interface between *Stereocaulon vesuvianum* and volcanic rock. *Clay Minerals* 32: 453-461.

Adamo, P., Vingiani S. and Violante, P. (2000) I licheni *Stereocaulon vesuvianum* Pers. e *Lecidea fuscoatra* (L.) Ach. ed il muschio *Grimmia pulvinata* agenti di bioalterazione di una tefrite fonolitica dell'Etna (Sicilia). In: Abstracts Convegno annuale della Società Lichenologica Italiana *La Lichenologia in Italia. Bilancio di fine secolo* (P., Adamo and G. Aprile, eds.), (Napoli, 22-24 Ottobre 1999). *Notiziario della Società Lichenologica Italiana* 13: 3-64.

4. Lichens and the Biodeterioration of Stonework

Alessi, P. and Visintin, D. (1988) Protective agents as a possible substrate for biogenic cycles. *Studia Geobotanica* 8: 99-112.

Atzeni, C., Cabiddu, M.G., Massidda, L., Sanna, U. and Sistu, G. (1994) Degradation and conservation of sandstone and pyroclastic rocks used in the prehistoric complex Genna Maria (Villanovaforru, Sardegna, Italy). In: Proceedings 3rd International Symposium *The Conservation of Monuments in the Mediterranean Basin* (V. Fassina V.H. Ott and F. Zezza eds.), (Venice, 22-25 June 1994): 533-539. Soprintendenza ai Beni Artistici e Etorici, Venezia.

Caneva, G. and Salvadori, O. (1988) Biodeterioration of stone. In: *The Deterioration and Conservation of Stone* (L. Lazzarini and R. Pieper, eds.). Studies and Documents on the Cultural Heritage, *UNESCO* 6: 182-234.

Caneva, G., Nugari, M. P., Ricci, S. and Salvadori, O. (1992) Pitting of marble in Roman monuments and the related microflora. In: Proceedings of the International Congress on *Deterioration and Conservation of Stone* (J. Delgado Rodrigues, F. Henriques and F. Telmo Jeremias, eds): 521-530. LNEC, Lisbon.

Caneva, G., Danin, A., Ricci, S. and Conti, C. (1994) The pitting of Trajan's column, Rome: an ecological model of its origin. *Contributi del Centro Linceo Interdisciplinare "Beniamino Segre"* 88: 77-102. Accademia Nazionale dei Lincei, Roma.

Danin, A. and Caneva, G. (1990) Deterioration of limestone walls in Jerusalem and marble monuments in Rome caused by cyanobacteria and cyanophilous lichens. *International Biodeterioration* 26: 397-417.

Del Monte, M. and Sabbioni, C. (1986) Chemical and biological weathering of an historical building: Reggio Emilia Cathedral. *The Science of the Total Environment* 50: 165-182.

Gorgoni, C., Lazzarini, L. and Salvadori, O. (1992) Minero-geochemical transformation induced by lichens in the biocalcarenite of the Selinuntine monuments. In: Proceedings of the International Congress on *Deterioration and Conservation of Stone* (J. Delgado Rodrigues, F. Henriques and F. Telmo Jeremias, eds.): 531-539. LNEC, Lisbon.

Modenesi, P. and Lajolo, L. (1988) Microscopical investigation on a marble encrusting lichen. *Studia Geobotanica* 8: 47-64.

Nimis, P.L. and Tretiach, M. (1996) Studies on the biodeterioration potential of lichens, with particular reference to endolithic forms. In: *Interactive physical weathering and bioreceptivity study on building stones, monitored by computerized X-ray tomography (CT) as a potential non-destructive research tool* (M. de Cleene ed.). European Commission Environment/Protection and Conservation of European Cultural Heritage, *Research Report* 2: 63-122. University of Ghent.

Pallecchi, P. and Pinna, D. (1988) Alteration of stone caused by lichen growth in the Roman Theatre of Fiesole (Firenze). In: Proceedings 6th International Congress *Deterioration and Conservation of Stone*: 30-47. Nicholas Copernicus University, Torun.

Pallecchi, P. and Pinna, D. (1988) Azione della crescita dei licheni sulla pietra nell'area archeologica di Fiesole. *Studia Geobotanica* 8: 113-124.

Piervittori, R., Gallo, L.M. and Laccisaglia, A. (1991) Analisi qualitativa dell' interfaccia lichene-substrato litico: metodologie con il microscopio polarizzatore. *Giornale Botanico Italiano* 125: 256.

Piervittori R., Salvadori O., Castelli S. and Favero-Longo S. (2002) Interazioni licheni-ofioliti in ambiente alpino. In: Abstracts del Convegno annuale della Società Lichenologica Italiana (G. Massari and S. Ravera, eds.), (Roma, 26-28 Ottobre 2002). *Notiziario della Società Lichenologica Italiana* 15: 38.

Pinna, D. and Salvadori, O. (1992) Effects of *Dirina massiliensis* and *stenhammari* growth on various substrata. In: Book of Abstracts, 2nd International Lichenological Symposium (IAL 2): 103. Bastad.

Roccardi A., Brunialti G., Modenesi P. and Senarega C. (2002) Studi preliminari sull'effetto della presenza di un lichene endolitico sulla permeabilità del travertino. In: Abstracts del Convegno annuale della Società Lichenologica Italiana (G. Massari and S. Ravera, eds.), (Roma, 26-28 Ottobre 2002). *Notiziario della Società Lichenologica Italiana* 15: 36-37.

Salvadori O. (2002) Colonizzazione dei manufatti lapidei da parte di organismi endolitici: un fenomeno sottostimato. In: Abstracts del Convegno annuale della Società Lichenologica Italiana (G. Massari and S. Ravera, eds.), (Roma, 26-28 Ottobre 2002). *Notiziario della Società Lichenologica Italiana* 15: 33.

Salvadori, O., Pinna, D. and Grillini, G.C. (1994) Lichen-induced deterioration on an ignimbrite of the Vulsini Complex (Central Italy). In: *Proceedings of the International Meeting, Lavas and Volcanic Tuffs* (A.E. Charola, R.J. Koestler and G. Lombardi, eds.), (Chile, 25-31 October 1990): 143-154. ICCROM, Roma.

Salvadori, O., Sorlini, C. and Zanardini, E. (1994) Microbiological and biochemical investigations on stone of the Ca' d'Oro facade (Venice). In: Proceedings of 3rd International Symposium *Conservation of Monuments in the Mediterranean Basin* (V. Fassina, H. Ott and F. Zezza, eds.): 343-347. Soprintendenza ai Beni Artistici e Storici, Venezia.

Salvadori, O., Appolonia, L. and Tretiach, M. (2000) Thallus-substratum interface of silicicolous lichens occurring on carbonatic rocks of the mediterranean regions. In: Book of Abstracts, Fourth IAL Symposium, *Progress and Problems in Lichenology at the turn of the Millenium*: 34. Universitat de Barcelona, Barcelona

Salvadori, O., Tretiach, M. and Appolonia, L. (2000) Relazione tallo-substrato in *Tephromela atra* (Norman) Hafellner v. *atra* e v. *calcarea* (Jatta) Clauz. In: Abstracts Convegno annuale della Società Lichenologica Italiana La Lichenologia in Italia. Bilancio di fine secolo (P., Adamo and G. Aprile, eds.), (Napoli, 22-24 Ottobre 1999). *Notiziario della Società Lichenologica Italiana* 13: 69.

Sampò, S. and Piervittori, R. (1990) Le malte come substrato elettivo per *Candelariella vitellina* (Ehrht.) Müll. Arg. In: Atti del Convegno Scienza e Beni Culturali, *Superfici dell'Architettura: Le Finiture* (G. Biscontin and S. Volpin eds.), (Bressanone, 26-29 Giugno 1990): 313-316. Libreria Progetto Editore, Padova.

Tretiach, M. (1995) Ecophysiology of calcicolous endolithic lichens: progress and problems. *Giornale Botanico Italiano* 129: 159-184.

Tretiach, M. and Geletti, A. (1997) CO_2 exchange of the endolithic lichen *Verrucaria baldensis* from karst habitats in Northern Italy. *Oecologia* 111: 515-522.

Tretiach, M. and Pecchiari, M. (1995) Gas exchange rates and chlorophyll content of epi- and endolithic lichens from the Trieste Karst (NE Italy). *New Phytologist* 130: 585-592.

3.4   Calcium oxalate films

The origin of these films is one of the most fascinating controversies in the field of biodeterioration. They are extended, uniform films, varying in color from yellow to brown, composed mainly of calcium oxalate, which is present in two hydrated forms: mono- (whewellite) and bi- (weddellite).

In the opinion of some authors, these films are derived from the transformation of organically-based substances which, in the past, were used for the protective and/or aesthetic treatment of stone surfaces. Others correlate the films with lichens, which favoured by an unpolluted atmosphere in the past, had colonized the surfaces of the monuments. According to others, the films could have been derived from the deposition of atmospheric particulates containing calcium oxalate. Finally, others hypothesize that the formation could, in some cases, be linked to the activity of microorganisms such as bacteria and fungi through the transformation of organic substances on the stonework which had previously been applied for protective purposes. Until a few decades ago, for example, calcium caseinate was widely used in the restoration of buildings.

Oxalic acid, produced by a large number of organisms, including lichens, can form calcium oxalate by chemical interaction with the calcium contained in the substratum. The production of oxalic acid by the mycobiont in lichens varies according to the kind of growth (epi- or endolithic). Generally epilithic lichens, after death, do not leave any obvious residue of their past metabolic activity. However, encrustations containing calcium oxalate have been found, for example, where *Dirina massiliensis* f. *sorediata* was present. The growth of crustose lichens occurs in a mosaic form and the ability to form oxalates varies according to the species. These two aspects contrast with the fact that the oxalate films are generally homogeneous and continuous.

Above and beyond the various lines of thought on the question of calcium oxalate films, it is correct to stress the need for further experimental studies: a) to obtain traces of the past metabolic action of lichens within the films using laboratory techniques, and b) to evaluate the possible relationship between oxalate films and lichen growth on natural outcrops.

Adamo, P., Terribile, F. and Violante, P. (1996) Micromorphological study of lichens activity on volcanic rock. In: Book of Abstracts, 10th International Working Meeting on *Soil Micromorphology*, (Moscow, 8-13 July 1996): 13. Moscow State University, Moscow.

Alessandrini, G. and Realini, M. (1990) Le pellicole ad ossalati: origine e significato nella conservazione. Note conclusive sul Convegno. *Arkos* 9/10: 19-36.

Alessandrini, G., Bonecchi, R., Peruzzi, R. and Toniolo, L. (1989) Caratteristiche composizionali e morfologiche di pellicole ad ossalato: studio comparato su substrati lapidei di diversa natura. In: Atti del Convegno *Scienza e Beni Culturali, Le pellicole ad ossalato: origine e significato nella conservazione delle opere d'arte* (V. Fassina, ed.): 137-150. Centro del C.N.R. "Gino Bozza", Milano.

Altieri, A., Laurenti, M.C. and Roccardi, A. (1999) The coinservation of archaeological sites: materials and techniques for short-term protection of archaeological remains. In: Proceedings of 6th International Conference *Non-destructive Testing and Microanalysis for the Diagnosticts and Conservation of the Cultural and Environmental Heritage*, (Rome, 17-19 May 1999): 673-687. Roma.

Alunno Rossetti, V. and Laurenzi Tabasso, M. (1973) Distribuzione degli ossalati di calcio $CaC_2O_4 H_2O$ e $CaC_2O_4 2H_2O$ nelle alterazioni delle pietre di monumenti esposti all'aperto. In: *Problemi di Conservazione* (G. Urbani, ed.): 375-386. Editrice Compositori, Bologna.

Appolonia, L., Grillini, G.C. and Pinna, D. (1996) Origin of oxalate films on stone monuments: I. Nature of films on unworked stone. In: Proceedings of 2nd International

Symposium *The oxalate films in the conservation of works of art* (M. Realini and L. Toniolo, eds.), (Milan, 25-27 March 1996): 257-268. Centro del C.N.R. "Gino Bozza", Milano.

Caneva, G. (1993) Ecological approach to the genesis of calcium oxalate patinas on stone monuments. *Aerobiologia* 9: 149-156.

Ciccarone, C. and Pinna, D. (1993) Calcium oxalate films on stone monuments. Microbiological observations. *Aerobiologia* 9: 33-37.

Cipriani, C. and Franchi, L. (1958) Sulla presenza di whewellite fra le croste di alterazione dei monumenti romani. *Bollettino del Servizio Geologico* 79: 555-564.

Del Monte, M. (1990) Microbioerosions and biodeposits on stone monuments: pitting and calcium oxalate patinas. In: Proceedings of the Advanced Workshop, *Analytical Methodologies for the Investigation of Damaged Stones* (F. Veniale and U. Zezza, eds.), (Pavia, 14-21 September 1990). Pavia.

Del Monte, M. and Sabbioni, C. (1983) Weddellite on limestone in the Venice environment. *Environmental Science and Technology* 17: 518-522.

Del Monte, M. and Sabbioni, C. (1987) A study of the patina called scialbatura on imperial Roman marbles. *Studies in Conservation* 32: 114-121.

Del Monte, M., Sabbioni, C. and Zappia G. (1987) The origin of calcium oxalates on historical buildings, monuments and natural outcrops. *The Science of the Total Environment* 67: 17-39.

Del Monte, M. and Ferrari, A. (1989) Patine di biointerazione alla luce delle superfici marmoree. In: Atti del Convegno *Scienza e Beni Culturali, Le pellicole ad ossalato: origine e significato nella conservazione delle opere d'arte* (V. Fassina, ed.): 171-182. Centro del C.N.R. "Gino Bozza", Milano.

Edwards, H.G.M. and Seaward, M.R.D. (1993) Raman microscopy of lichen-substratum interfaces. *Journal of the Hattori Botanical Laboratory* 74: 303-316.

Edwards, H.G.M., Farwell, D.W. and Seaward, M.R.D. (1991) Preliminary Raman microscopic analyses of a lichen encrustation involved in the biodeterioration of Renaissance frescoes in central Italy. *International Biodeterioration* 27: 1-9.

Edwards, H. G. M., Farwell, D. W., Seaward, M. R. D. and Giacobini, C. (1991) Preliminary Raman microscopic analysis of a lichen encrustation involved in the biodeterioration of Renaissance frescoes in central Italy. *International Biodeterioration* 27: 1-9.

Edwards, H.G.M., Farwell, D.W., Giacobini, C., Lewis, I.R. and Seaward, M.R.D. (1994) Encrustations of the lichen *Dirina massiliensis* forma *sorediata* on Renaissance frescoes: an FT-Raman spectroscopic study. In: Proceedings of XIVth International Conference on *Raman Spectroscopy* (N-T. Yu and X-Y. Li, eds.), (Hong Kong, August 1994): 896-897. J. Wiley, Chichester.

Edwards, H.G.M., Farwell, D.W. and Seaward M.R.D. (1997) Raman spectroscopy of *Dirina massiliensis* f. *sorediata* encrustations growing on diverse substrata. *The Lichenologist* 29: 83-90.

Edwards, H.G.M., Gwyer, E.R. and Tait, J.K.F. (1997) Fourier transform Raman analysis of paint fragments from biodeteriorated Renaissance frescoes. *Journal of Raman Spectroscopy* 28: 677-684.

Favali, M.A., Fossati, F., Mioni, A. and Realini M. (1995) Biodeterioramento da licheni crostosi dei calcari selciferi lombardi. In: Atti del Convegno *Scienza e Beni Culturali, La pulitura delle superfici dell'architettura* (G. Biscontin and G. Driussi, eds.): 201-209. Libreria Progetto Editore, Padova.

Favali, M.A., Fossati, F. and Realini, M. (1989) Studio della natura delle pellicole osservate sul Duomo e sul Battistero di Parma. In: Atti del Convegno *Scienza e Beni Culturali, Le*

4. Lichens and the Biodeterioration of Stonework

pellicole ad ossalato: origine e significato nella conservazione delle opere d'arte (V. Fassina, ed.): 261-270. Centro del C.N.R. "Gino Bozza", Milano.

Franceschi, V.R. and Horner, H.T. (1980) Calcium oxalate crystal in plants. *Botanical Reviews* 46: 361-427.

Franzini M., Gratzui, C. and Wicks, E. (1984) Patine ad ossalato di calcio sui monumenti marmorei. *Rendiconti Società Italiana di Mineralogia e Petrografia* 39: 59-70.

Lazzarini, L. and Salvadori, O. (1989) A reassessment of the formation of the patina called scialbatura. *Studies in Conservation* 34: 20-26.

Matteini, M. and Moles, A. (1986) Le patine di ossalato sui manufatti in marmo. In: *Restauro del Marmo: Opere e Problemi.* Quaderni dell'Opificio delle Pietre Dure e Laboratori di Restauro di Firenze: 65-73. Opus Libri, Firenze.

Mattirolo, O. (1917) Sulla natura della colorazione rosea della calce dei muri vetusti e sui vegetali inferiori che danneggiano i monumenti e le opere d'arte. *Rivista Archeologica della Provincia e Antica Diocesi di Como* 73/75: 1-19.

Modenesi, P., Canepa, R. and Tafanelli, A. (1996) New hypotheses on the role of calcium oxalates in epiphytic lichens. In: Book of Abstracts, 3rd IAL Symposium *Progress and Problems in Lichenology in the Nineties* (Salzburg, 1-7 September 1996): 28.

Modenesi, P., Giordani, P. and Brunialti, G. (2000) Factors determining the formation of weddellite and whewellite in lichens. In: Book of Abstracts of 4th IAL Symposium on *Progress and Problems in Lichenology at the Turn of the Millenium*, (Barcelona, 3-8 September 2000): 37-38. Universitat de Barcelona, Barcelona.

Pinna, D. (1993) Fungal physiology and the formation of calcium oxalate films stone monuments. *Aerobiologia* 9: 157-167.

Pinna, D., Salvadori O. and Tretiach, M. (1998) An anatomical investigation of calcicolous endolithic lichens from the Trieste Karst (NE Italy). *Plant Biosystems* 132. 183-195.

Rossi Manaresi, R., Grillini, G.C., Pinna, D. and Tucci, A. (1989) La formazione di ossalati di calcio su superfici monumentali: genesi biologica o da trattamenti. In: Atti del Convegno *Scienza e Beni Culturali, Le pellicole ad ossalato: origine e significato nella conservazione delle opere d'arte* (V. Fassina, ed.): 113-125. Centro del C.N.R. "Gino Bozza", Milano.

Rossi Manaresi, R., Grillini, G.C., Pinna, D. and Tucci, A. (1989) Presenza di ossalati di calcio su superfici lapidee esposte all'aperto. In: Atti del Convegno *Scienza e Beni Culturali, Le pellicole ad ossalato: origine e significato nella conservazione delle opere d'arte* (V. Fassina ed.): 195-205. Centro del C.N.R. "Gino Bozza", Milano.

Sabbioni, C. and Zappia, G. (1991) Oxalate patinas on ancient monuments: the biological hypothesis. *Aerobiologia* 7: 31-37.

Salvadori, O. and Zitelli, A. (1981) Monohydrate and dihydrate calcium oxalate in living lichen incrustation biodeteriorating marble columns of the basilica of S. Maria Assunta on the island of Torecello (Venice). In: Proceedings of 2nd International Symposium on *The Conservation of Stone II*, (R. Rossi Manaresi, ed): 759-767. Centro per la Conservazione delle Sculture all'Aperto, Bologna.

Salvadori, O., Appolonia, L. and Tretiach, M. (2000) Thallus-substratum interface of silicicolous lichens occurring on carbonatic rocks of the Mediterranean regions. In: Book of Abstracts, Fourth IAL Symposium *Progress and Problems in Lichenology at the Turn of the Millennium*, (Barcelona, 3-8 September 2000): 34. Universitat de Barcelona.

Seaward, M.R.D. and Giacobini, C. (1989) Oxalate encrustation by the lichen *Dirina massiliensis* forma *sorediata* and its role in the deterioration of works of art. In: Atti del Convegno *Scienza e Beni Culturali, Le pellicole ad ossalato: origine e significato nella conservazione delle opere d'arte* (V. Fassina, ed.): 215-219. Centro del C.N.R. "Gino Bozza", Milano.

3.5 Methods of prevention and control

When planning a restoration work, it is advisable, before totally eliminating the thalli, which is not always necessary, to conduct:

- a qualitative and quantitative analysis of the lichen flora present;
- a careful appraisal of the kind of damage caused to the stone;
- a search for the causes which might have favoured the lichen colonization.

Only in this way is it possible to plan correctly the action to be taken, including preventive measures for controlling those factors which could favour the reappearance of the lichen thalli.

Investigations have been conducted on the different sensitivity of the various forms of lichen growth to biocides and the methods of removing thalli from the substratum. A correct knowledge of the reproductive methods of lichens facilitates the choice of the technique to be adopted. Thus in the presence of soredium-bearing thalli it is not advisable to remove lichen solely by mechanical means without first treating with a biocide, as this would favour the dipersion in the air of the countless vegetative propagules.

In the choice of biocides one should not only evaluate the effectiveness of the product against biodeteriogens, but also consider other aspects which if underestimated could set off processes which are more damaging than the bioalteration. When making the choice of the biocide treatment to be used, for example, one must bear in mind that this could interfere chemically or physically with the substratum. However, the experimental tests conducted so far are still insufficient to provide precise indications on the effects that a certain product can produce on the substratum. Another side effect which must not be overlooked is the possibility of altering the balance of the ecosystem favouring, for example, the spread of those species which are more resistant and aggressive for the stone.

At present, since specific products for the elimination of lichens (lichenocides) do not exist, there is a tendency to advise the use of products belonging to toxicological classes III and IV (quaternary ammonium salts) and urea derivatives. Good results have been obtained recently from using proteolithic enzymes, although their use is more problematic over large areas.

The products are applied in three ways: spraying (the most common), with a brush, or with a compress.

4. Lichens and the Biodeterioration of Stonework

Affini, A., Favali, M.A., Pedrazzini, R. and Fossati, F. (1996) Metodologie di intervento e di conservazione dei monumenti del Giardino Ducale di Parma. *Giornale Botanico Italiano* 130: 430.

Amadori, L., Mecchi, A.M., Monte, M., Musco, S. and Salvatori, A. (1989) La conoscenza dei materiali e delle strutture per un progetto di restauro nel Parco Archeologico di Gabii. In: Atti del *Convegno Scienza e Beni Culturali, Il Cantiere della conoscenza, il Cantiere del Restauro* (G. Biscontin, M. Dal Col and S. Volpin, eds.): 295-308. Libreria Progetto Editrice, Padova.

Altieri, A., Giuliani, M.R., Nugari, M.P., Pietrini, A.M., Ricci, S. and Roccardi, A. (1998) Il degrado di origine biologica delle opere d'arte: diagnosi e interventi. In: *Diagnosi e progetto per la conservazione dei materiali dell'architettura* (Istituto Centrale per il Restauro, ed.):125-137. De Luca Edizioni, Roma.

Basile, G., Chilosi, M.G. and Martellotti, G. (1987) La facciata della cattedrale di Termoli: un esempio di manutenzione programmata. *Bollettino d'Arte*, 41: 283-30.

Bernardini, C. (1993) Biocidi e prevenzione microbiologica: alcune osservazioni in cantiere. *Kermes* 16: 12-19.

Bettini, C. (1984) Gli interventi di restauro sulla decorazione della Villa di papa Giulio II. *Bollettino d'Arte* 27: 127-131.

Bettini, C. and Villa, A. (1981) Description of a method for cleaning tombstones. In: Proceedings of 2nd International Symposium *The Conservation of Stone* (R. Rossi Manaresi, ed.): 523-534. Centro per la Conservazione delle Sculture all'Aperto, Bologna.

Biscontin, G. (1983) La conservazione dei materiali lapidei: trattamenti conservativi. In: Atti del Convegno Internazionale *La Pietra: interventi, conservazione, restauro* (A. Cassiano, O. Curti and G. Delli Ponti, eds.), (Lecce, 6-8 Novembre 1981): 111-120. Congedo Editore, Galatina (Lecce).

Caneva, G. and Salvadori, O. (1988) I pesticidi nel diserbo dei monumenti. *Verde Ambiente* 1: 79-84.

Caneva, G., Nugari, M.P., Pinna, D. and Salvadori, O. (1996) *Il Controllo del Degrado Biologico. I biocidi nel restauro dei materiali lapidei.* Nardini Editore, Firenze.

Caneva, G. and Salvadori, O. (1987) I pesticidi nel controllo del biodeterioramento dei monumenti: problemi tecnici e sanitari. In: Atti del Convegno *Inquinamento in ambienti di vita e di lavoro: esperienze e linee di intervento* (F. Candura and A. Messineo, eds.): 81-91. Edizioni Acta Medica, Roma.

Capponi, G. and Meucci, C. (1987) Il restauro del paramento lapideo della facciata della Chiesa di S. Croce a Lecce. *Bollettino d'Arte* 41: 263-282.

Cardilli Aloisi, L. (1985) La restauration de la Porta del Popolo a Rome. In: Proceedings of the 5th International Congress *Deterioration and Conservation of Stone* (G. Félix, ed.): 1083-1091. Presses Polytechniques Romandes, Lausanne.

Cavaletti, R., Strazzabosco, G., Manoli, N. and Toson P. (1990) Relazione tecnica di restauro della statua n. 85 "Andrea Briosco". In: Atti della Giornata di Studio *Il Prato della Valle e le opere di pietra calcarea collocate all'aperto* (S. Borsella, V. Fassina and A.M. Spiazzi, eds.), (Padova, 6 Aprile 1990): 313-331. Libreria Progetto Editore, Padova.

Cherido, M.M. (1990) Relazione tecnica di progetto per il restauro della statua n. 41 "Cesare Piovene". In: Atti Giornata di Studio, *Il Prato della Valle e le opere di pietra calcarea collocate all'aperto* (S. Borsella, V. Fassina and A.M. Spiazzi, eds.), (Padova, 6 Aprile 1990): 355-359. Libreria Progetto Editore, Padova.

Curri, S. (1986) Processi litoclastici di origine biologica: diagnosi e trattamenti. In: *Tecniche della Conservazione* (A. Bellini, ed.): 144-206. Franco Angeli, Milano.

Enteco, S.R.L. (1990) Relazione tecnica di restauro della statua n. 86 "Albertino Papafava di Carrarresi". In: Atti della Giornata di Studio *Il Prato della Valle e le opere di pietra calcarea collocate all'aperto* (S. Borsella, V. Fassina and A.M. Spiazzi, eds.), (Padova, 6 Aprile 1990): 299-311. Libreria Progetto Editore, Padova.

Gabrielli, N. (1986) Restauro di opere lapidee. *Rassegna dei Beni Culturali* 8/9: 44-50.

Giacobini, C. and Bettini, C. (1978) Traitements des vestiges archeologiques détériorés par les lichens et les algues. In: Proceedings International Symposium *Deterioration and Protection of Stone Monuments*: 4-3. UNESCO/RILEM, Paris.

Giacobini, C., Bettini, C. and Villa, A. (1979) Il controllo dei licheni, alghe e muschi. In: Proceedings of the 3rd International Symposium on the *Deterioration and Preservation of Stones*, (Venice, 24-27 October 1979): 305-312. Venezia.

Giacobini, C., Roccardi, A. and Tigliè, I. (1986) Ricerche sul biodeterioramento. In: Atti del Convegno *Scienza e Beni Culturali, Manutenzione e conservazione del costruito tra tradizione e innovazione* (G. Biscontin, ed.), (Bressanone, 24-27 Giugno 1986): 687-705. Libreria Progetto Editore, Padova.

Lazzarini, L. and Laurenzi Tabasso, M. (1986) *Il restauro della pietra*. Cedam, Padova.

Marchese, E.P., Razzara, S., Grillo, M. and Galesi, R. (1990) Indagine floristica e restauro conservativo dell'Abbazia di San Nicolò l'Arena di Nicolosi (Etna). *Bollettino Accademia Gioenia Scienze Naturali* 23: 707-720.

Micheli, M.P. and Napoli, B. (1987) Interventi conservativi sulla facciata della Chiesa di Santa Maria Assunta a Ponte di Cerreto. *Bollettino d'Arte* 41: 247-262.

Monte Sila, M. (1986) Biodeterioramento dei materiali musivi e proposte di intervento. In: Atti del II Seminario di Studi, *Metodologia e prassi della conservazione musiva*, (Ravenna, 22-23 Gennaio 1986): 345-51. Longo Editore, Ravenna.

Monte, M. and Nichi, D. (1997) Azione di due biocidi nell'eliminazione di licheni dai monumenti in pietra. In: Abstracts del Convegno annuale della Società Lichenologica Italiana *Licheni e Ambiente* (D. Ottonello, ed.), (Palermo, 9-12 Dicembre 1995). *Notiziario della Società Lichenologica Italiana* 10: 83.

Monte, M. and Nichi, D. (1997) Effects of two biocides in the elimination of lichens from stone monuments: preliminary findings. *Science and Technology for Cultural Heritage* 6: 209-216.

Narduzzi, P.A. and Soccal, M. (1990) Relazione tecnica di cantiere: il restauro della statua n.36 "Galileo Galilei". In: Atti della Giornata di Studio, *Il Preato della Valle le opere di pietra calcarea collocate all'aperto* (S. Borsella, V. Fassina and A.M. Spiazzi, eds.), (Padova, 6 Aprile 1990): 337-346. Libreria Progetto Editore, Padova.

Nimis, P.L. and Salvadori, O. (1997) La crescita dei licheni sui monumenti di un parco. Uno studio pilota a villa Manin. In: *Restauro delle sculture lapidee nel parco di Villa Manin a Passariano. Il viale delle Erme* (E. Accornero, ed.): 109-142. Regione Autonoma Friuli-Venezia Giulia, Centro di Catalogazione e Restauro dei Beni Culturali, Venezia.

Nugari, M.P., D'Urbano, M.S. and Salvadori, O. (1993) Test methods for comparative evaluation of biocide treatments. In: Proceedings of the International RILEM/UNESCO Congress *Conservation of Stone and Other Material*: 565-572. Spon, London.

Nugari, M.P., Pallecchi, P. and Pinna, D. (1993) Methodological evaluation of biocidal interference with stone materials. Preliminary laboratory tests. In: Proceedings of the International RILEM/UNESCO Congress *Conservation of stone and other materials*: 295-302. Spon, London.

Paleni, A. and Curri, S. (1972) La pulitura delle sculture all'aperto. *Petrolieri d'Italia*: 5.

Paleni, A. and Curri, S. (1972) The attack of algae and lichens on stone and means of their control. In: Proceedings of International Symposium *Deterioration of Building Stone*: 157-166. La Rochelle, France.
Piervittori, R. and Caramiello, R. (2002) Importance of biological elements in conservation of stonework: a case study on a Romanesque church (Cortazzone, N. Italy). In: Proceedings 3rd International Congress *Science and Technology for the Safeguard of Cultural Heritage in the Mediterranean Basin* (A. Guarino, ed.), (Alcalá de Henares-Spain, July 9-14 2001): 891-894. Universidad de Alcalá, Spain.
Piervittori, R. and Salvadori, O. (2002) Il contributo della Lichenologia alla conoscenza e conservazione dei Beni Culturali: esperienze di studio e proposta di protocollo metodologico. In: Book of Abstracts, 97° Congresso Nazionale della Società Botanica Italiana (Lecce, 24-27 Settembre 2002): 9. Edizioni del Grifo, Lecce.
Pinna, D. (2002) Crescita biologica su monumenti lapidei trattati con protettivi e consolidanti. In: Abstracts del Convegno annuale della Società Lichenologica Italiana (G. Massari and S. Ravera, eds.), (Roma, 26-28 Ottobre 2002). *Notiziario della Società Lichenologica Italiana* 15: 34-35.
Pinna, D., Biscontin, G. and Driussi, G. (1995) La pulitura e il controllo della crescita biologica sui materiali lapidei. In: Atti del Convegno di Studi *Scienza e Beni Culturali, La pulitura delle superfici dell'architettura* (G. Biscontin and G. Driussi, eds.): 619-624. Libreria Progetto Editore, Padova.
Tiano, P. (1998) Biodeterioration of monumental rocks: decay mechanisms and control methods. *Science and Technology for Cultural Heritage* 7: 19-38.

4. NEW LINES OF RESEARCH

4.1 Biomonitoring and heritage buildings

Stone materials also undergo deterioration due to atmospheric pollutants. While atmospheric pollution is a limiting factor for epiphytic lichens, the same cannot be said for saxicolous lichens. Polluting substances condition the presence, diffusion and distribution of lichens; for example, the use of fertilizers in agricultural areas can favour the growth of nitrophilous lichens on monumental buildings.

Lichens not only play a role as agents of biodeterioration but also as valid indicators of environmental quality. Floristic and vegetational studies that have to be made on stone materials also allow us to ascertain whether the pollution-tolerant species are more frequent than the poleophobic ones.

Camuffo, D. (1993) Reconstructing the climate and air pollution of Rome during the life of the Trajan column. *The Science of the Total Environment* 128: 205-226.
Caneva, G., Gori, E. and Danin, A. (1992) Incident rainfall in Rome and its relation to biodeterioration of buildings. *Atmospheric Environment* 26B: 255-259.
Monte, M. (1991). Lichens on monuments: environmental bioindicators. In: *Science, Technology and European Cultural Heritage* (N.S. Baer, C. Sabbioni and A.I. Sors, eds.): 355-359. Butterworth-Heinemann, London.

Monte, M. (1994) Licheni come bioindicatori: analisi dell'ambiente e del degrado dei monumenti. In: *Studi e Ricerche sulla Conservazione delle Opere d'Arte alla Memoria di Marcello Paribeni*: 211-221. C.N.R., Roma.

4.2 Aerobiology and heritage buildings

Aerobiology is the study of the release, dispersion and transport in the air of particles of biological origin and the effects they have on the impact surface. It is a discipline which for long had its main applications in the medical sector; only in the last few years has it also been used in that of the conservation of heritage buildings. A new impetus has been given to this research sector by an interdisciplinary working group set up in the early 1990s, within the Associazione Italiana di Aerobiologia with the aim of:

- improving knowledge on the relationships between
- air-dispersed microflora and the phenomenon of the
- deterioration of materials used in works of art;
- trying to standardize sampling methods;
- determining the risk threshold values concerning biodeteriogens involved in the phenomenon of deterioration.

The objectives of aerobiological monitoring applied to the conservation of exposed heritage buildings, for which so far little has been done, can be multifold, involving various monitoring and data analysis methods. Normally no particular attention is paid to phenomena of dispersion in the air of organisms which are pathogenic for materials and the analyses conducted start from an *in situ* inspection of the phenomenon and, above all in the case of external ambients, the studies made have been more of interest to theoretical research than to routine applications. There is, however, a close correlation between Aerobiology and the biodeterioration of works of art if we consider that air is the main means of transport of biological particles. By "bioaerosol" pollution we mean the contamination of air by organisms and microorganisms, or parts of them which may be freely present or transported in the air by particles of various kinds.

Thus air is one of the agents which facilitates the transport of spores and/or vegetative lichen propagules, contributing to their dispersion, and favouring their establishment on stonework and in many cases contributing to the deterioration of the latter. This phenomenon is found essentially in open environments, to a lesser extent in semi-closed ones, and absent in closed ones.

Stonework of historical and artistic interest such as monuments, churches and archaeological sites make up the substrata most exposed to this kind of biological attack. Once the colonization mechanism has been started, suitable analyses always need to be conducted to verify the causes and evaluate the effects of a lichen-

4. Lichens and the Biodeterioration of Stonework

induced alteration. This needs to be done in order to adopt, in the restoration phase, suitable measures for the prevention and control of new biodeterioration phenomena.

Lichen sampling has so far been performed on the substratum on which the lichens develop and never on the air which may transport them. Experimentation and the development of new techniques are under way for: a) direct sampling; b) analysis of collected material; and c) the evaluation of the correlation between the air-dispersed lichen component and the colonization present on the stonework.

Caneva, G., Piervittori, R. and Roccardi, A. (1998) Ambienti esterni: problematiche specifiche. In: *Aerobiologia e Beni culturali. Metodologie e tecniche di misura* (P. Mandrioli and G. Caneva, eds.): 247-251. Nardini Editore, Firenze.

Caramiello, R., Piervittori, R., Papa, G. and Fossa, V. (1991) Estrazione di pollini e spore da talli lichenici. *Giornale Botanico Italiano* 125: 331.

Caramiello, R., Siniscalco, C. and Piervittori, R. (1991) The relationship between vegetation and pollen deposition in soil and in biological traps. *Grana* 30: 291-300.

Maggi, O., Pietrini, A.M., Piervittori, R., Ricci, S. and Roccardi, A. (1998) L'aerobiologia applicata ai beni culturali. In: *Areobiologia e Beni culturali. Metodologie e tecniche di misura* (P. Mandrioli and G. Caneva, eds.): 32-36. Nardini Editore, Firenze.

Piervittori, R. and Laccisaglia, A. (1993) Lichens as biodeterioration agents and biomonitors. *Aerobiologia* 9: 181-186.

Piervittori, R. and Roccardi, A. (1998) Licheni. In: *Aerobiologia e Beni culturali. Metodologie e tecniche di misura* (P. Mandrioli and G. Caneva, eds.): 179-183. Nardini Editore, Firenze.

Piervittori, R. and Roccardi, A. (2002) Indagini aerobiologiche in ambienti esterni: valutazione della componente lichenica. In: Book of Abstracts, X Convegno Nazionale di Aerobiologia Aria e Salute, Sezione Beni Culturali (Bologna, 13-15 Novembre 2002): 70. Associazione Italiana di Aerobiologia, Bologna.

Piervittori, R., Roccardi, A. and Isocrono, D. (2002) Aspetti della colonizzazione lichenica sui monumenti. In: Abstracts del Convegno annuale della Società Lichenologica Italiana (G. Massari and S. Ravera, eds.), (Roma, 26-28 Ottobre 2002). *Notiziario della Società Lichenologica Italiana* 15: 71-72.

Roccardi, A. and Piervittori, R. (1998) The aerodiffused lichen-component: problems and methods. In: 6th International Congress on *Aerobiology*, Book of Abstracts, (Perugia, 31 Agosto-5 Settembre 1998): 268. Università di Perugia.

Roccardi, A. and Piervittori, R. (2000) Aerobiologia e Beni Culturali: la componente lichenica aerodiffusa. In: Abstracts del Convegno annuale della Società Lichenologica Italiana *La Lichenologia in Italia. Bilancio di fine secolo* (P., Adamo and G. Aprile, eds.), (Napoli, 22-24 Ottobre 1999). *Notiziario della Società Lichenologica Italiana* 13: 67-68.

5. CONCLUSION

This review summarizes the current state of knowledge of the various aspects of lichen colonization on stonework of artistic value in an Italian context. Further information can be found in specific critical bibliographies (see below) and in

Chapter 14 of this volume, and on the home-page of the Italian Lichen Society (http://dbiodbs.univ.trieste.it/sli/home.html).

Piervittori, R., Salvadori, O. and Laccisaglia, A. (1994) Literature on lichens and biodeterioration of stonework I. *The Lichenologist* 26: 171-192.
Piervittori, R., Salvadori, O. and Laccisaglia, A. (1996) Literature on lichens and biodeterioration of stonework II. *The Lichenologist* 28, 471-483.
Piervittori, R., Salvadori, O. and Isocrono, D., (1998) Literature on lichens and biodeterioration of stonework III. *The Lichenologist* 30: 263-277.
Piervittori, R., Salvadori, O. and Isocrono, D. (2004). Literature on lichens and biodeterioration of stonework. IV. *The Lichenologist* 36: 145-157.
Piervittori, R., Valcuvia-Passadore, M. and Nola, P. (1990/91) Italian lichenological bibliography: 1568-1989. *Allionia* 30: 99-169.
Piervittori, R., Valcuvia-Passadore, M. and Laccisaglia, A. (1995) Italian lichenological bibliography. First update (1989-1994) and addenda. *Allionia* 33: 153-179.
Piervittori, R., Valcuvia-Passadore, M. and Isocrono, D. (1998) Italian lichenological bibliography. Second update (1995-1998) and addenda. *Allionia* 36: 67-88.
Piervittori, R., Valcuvia-Passadore, M. and Isocrono, D. (2001) Italian lichenological bibliography. Third update (1999-2001) and addenda. *Allionia* 38: 81-94.

ACKNOWLEDGEMENTS

I am greatly indebted to Prof. Larry St. Clair and Prof. Mark Seaward for their courtesy and particular thanks go my Italian lichenologist friends: Paola Adamo (Napoli), Giulia Caneva (Roma), Deborah Isocrono (Torino), Paolo Modenesi (Genova), Pier Luigi Nimis (Trieste), Domenico Ottonello (Palermo), Daniela Pinna (Bologna), Ada Roccardi (Roma), Ornella Salvadori (Venezia) and Mauro Tretiach (Trieste). I am also grateful to Mauro Guolo, Franco Estivi and Marina Galimberti (Library of Department of Plant Biology, University of Turin) for their collaboration. This study was supported by M.I.U.R. funds.

Chapter 5

DETERIORATIVE EFFECTS OF LICHENS ON GRANITE MONUMENTS

B. SILVA and B. PRIETO
Departamento Edafología y Química Agrícola, Facultad Farmacia, Univ. Santiago de Compostela, 15782-Santiago de Compostela, Spain

Abstract: The contribution of lichens to the detrioration of granite monuments has been studied from ecological, physical and mineralogical points of view. The results obtained from more than 100 samples demonstrate the importance of lichens in the weathering of granitic rocks since, among other effects, they give rise to (a) intense disaggregation of the substratum in contact with them, (b) dissolution of minerals and transformation of primary minerals into others, namely biotite and hydroxyaluminum-vermiculite, and (c) neoformation of calcium minerals, such as calcium carbonate, calcium oxalates and gypsum. All these data were very useful in research aimed at testing methods for the conservation and restoration of monuments.

Keywords: granite; lichens; monuments; weathering; Galicia

1. INTRODUCTION

Interest in the biodeterioration of ancient buildings and monuments in Galicia (NW Spain) stems from the fact that the mild oceanic climate strongly favors biological colonization of these structures, most of them built of granite (the main geological substrate of the region). The most obvious, if not the most abundant, colonizing organisms are lichens, the deteriorative effects of which have been investigated in a study in which more than 100 samples from granite monuments were examined. The main aims of the study were to identify the lichens colonizing the monuments and to analyze their effects in terms of weathering of the rock, with a view to preventing biodeterioration or even colonization itself.

For this, as well as analysing the importance of the role of lichens in physical, chemical and mineralogical weathering of granite, an ecological study was carried out of the lichen flora colonizing 25 granite monuments. The colonizing species were identified and the main factors favouring their presence and distribution were established. Different biocides were also tested to determine their efficiency in destroying lichens and their effect on the substratum.

The importance of this study lies in the fact that, as well as being the only study of its kind carried out in Galicia, it also has practical applications. The restoration and conservation of works of art should only be carried out after identifying the agents causing the biodeterioration, the conditions favouring their growth, the importance of their role in the deterioration processes and the best methods of removing them and avoiding recolonization. As a result of this wide-ranging study, we have published 15 articles describing different aspects of the weathering of granite by lichens in Galicia (Carballal *et al.*, 2001; Prieto *et al.* 1994, 1995a, 1995b, 1995c, 1996, 1997, 1999a, 1999b, 1999c, 2000, 2002a, 2002b; Silva *et al.* 1997, 1999). The following overview summarizes some of the most interesting aspects of these studies.

2. ECOLOGY OF THE LICHEN FLORA COLONIZING GRANITE MONUMENTS

In order to investigate the ecology of the lichens that colonize granite monuments, lichens growing on 7 dolmens and 20 churches were studied. From the results obtained it was concluded that the monuments constitute very specific ecosystems since 11 species, 1 subspecies and 1 variety of the 96 taxa identified were recorded for the first time from Galicia (Table 5-1). Such a study is of great interest in terms of improving our knowledge of not only the region's lichen flora, but also of how to conserve and restore our monuments, since it is impossible to extrapolate information obtained from natural outcrops or other habitats when considering lichen removal.

Table 5-1. List of lichens species reported for the first time in Galicia (Prieto et al., 1994, 1995a).

Lichens	Location
Aspicilia contorta (Hoffm.) Krempelh.	Argalo Dolmen
Buellia lusitanica Steiner	Axeitos Dolmen
Buellia sejuncta Steiner	San Colmado Dolmen
Lecanora cenisia Ach.	Casa dos Mouros Dolmen
Lecanora muralis (Schreber) Rabenh. ssp *dubyi* (Müll. Arg.) Poelt	San Colmado Dolmen
Lecidella subincongrua (Nyl.) Hertel & Leuckert	Casa dos Mouros and Axeitos Dolmens
Pertusaria leucosora Nyl.	Various churches

5. Deteriorative Effects of Lichens on Granite Monuments

Table 5-1.1 (Continued) List of lichens species reported for the first time in Galicia (Prieto et al., 1994, 1995a).

Lichens	Location
Micarea leprosula (Th. Fr.) Coppins & A. Fletcher	Candean Dolmen
Pertusaria excludens Nyl.	Argalo Dolmen
Pertusaria mammosa Harm.	Argalo Dolmen
Rinodina teichophila (Nyl.) Arnold	Axeitos Dolmen
Scoliciosporum umbrinum (Ach.) Arnold var. *compactum* (Körber) Clauz. et Roux	Casa dos Mouros and Argalo Dolmens
Verrucaria macrostoma Dufour ex DC:	Various churches

The study also demonstrated the importance of lichen colonization in terms of cover (with more than 70% of the external surface was colonized) and in terms of the variety of species that colonize each monument (with, on average, 27 species present on each monument). However, overall analysis of the results showed that colonization was fairly uniform, as many of the same species were present on many of the buildings. Furthermore, there were differences in the colonization of dolmens and churches, a greater variety of flora being present on the former, with, in total, 33 and 65 lichen taxa respectively found on the churches and dolmens.

Differences in the composition of the lichen floras were confirmed and explained by application of the ecological indices proposed by Wirth (1980) and Nimis *et al.* (1987, 1992) and by statistical analysis of the results. These analytical methods revealed that the differences between dolmens and churches in terms of the colonizing flora were due to the higher pH values of the church walls resulting from the use of mortars containing lime and to the higher levels of nitrogen on the church walls caused by the presence of humans and animals, especially birds.

Statistical treatment allowed us to identify a group of lichen species which typically colonize granite monuments, namely: Aspicilia caesiocinerea, *A. cinerea, Caloplaca citrina, C. crenularia, Candelariella vitellina, Dirina massiliensis, Lasallia pustulata, Lecanora campestris, L. muralis ssp. dubyi, L. polytropa, Lecidea fuscoatra, Ochrolechia parella, Parmelia conspersa, Pertusaria coccodes, P. leucosora, P. pseudocorallina, Rhizocarpon geographicum, Sarcogyne clavus, Tephromela atra, Verrucaria macrostoma* and *Xanthoria parietina.*

Of these, *Verrucaria macrostoma, Sarcogyne clavus* and *Caloplaca citrina* were only present on churches. These species can thus be considered characteristic of this type of building and therefore as bioindicators of the factors previously mentioned as causing the differences between colonization of dolmens and churches. For the same reasons, *Lasallia pustulata* is considered to be characteristic of dolmens, since it only colonizes these monuments; it is therefore a bioindicator of very rural and unpolluted environments, of simple construction (i.e. absence of materials other than granite), low pH and absence of nitrogen.

Taking the above into account with the results of the statistical analysis, the group of species which typically colonize Galician granite churches comprises: *Caloplaca citrina, Caloplaca crenularia, Lecanora campestris, Pertusaria leucosora, Sarcogyne clavus* and *Verrucaria macrostoma*, whereas the group of lichens characteristic of granite dolmens comprises: *Aspicilia cinerea, Diploschistes scruposus, Lasallia pustulata, Parmelia conspersa, Rhizocarpon geographicum, Lecidea fuscoatra, Parmelia revoluta, Pertusaria coccodes, Rhizocarpon obscuratum, Rinodina confragosa, Trapelia coarctata* and *Lecanora polytropa*.

3. THE IMPORTANCE OF LICHENS IN THE WEATHERING OF GRANITE

Of all the species of lichens identified from the monuments studied, 13 were chosen to study the importance of the role of lichens in granite weathering processes. Because of the difficulty in collecting samples directly from the monuments, the lichens were sometimes collected from nearby outcrops of the same material as the dolmens and churches studied.

Micromorphological examination of the samples showed the importance of the physical deterioration caused by the lichens, the hyphae penetrating into the rock to a depth which depends on the physical characteristics of the granite (in particular the porosity) as well as the morphology of the thalli, the crustose lichens being the most penetrative. Although the hyphae mainly penetrate the spaces between rock grains (Fig. 5-1a and 5-1b), it has also been observed that the lichens are capable of creating new routes of access along the mineral exfoliation planes (Fig. 5-2a).

Penetration of the hyphae results in intense disaggregation in the area in contact with the lichen (Fig. 5-2b) and most of the detached minerals are retained within the thallus. There is also an associated increase in fissuration and a decrease in structural stability of the affected area. This physical deterioration increases the area of contact between the lichen hyphae and the rock, catalysing further weathering reactions produced by both the lichens themselves and other agents.

5. Deteriorative Effects of Lichens on Granite Monuments

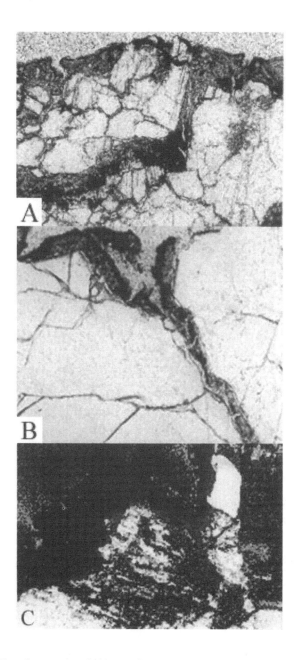

Figure 5-1. Microphotograph of thin sections of lichen-colonized granite samples cut perpendicular to the surface, showing occupation of intergranular voids by hyphae (a and b) and dissolution of minerals (c)

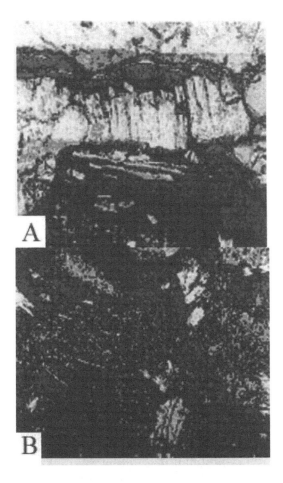

Figure 5-2. Scanning electron microphotograph showing indentations in a feldspar crystal associated with penetration of the granite by hyphae along the exfoliation planes (a) and powdering of grains of an orthoclase (b)

The active role of lichens in mineralogical weathering of granite has been demonstrated by:
a) The contribution to the transformation of biotite to hydroxyaluminic vermiculite as a result of the removal of potassium ions from the interlayer due to the effect of the organic acids excreted.
b) The excretion of oxalic acid, giving rise to neoformation of calcium oxalate in both its monohydrate and dihydrate forms (whewellite and wedellite) through chemical weathering of plagioclases containing calcium.
c) Occasional neoformation of calcium carbonate crystals due to a localized increase in pH caused by weathering of plagioclase. This indicates that many of

5. Deteriorative Effects of Lichens on Granite Monuments

the weathering reactions occur at the microsite level and therefore studies must be carried out at this level.

d) Neoformation of gypsum, probably from atmospheric SO_2 and calcium from the rock (due to weathering of plagioclase).

e) The presence of silica gels at the lichen-rock interface and in the lichen thalli, which provides evidence of the importance of biological breakdown of the minerals.

f) Indirect intervention in the formation of kaolinite, by increasing the time of contact between the rock and the attacking solutions and avoiding erosion of kaolinized minerals.

The chemical action of the lichens on the granite causes dissolution of the minerals (Fig. 5-1c) by production of organic acids as well as by an increase in the time of contact between the rock and the attacking solution.

Differences were found in the concentration of elements at the surface of the rock in direct contact with the lichen (interface) and in the interior of the rock, indicating the capacity of the lichen to take up elements from the rock (Fig. 5-3).

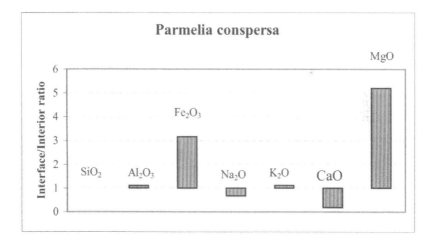

Figure 5-3. Ratio of the concentration of each element at the rock-lichen interface to its concentration in the corresponding interior rock sample (the latter being taken to be the reference value for rock unaffected by colonization)

Laboratory experiments in which samples of granite and isolated granite minerals were placed in contact with lichenic substances extracted from *Xanthoria parietina* and *Lasallia pustulata* and also with commercially available oxalic and stictic and usnic acids have demonstrated the direct action of these acids in chemical weathering; all the extracts used increased the mobility of the elements to a greater or lesser degree. The fact that some of these acids are very active in extracting potassium supports the previously mentioned finding that lichens contribute to the transformation of biotite into hydroxyaluminum-vermiculite.

The importance of the chemical weathering caused by lichenic acids lies in the speed of the action and the capacity of oxalic acid to complex Fe and Al, thereby causing the destruction of the structure of the minerals. The action of the other acids is also important because of their capacity to solubilize alkaline and alkaline-earth cations, especially potassium; they are therefore particularly important in the weathering of micaceous minerals.

4. CONTROL AND PREVENTION: BIOCIDE TESTING

The importance of the effects cited above often necessitates the remove of lichens from monuments. To determine the effectiveness, persistence and possible interaction of different biocides with the granite, some of the biocides most commonly used in restoration work (AB-57), Neodesogen, Hyvar-X, Sanit-S, Paragón invisible) were selected for laboratory and field tests on a granite building.

With the exception of Paragón invisible, all of the products were effective in removing lichens from the rock surfaces and the effect lasted for a considerable time (more than two years). However, study of samples taken from the interior of the rock revealed, in some cases, remnants of lichens and algae between the grains of quartz and feldspar persisted after biocidal treatment.

In the study of the interaction of the biocides with the substratum, no mineralogical changes were observed, the only effect being a notable change of color caused by the least effective biocide. Analysis of the treated samples revealed the presence of soluble salts. However, the quantities found were insignificant, and it was concluded that there was no risk of the biocides producing saline residues if they were rinsed with a sufficient amount of water after application.

To summarize, of the biocides tested, those containing quaternary ammonium salt were the most effective and also had the longest lasting effects and did not interact with the substratum.

ACKNOWLEDGEMENT

This work has been partially supported by Ministerio de Ciencia y Tecnologia (Programa Ramon y Cajal).

REFERENCES

Carballal, R., Paz-Bermudez, G., Sánchez-Biezma, M.J. and Prieto, B. (2001) Lichen colonization of coastal churches in Galicia: Biodeterioration implications. *International Biodeterioration and Biodegradation* 47: 157-163.

Nimis, P. L., Monte, M. and Tretiach, M. (1987) Flora e vegetaciones licencia di aree archeologiche del Lazio. *Studia Geobotánica* 7: 3-161.

Nimis P. L., Pinna, D. and Salvadori, O. (1992) *Licheni e Conservazione dei Monumenti*. Cooperative Libraria Universitaria Editrice Bologna.

Prieto, B., Rivas, T., Silva, B., Carballal, R. and Lopéz de Silanes, M.E. (1994) Colonization by lichens of granitic dolmens in Galicia (N.W. Spain). *International Biodeterioration and Biodegradation*, 34: 47-60.

Prieto, B., Rivas, T., Silva, B., Carballal. R. and Sánchez-Biezma, M.J. (1995a) Etude écologique de la colonisation lichénique des églises des environs de Saint-Jacques de Compostelle (NW Spain). *Cryptogamie, Bryologie, Lichénologie*, 16: 219-228.

Prieto Lamas, B., Rivas Brea, M.T. and Silva Hermo, B.M. (1995b) Colonization by lichens of granite churches in Galicia (Northwest Spain). *The Science of the Total Environment* 167: 343-351.

Prieto, B., Rivas, T. and Silva, B (1995c) The effect of selected biocides on granites colonized by lichens. In: *Biodeterioration and Biodegradation 9* (A. Bousher, M. Chandra & I.R. Edyvean eds): 204-209.

Prieto, B., Rivas, M.T., Silva, B.M. and López de Silanes, M.E. (1996) Ecological characteristics of lichens colonizing granite monuments in Galicia (Northwest Spain). In: *Degradation and Conservation of Granitic Rocks in Monuments*. (M.A. Vicente, J. Delgado & J. Acevedo eds): 295-300. European Commission DG XII, Brussels.

Prieto, B., Silva, B., Rivas, T., Wierzchos, J. and Ascaso, C. (1997) Mineralogical transformation and neoformation in granite caused by the lichens *Tephromela atra* and *Ochrolechia parella*. *International Biodeterioration and Biodegradation* 40: 191-199.

Prieto, B., Seaward, M.R.D., Edwards, H.G.M., Rivas, T. and Silva, B. (1999a).An FT-Raman spectroscopic study of gypsum neoformation by lichens growing on granitic rocks. *Spectrochimica Acta* 55A: 211-217.

Prieto, B., Seaward, M.R.D., Edwards, H.G.M., Rivas, T. and Silva, B. (1999b) Biodeterioration of granite monuments by *Ochrolechia parella* (L.). Mass.: an FT Raman spectroscopy study. *Bioespectroscopy* 5: 53-59.

Prieto, B., Rivas, T. and Silva, B. (1999c) Environmental factors affecting the distribution of lichens on granitic monuments. *The Lichenologist* 31: 291-305.

Prieto, B., Edwards, H.G.M. and Seaward, M.R.D. (2000).A Fourier Transform-Raman spectroscopic study of lichen strategies on granite monuments. *Geomicrobiology Journal* 17: 55-60.

Prieto, B., Rivas, T. and Silva, B (2002a) Alteración del granito por acción de los líquenes. Aspectos biogeoquímicos y biogeofísicos. In: *Biodeterioro de Monumentos Históricos de Iberoamérica*, Vol. II. (H. Videla & C. Saiz-Jimenez eds). CYTED, La Plata.

Prieto, B., Rivas, T. and Silva, B. (2002b) Rapid quantification of phototrophic microorganisms and their physiological state through their color. *Biofouling* 18: 229-236.

Silva, B., Prieto, B., Rivas, T., Sánchez-Biezma, M.J., Paz, G. and Carballal, R. (1997) Rapid biological colonization of a granitic building by lichens. *International Biodeterioration and Biodegradation* 40: 263-267.

Silva, B., Rivas, T. and Prieto, B. (1999) Effects of lichens on the geochemical weathering of granitic rocks. *Chemosphere* 39: 379-388.

Wirth, V. (1980). *Flechtenflora*. Ulmer, Stuttgart.

Chapter 6

MICROBIAL BIOFILMS ON CARBONATE ROCKS FROM A QUARRY AND MONUMENTS IN NOVELDA (ALICANTE, SPAIN)

C. ASCASO, M.A.[1], GARCIÁ DEL CURA[2] and A. DE LOS RÍOS[1]
[1]*Departamento de Biología Ambiental y Servicio de Microscopía Electrónica, Centro de Ciencias Mediambientales CSIC, 28006 Madrid, Spain* [2]*Instituto de Geología Económica CSIC-UCM and Laboratorio de Petrología Aplicada, Unidad Asociada CSIC –UA, Alicante, Spain*

Abstract:	Dolomite complex rock of the Upper Jurassic and biocalcarenite (Bateig stone) used for monuments in Novelda (Alicante, Spain) support a considerable biocomplexity of biofilms. The climate of the Novelda region is described as Iberolevantine-mesomediterranean. The La Mola quarry, from which samples were obtained for the present study, was opened at the end of the 19th century and remained in use until the end of the 1970s. The Novelda monuments studied were constructed at the end of the 19th century. In the quarry boundary zone there is an abundance of lichens, such as *Collema*, *Catapyrenium* and *Lecidea*, mosses and chroococcoid cyanobacteria (*Gloeocapsa* sp. and *Chroococcus* sp.). Fragments of quarry rock and monument stone were investigated by SEM-BSE and/or EDS analysis according to a new methodology. In the Mola quarry, the combined action of epilithic and endolithic microorganisms (mostly cyanobacteria) degrade the rock, becoming truly euendolithic since they bore into the rock. Alteration of the biocalcarenite by lichen thalli and microorganisms can be modelled: the mycobiont of the algal zone and upper cortex harbor biomineral fragments of the substratum including calcium oxalate. A second layer lying close to the lithic substratum is composed on the mycobiont intermixed with the bioclasts and minerals detached from it, resulting in significant mechanical alteration. Within the rock, microorganisms are present between the hollows and cavities: here the bioclasts and minerals shelter the microorganisms and important element biomobilization phenomena take place.
Keywords:	cyanobacteria; EDS; SEM-BSE; endolithic; epilithic; euendolithic; microbial biofilms; lichens; biocalcarenite; Alicante

1. INTRODUCTION

The complex interactions between lithobionts and their mineral substrata are currently of interest due to their implications for the bioweathering of stone monuments. The term biocomplexity has recently been applied to the study of ecosystems at the highest level. Whether analysing an ecosystem or a lithic microecosystem on a scale of only hundreds of cubic microns, its components and dynamic processes need to be established and understood.

Brock (1987) defined microbial ecology as the study of microorganisms in their natural environment, and emphasized that the researcher should examine the microorganisms *in situ*. We interpret this as a need to avoid disturbing the epilithic and endolithic microhabitat. When investigating the biodeterioration that occurs in stone monuments, the researcher needs to adhere to the strict rules of microbial ecology of lithic substrata. To this end, the phenomena of interrelationships among microorganisms or interactions between microorganisms and the microclimate and the mineral substrate need to be examined *in situ*. Moreover, detailed knowledge of bioweathering mechanisms may help our understanding of the relative effects of physical and chemical changes in the substrate, such that treatments aimed at resolving the problems associated with the presence of the lithobiontic community in the stone may be appropriately selected.

Lichens, fungi and cyanobacteria are particularly involved in a range of effects on stone buildings. The effects of lichens on monuments were first explored taxonomically and autecologically (Seaward, 1988); in some cases, these data were precisely related to ecological factors (Nimis *et al.*, 1992). The ecology of the communities concerned has also been investigated through multivariate analysis aimed at the conservation of monuments (Monte, 1991). According to Nimis and Salvadori (1997), taxonomic identification is always necessary, since each species may contribute to the process of degradation in a different way.

Ascaso and Ollacarizqueta (1991), through observations based at the square-micron scale, tried to determine the true effects of epilithic and endolithic microorganisms, in this particular case, of lichen symbionts and free-living fungi inhabiting the building stone of the Monasterio de Silos (Burgos, Spain). At that time there was considerable debate as to the effects of lichens and other microorganisms on stone, and some authors suggested a protective role for lichen thalli (Garg *et al.*, 1988; Ariño *et al.*, 1995).

This controversy had already existed earlier: Bech-Anderson (1983) stated that since the application of light microscopy methods to stone weathering studies, there was evidence that lichens did cause significant alteration to the rock substratum. In fact, there were hardly any studies that demonstrated what really occurs in the biofilm, with its lichens and free-living microorganisms, and the lithic substratum. This occurred since, traditionally, the biological components of the mineral substratum had been explored using microbiological techniques based mainly on the

6. Microbial Biofilms on Carbonate Rocks

identification of microorganisms by Petri dish culture. This technique was subsequently questioned by Amann *et al.* (1995), since many of the microorganisms inhabiting the rock substrate cannot be isolated or cultured. These microorganisms were referred to as viable but non-culturable.

It is for this reason that techniques of molecular biology have recently been applied to establish the taxonomy of the microorganisms that make up the biofilm found on paintings (Rolleke *et al.*, 1996, 1998). Molecular taxonomies were inferred from microorganisms previously isolated from their substratum, in this case paint. However, the difficulty of isolating all the microorganisms inhabiting a lithic substrate means that it would be most appropriate to apply molecular techniques on microorganisms without extracting them from the stone interior. Furthermore, Koestler and Salvadori (1996) encountered problems related to the evaluation of the efficiency of biocides, when using cultures of microorganisms removed from their natural environment. Recent studies have therefore opted for the use of fluorescence techniques such as measurement of the loss of chlorophyll from the algal layer to evaluate the *in situ* efficiency of a biocide.

However, neither the widely applied plate culture techniques and molecular techniques with their associated inherent limitations described above nor the fluorescence procedures can provide us with information as essential as the morphological and structural characteristics of the biological components (dead and alive) of the biofilms. Neither are these methods useful for revealing the morphological and geophysico-chemical features of the altered minerals or the extent of their bioalteration, nor do they provide information upon which different biocides act most efficiently on microorganisms deep within the rock substratum. Hence, as for investigations of the ecology of lithobiontic microorganisms in any lithic substrate (once again keeping in mind Brock's recommendations of 1987), in biodeterioration studies our attention should be focussed on *in situ* evaluations of the microbiota of monument stone.

For *in situ* analysis of the microhabitats of monument stone, special scanning electron microscopy applications were developed at the start of the 1990s, enabling zones of up to several square centimeters of the lithic microbial biofilm and the organomineral interface, which the microorganisms (photobiont and mycobiont) of the lichen thallus and adjacent microorganisms form with the mineral substratum, to be observed in great detail (Wierzchos and Ascaso, 1994). The method that we denote SEM-BSE (scanning electron microscopy with backscattered electron imaging) involves two main steps. The first corresponds to the sample preparation procedure and combines glutaraldehyde fixing with postfixing in osmium tetroxide and preparing polished resin blocks that are finally coated with carbon (Wierzchos and Ascaso, 1994). The second step involves the observation of the polished carbon-coated surfaces in the scanning electron microscope using the back-scattered electron (BSE) detector. The BSE signal is strongly dependent on the mean atomic number of the object bombarded with electrons. Thus, the SEM-BSE allows the ultrastructural elements of the biological components to be identified, along with the simultaneous observation of the microhabitat's inorganic components, among which are the minerals that surround the microorganisms (Ascaso and Wierzchos, 1994, 1995; Sanders *et al.*, 1994; Ascaso *et al.*, 1995). Energy dispersive spectroscopy (EDS), a microanalytical technique, allows these mineral components to be characterized (qualitative/quantitative determination and spatial distribution of elements) (Ascaso *et al.*, 1998a; Wierzchos and Ascaso, 1996, 1998).

This chapter describes evaluations performed on the biocolonization by lichens and free-living microorganisms on carbonate rocks from a quarry located 5 km from Novelda (Alicante-Spain) and on monuments constructed of the same material in the city.

2. MATERIALS AND METHODS

2.1 Materials

The climate of the Novelda region (Alicante) is Iberolevantine-mesomediterranean. Annual rainfall ranges from 250 to 300 mm, with a mean of 300 mm recorded for the past 35 years and a winter peak and scarce rainfall in the winter and spring (semiarid climate). Mean annual temperature ranges from 13 to 17°C. The Novelda region is well known for its quarries, which have supplied good quality stone for the construction of monuments in many areas of Spain. The quarry of La Mola lies 5 km outside the city of Novelda. There is a chronological record of the start of biocolonization both of the quarry stone and that of the city's monuments.

The La Mola quarry, from which samples were obtained for the present study, was opened at the end of the 19th century and remained in use until the end of the 1970s (Monzó and García del Cura, 1999, 2000). The quarried stone is comprised of a dolomite complex of the Upper Jurassic of different size crystals (5 to 200 microns) and different degrees of dedolomitization and jointing. The presence of calcite cement may be noted on polarized light microscopy with the help of Alizarin red S staining. In the quarry boundary zone, there is an abundance of lichens, mosses and cyanobacteria (*Gloeocapsa* sp. and *Chroococcus* sp.) The commonest genera of lichens are *Collema*, *Catapyrenium* and *Lecidea*. In the exploited zone of the quarry abandoned at the end of the 1970s, there is less colonization and the initial stages of microbial invasion can be observed.

The monuments of Novelda studied were built at the end of the 19th century, the rocks (biocalcarenites) used in their construction being identified as Novelda stone, now known as Bateig stone (Ordóñez *et al.*, 1994, 1997a, 1997b). This stone is a sandy biocalcarenite with a micritic matrix composed mainly of foraminifera with a few mollusk, bryozoan and echinoderm fragments. Its terrigenous components are quartz, some feldspar and fragments of rock. These fragments are polymictic and are

6. *Microbial Biofilms on Carbonate Rocks*　　　　　　　　　　　　　　　　　　　　　83

mainly comprised of slates, metaquartzites, limestones and dolomites, together with authigenic minerals (glauconite and microcrystalline quartz) and occasionally chalcedony.

2.2　Methods

Fragments of quarry rock and monument stone were processed for SEM-BSE and/or EDS analysis according to a method described elsewhere (Wierzchos and Ascaso, 1994). The stone fragments, once fixed (3.25% glutaraldehyde followed by 1% OsO_4) and dehydrated in an ethanol series, were embedded in epoxy resin. After polymerization, the blocks were cut and finely polished. Transverse sections of polished surfaces of the stone were stained with uranyl acetate followed by lead citrate, carbon-coated and examined using a DSM 940 A Zeiss and a DSM 960 A Zeiss microscope (both equipped with a four-diode, semiconductor BSE detector and a Link ISIS microanalytical EDS system). BSE and EDS examinations of the samples were simultaneously performed. The microscope operating conditions were as follows: 0° tilt angle, 35° take-off angle, 15 kV acceleration potential, 6 or 25 mm working distance and 1-5 nA specimen current.

3.　RESULTS

3.1　La Mola quarry

The limiting zone of the quarry shows a high degree of biological alteration (Fig. 6-1), involving significant biomobilization processes provoked by endolithic microorganisms (arrows). The combined action of epilithic (arrowhead in Fig. 6-2) and endolithic (arrow in Fig. 6-2) microorganisms has the final consequence of degrading the rock. Endolithic microorganisms are truly euendolithic; they penetrate the rock and establish themselves in spaces that present the shape of the contours of cells that colonize them (arrow in Fig. 6-2). As a consequence of the chemical changes produced by microorganisms, in the case of cyanobacteria, possibly *Gloeocapsa* (arrowhead in Fig. 6-3), a mechanical alteration effect is produced that disaggregates the substrate forming microparticles (arrows in Fig. 6-3). These may become trapped among the cyanobacterial colonies. In neighbouring areas (Fig. 6-4), it is possible to observe large numbers of bacterial colonies that share their microhabitat with protolichens (arrow in Fig. 6-4).

Figure 6-5 shows a detail of Figure 6-4 in which, as well as the bacteria (arrows), it is possible to observe fungi and algae organized in the stage of protolichen (arrowheads mark the contours of the protolichen). In this case, the epilithic bacteria and protolichen appear to have little effect on the rhombohedral dolomite over which they lie (Fig. 6-4). EDS mapping of the dolomite rhombus mainly showed Mg and Ca with no loss of these elements in the zone close to the microorganisms. In internal sites, some cryptoendolithic microorganisms can be observed (curved arrow in Fig. 6-4), but only at a distance from the zone of the dolomite rhombus.

In the area of the quarry abandoned 25 years ago, cyanobacteria in the form of colonies are the dominant microorganisms. Fragments of minerals (curved arrow in Fig. 6-6) are seen to be detached from the substratum during the growth of cyanobacteria. These *Chroococcus*-type cyanobacteria occupy epilithic zones and, as may be observed in Figure 6-7, there is no substantial endolithic colonization (neither by cyanobacteria nor by other microorganisms). This stone, from a quarry abandoned 25 years ago, shows the initial stages of colonization.

3.2 Novelda monument stone

Considerable lichen colonization was observed over vertical (north-facing) and horizontal surfaces of the stones of the monument constructed in 1886, producing intense mechanical and chemical alteration to the substrate (Fig. 6-8). These effects are mainly exerted by the mycobiont of the lichen thallus (arrowheads A and B in Fig. 6-8), particularly on bioclasts (red algae, bryozoans foraminifera) (black arrow in Fig. 6-8), limestone clasts (star in Fig. 6-8) and in smaller measure on the quartz (white arrow in Fig. 6-8). Quartz, dolomite and calcite appear as microdivided minerals within the zone indicated in Figure 6-8 by arrowhead B, i.e. the part of the thallus that does not produce the apothecium. Figure 6-9 is a scan of the zone marked with a line in Figure 6-8. This scan line shows the elements Mg, Al, Si and Ca at the points of the sample that are crossed by the transect.

Figure 6-10 shows an enlarged view of the zone marked by the arrowhead A in Figure 6-8. Being an apothecium, it is possible to observe the asci (arrows in Fig. 6-10) and paraphyses. In zones where there is an apothecium, the lower part of the thallus shows the largest amounts of highly microdivided minerals, a product of the mechanical alteration of the rock components such as quartz, fragments of carbonate rock and the bioclasts, *Globigerina* and red algae (Fig. 6-11). Mapping the distribution of chemical elements in the zone indicated the nature of the components of the substrate underlying the thallus.

Figure 6-12 shows a fragment of dolomitic rock in detail and demonstrates the effects of a disaggregation process. Microdivision is apparent in the upper part of the fragment (white arrows in Fig. 6-12), while the lower section remains more compact. In the space left between the fragment and substratum in the shape of a fissure, a set of microorganisms (black arrow in Fig. 6-12) can be observed.

Figure 6-13 shows the accumulation of microfragmented minerals (arrows in Fig. 6-13) among the hyphae of the upper region of the thallus.

6. Microbial Biofilms on Carbonate Rocks 85

Figure 6-14 shows how mineral fragments (star in Fig. 6-14) become detached from the substrate by the wedge-like action of the fungal hyphae of the lower part of the lichen thallus. A fragment of red algae (black arrows in Fig. 6-14) located next to the substrate's detached mineral has also undergone mechanical alteration (white arrows in Fig. 6-14), breaking up into fragments.

Figure 6-15 shows an ascocarp with asci and ascospores (arrowheads in Fig. 6-15); at the interface between the thallus and substrate, mechanical alteration leads to its disintegration (black arrows in Fig. 6-15). Figure 6-16 shows a soredium (white arrows in Fig. 6-16) enveloped by microdivided minerals derived from the mechanically altered substratum, in which it was possible to observe calcite with rhombi of dolomite (white arrowhead in Fig. 6-16) and glauconite (black arrowhead in Fig. 6-16).

Figure 6-17 shows a calcium map corresponding to the medullary zone of the thallus shown in Figure 6-13. This element is shown in the fragmented minerals, which become adhered to the hyphae (pentagonal arrows), and in the neoformed calcium oxalate crystals (biomineralization) (arrows).

4. DISCUSSION

Owing to the crustose morphology of the lichens colonizing the monument's biocalcarenite, the species examined here are mainly involved in geophysical processes. Alteration of the lithic substratum by lichen thalli and microorganisms has recently been modelled in alumino-silicate rocks by Banfield *et al.* (1999) and De los Ríos *et al.* (2002).

According to these models, we can also distinguish different layers in the microbial biofilms examined here. The first layer, consisting of the upper part of the lichen thallus, is practically devoid of minerals when an ascocarp is present. However, in the absence of the latter, many microdivided or neoformed (biominerals) minerals are able to settle in the thallus medulla and may even reach the algal zone (Fig. 6-8) and upper cortex (Fig. 6-13). The microorganisms (mainly the mycobiont) harbour the minerals in this layer. This zone shows a substantial presence of the biomineral calcium oxalate (Fig. 6-17). The second layer lies closest to the lithic substrate, where the symbionts of the thallus, generally the mycobiont, closely intermix with the minerals detached from the substratum, producing microfractures. It is sometimes also possible to observe minerals in an intermediate state, i.e. part of the mineral remains intact, adhering to the substratum, while another part is totally microfragmented (see clast of carbonate rock in Fig. 6-12). This is the area where the thallus makes most contact with the lithic substrate and where geophysical alteration phenomena take place.

These biophysical phenomena are accompanied by the biomobilization of elements such as calcium that subsequently forms biominerals in the form of oxalates in the first layer. Finally, within the substratum, some microorganisms can be seen to have settled between the hollows and cavities formed by the persistent mechanical alteration of the minerals (fissured zone marked by the white and black arrows in Fig. 6-8, zone marked by the black arrow in Fig. 6-12 and lower part of the mineral fragment indicated by a white star in Fig. 6-14). In this case, it is the minerals that harbour or shelter the microorganisms.

In summary, the first zone contains the microfragmented minerals and sustains biomineralization phenomena that give rise to compounds such as oxalates. However, it is in the second zone where obvious mechanical alteration of the substrate takes place, causing mineral microfragmentation. In the third zone, microorganisms settle in a pure mineral microhabitat where element biomobilization phenomena occur. Hence in the biocalcarenites harbouring the lichen thalli, biogeophysical and biomineralization phenomena can be observed, indicating the occurrence of biogeochemical biomobilization processes. In the same way, in the case of the calcareous litharenite forming the façades of the Jaca cathedral (Huesca, Spain), it is the mycobiont which is mainly responsible for the bioalteration of the substratum (Ascaso et al., 1998b). Hirsch et al. (1995) described the ecological advantages of fungi over bacteria and algae as lithobionts despite poor nutrient concentrations.

Biogeochemical processes are also clearly visible in the bioalteration of the boundary of La Mola quarry, where cyanobacteria induce biomobilization processes, behaving as euendolithic microorganisms according to the classification of Golubic et al. (1981). An indication of the chemical alteration of the calcareous substrate is that the spaces around the microorganisms have the same shape as their contours. Biogeochemical action gives rise to an increase in the rugosity and porosity of the stone surface, as had already been noted by SEM-BSE in the façades of the Torre de Belem in Lisbon (Portugal) (Ascaso et al., 1998c). At Torre de Belem, calcium deposits were observed coating the cells that produce the spaces in which they lay.

Furthermore, biomineralization due to the calcification of cyanobacteria was reported by Pentecost (1991). In the Monasterio de los Jeronimos, the cyanobacteria are found in hollows that define the shape of the organism, as occurs in the present study, and calcium precipitation around cyanobacteria was revealed by digital mapping (Ascaso et al., 2002). All these cases suggest calcium biomobilization and euendolithic behavior. Extracellular polymeric substances (EPS) were observed by low temperature scanning electron microscopy (LTSEM) around the cyanobacteria in the stone of the Monasterio de los Jeronimos (Lisbon) (Ascaso et al., 2002). This is consistent with the capacity of EPS to provide nucleation sites for the formation of minerals, especially in zones of high calcium content (Barker and Banfield, 1996; Barker et al. 1997).

In contrast, in the area of the quarry that had been in use up to the end of the 1970s and then abandoned, only the first stages of colonization by cyanobacteria that rarely behave as euendoliths could be observed. One of the images of this area of the quarry (Fig. 6-5) is, in our opinion, of great relevance, since it clearly shows the structure of a biofilm with a large proportion of bacteria. Once bacteria had been acknowledged as exhibiting phenotypic plasticity, Boivin and Costerton (1990) applied morphological methods to examine bacteria *in situ*. Biofilm formation is a *sine qua non* in the biodeterioration of complex substrata according to these authors, who emphasize that "we must understand biofilms if we are to understand and control biodeterioration". In the biofilms observed on these rocks, the bacterial colonies remained at the surface along with the microorganisms comprising the protolichen and together exerted a disaggegating effect on the substrate.

In the present study, the great biocomplexity of the biofilms found on carbonate rocks was revealed. This biocomplexity needs to be taken into account when designing strategies aimed at controlling or eliminating biodeterioration in these lithic substrates. The efficiency of several biocides on lithobionts embedded in the wall stone of the Monasterio de los Jeronimos was recently tested (Ascaso *et al.*, 2002), highlighting the importance of this type of *in situ* analysis. It is thus recommended that both biofilm/substrate characterization and the evaluation of treatments directed towards reducing biodeterioration phenomena should be performed without removing the biofilm from its rock substratum.

ACKNOWLEDGEMENTS

This study was funded by the PGC project BOSS-2003-02418. The authors thank Jose Carlos Monzó for sample collection, Virginia Souza-Egipsy and Jacek Wierzchos for their help in sample preparation, and Fernando Pinto and Rosario Santos of the Microscopy Service for technical assistance.

REFERENCES

Amman, R.I., Ludwig, W. and Schleifer, K.H. (1995) Phylogenetic identification and *in situ* detection of individual microbial cells without cultivation. *Microbiological Reviews* 59: 143-169.

Ariño, X., Ortega-Calvo, J.J., Gomez-Bolea, A. and Saiz-Jimenez, C. (1995) Lichen colonization of the Roman pavement at Baelo Claudia (Cadiz, Spain): biodeterioration vs. bioprotection. *The Science of the Total Environment* 167: 353-363.

Ascaso, C. and Ollacarizqueta, M.A. (1991) Structural relationship between lichen and carved stonework of Silos Monastery, Burgos, Spain. *International Biodeterioration* 27: 337-349.

Ascaso, C. and Wierzchos, J. (1994) Structural aspects of the lichen-rock interface using back-scattered electron imaging. *Botanica Acta* 107: 251-256.

Ascaso, C. and Wierzchos, J. (1995) Study of the biodeterioration zone between the lichen thallus and the substrate. *Cryptogamic Botany* 5: 270-281.

Ascaso, C., Wierzchos, J. and de los Ríos, A. (1995) Cytological investigations of lithobiontic microorganisms in granitic rocks. *Botanica Acta* 108: 474-481.

Ascaso, C., Wierzchos, J. and de los Ríos, A. (1998a) *In situ* cellular and enzymatic investigations of saxicolous lichens using correlative microscopical and microanalytical techniques. *Symbiosis* 24: 221-234.

Ascaso, C., Wierzchos, J. and Castello, R. (1998b) Study of the biogenic weathering of calcareous litharenite stones caused by lichen and endolithic microorganisms. *International Biodeterioration and Biodegradation* 42: 29-38.

Ascaso, C., Wierzchos, J., Delgado, J., Aires-Barros, L., Henriquez, F.M.A. and Charola, A.E. (1998c) Endolithic microorganisms in the biodeterioration of the Belem Tower. Intern. *Zeitschrift für Bauinstandsetzen* 4: 627-640.

Ascaso, C., Wierzchos, J., Souza-Egipsy, V., De los Rios, A., and Delgado-Rodriguez, J. (2002) *In situ* evaluation of the biodeteriorating action of microorganisms and the effects of biocides on carbonate rock of the Jerónimos Monastery (Lisbon). *International Biodeterioration and Biodegradation* 49: 1-12.

Banfield, J.F., Barker, W.W., Welch, S.A., and Taunton, A. (1999) Biological impact on mineral dissolution: application on the lichen model to understand mineral weathering in the rhizosphere. *Proceedings of the National Academy of Sciences* 96: 3404-3411.

Barker, W.W. and Banfield, J.F. (1996) Biologically versus inorganically mediated weathering reactions: relationship between minerals and extracellular microbial polymers in lithobiontic communities. *Chemical Geology* 132: 55-69.

Barker, W.W., Welch, S.A. and Banfield, J.F. (1997) Biochemical weathering of silicate minerals. In: *Geomicrobiology: interactions between microbes and minerals*. (J. F. Banfield and K. H. Nealson ed): 391-428. Mineralogical Society of America, Washington.

Bech-Andersen, J.C.P. (1983) Studies of lichen growth and deterioration of rocks and building materials using optical methods. *Biodeterioration* (T.A.Oxley and S.Barry ed) 5: 568-572. Wiley, London.

Boivin J. and Costerton J.W. (1990) Biofilms and biodeterioration. In: *Biodeterioration and Biodegradation 8*. (H.W. Rassmon ed): 53-62. Elsevier Applied Science, London.

Brock, T.D. (1987) The study of the microorganisms *in situ*: progress and problems. In: *Ecology of Microbial Communities* (M. Fletcher, T.R.G. Gray and J. G. Jones ed): 1-17. Society for General Microbiology, Cambridge.

De los Rios, A., Wierzchos, J. and Ascaso, C. (2002) Microhabitats and chemical microenvironments under saxicolous lichens growing on granite. *Microbial Ecology* 43: 181-188.

Garg, K.L., Dhawan, S. and Agrawal, O.P. (1988) Deterioration of stone and building materials by algae and lichens: a review. In: *Preprints of the International Conference on Biodeterioration of Cultural Property*: 1-43. National Research Laboratory for the Conservation of Cultural Property, Lucknow.

Golubic, S., Friedmann, I., and Schneider, J. (1981) The lithobiontic ecological niche, with special reference to microorganisms. *Journal of Sedimentary Petrology* 51: 475-478.

Hirsch, P., Eckhardt, F.E.W. and Palmer, R.J. (1995) Fungi active in weathering of rock and stone monuments. *Canadian Journal of Botany* 73: 1384-1390.

Koestler, R.J. and Salvadori, O. (1996) Methods of evaluating biocides for the conservation of porous building materials. *Science and Technology for Cultural Heritage* 5: 63-68.

Monte, M. (1991) Multivariate analysis applied to the conservation of monuments: lichens on the Roman aqueduct Anio Vetus in S. Gregorio. *International Biodeterioration* 28: 133-150.

Monzó Giménez, J.C. and García del Cura, M.A. (1999) Efecto de la bioalteración en la corrección de los impactos visuales cromáticos de canteras de rocas carbonáticas: El Cerro de la Mola, Novelda (Alicante). *Ingeopres* 77: 76-86.

Monzó Giménez, J.C. and García del Cura, M.A. (2000) Bioalteración de rocas dolomíticas en clima mediterráneo semiárido: la cantera del Cerro de la Mola. *Geotemas* 1: 267-272.

Nimis, P.L., Pinna, D. and Salvadori, O. (1992). *Licheni e Conservazione dei Monumenti*. Cooperativa Libraria Universitaria, Bologna.

Nimis, P.L. and Salvadori, O. (1997) La crescita dei licheni sui monumenti di un parco. Uno studio pilota a Villa Manin. In: *Il restauro delle sculture lapidee nel parco di Villa Manin a passariano il viale delle erme*. (E.Accornero ed) 4: 109-142. Centro di Catalogazione Restauro dei Beni Culturali Villa Manim di Passatiano, Venezia.

Ordóñez, S., García del Cura, M.A., Fort, R., Louis, M., Lopez De Azcona, M.C. and Mingarro, F. (1994) Physical properties and petrographic characteristics of some Bateig stone varieties. *7th Congress of the International Association of Engineering Geology*. (R.Oliveira, A.P.Cunha, A.G.Coelho and L.F.Rodríguez eds): 3595-3603. Balkema, Rótterdam.

Ordóñez, S., Fort, R. and García del Cura, M.A. (1997a) Pore size distribution in durability evaluation of porous limestones: Bateig Stone (Alicante, Spain). *The Quarterly Journal of Engineering Geology* 30: 221-230.

Ordóñez, S., García del Cura, M.A., Bernabéu, A. and Rodríguez, M.A. (1997b) Rocas ornamentales porosas del Mioceno marino de Levante (Alicante-Murcia-Albacete) In: *Avances en el conocimiento del Terciario Ibérico* (J.P. Calvo and J. Morales ed): 141-144. Universidad Complutense. Madrid.

Pentecost, A. (1991) Calcification processes in algae and cyanobacteria. In: *Calcareous Algae and Stromatolites*. (R. Riding ed): 1-20. Springer-Verlag, New York.

Rolleke, S., Muyzer, G., Wawer, C., Wanner, G. and Lubitz, W. (1996) Identification of bacteria in a biodegraded wall painting by denaturing gradient gel electrophoresis of PCR-amplified gene fragments coding for 16S rRNA. *Applied Environmental Microbiology* 62: 2059-2065.

Rolleke, S., Witte, A., Wanner, G. and Lubitz, W. (1998) Medieval wall painting – an habitat for *Archaea*: identification of *Archaea* by denaturing gradient gel electrophoresis (DGGE) of PCR-amplified gene fragments coding 16S rRNA in a medieval painting. *International Biodeterioration and Biodegradation* 41: 85- 92.

Sanders, W.B., Ascaso, C. and Wierzchos, J. (1994) Physical interactions of two rhizomorph-forming lichens with their rock substrate. *Botanica Acta* 107: 432- 439.

Seaward, M.R.D. (1988) Lichen damage to ancient monuments: a case study. *The Lichenologist* 20: 291-294.

Wierzchos. J. and Ascaso, C. (1994) Application of back-scattered electron imaging to the study of the lichen-rock interface. *Journal of Microscopy* 175: 54-59.

Wierzchos, J. and Ascaso, C. (1996) Morphological and chemical features of bioweathered granitic biotite induced by lichen activity. *Clays and Clay Minerals* 44: 652-657.

Wierzchos, J. and Ascaso, C. (1998) Mineralogical transformation of bioweathered granitic biotite studied by HRTEM: evidence for a new pathway in lichen activity. *Clays and Clay Minerals* 46: 446-452.

SEM-BSE images of dolostone (sedimentary rock formed by dolomite mesocrystals); asterisks point to the lithic substratum.

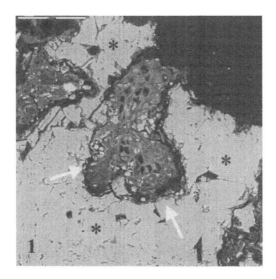

Figure 6-1. Endolithic microorganisms (arrows)

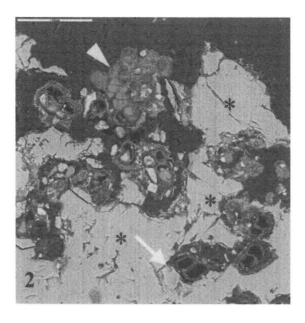

Figure 6-2. Epilithic (arrowhead) and endolithic (arrow) microorganisms. Rock disaggregation is revealed by the presence of rock fragments marked with asterisks in image

6. Microbial Biofilms on Carbonate Rocks

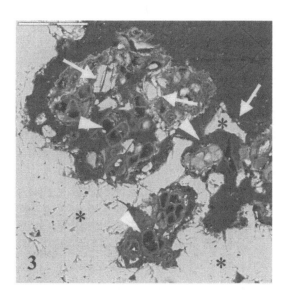

Figure 6-3. Euendolithic microorganisms (indicated by an arrowhead) and microparticles of minerals detached from the substrate (arrows)

Figure 6-4. Lithobiontic microorganisms in close proximity to rhombohedral dolomite (asterisks). Quartz (arrowheads) and limestone (stars) fragments can also be seen in the zone. In the dolomite rhombus in the upper part of the figure it may be seen how bacterial colonies share their habitat with protolichens (arrow). Areas occupied by cryptoendolithic microorganisms appear under the rhombi (curved arrow)

92   *Biodeterioration of Rock Surfaces*

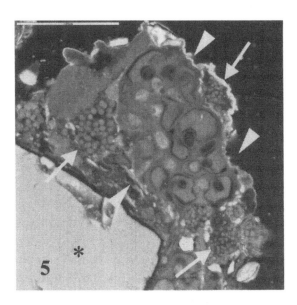

Figure 6-5. Detail of Figure 4. Sample from the boundary zone of the Mola Quarry: colonies of bacteria (arrows) and a protolichen with its photobiont and mycobiont. The area occupied by the protolichen is defined by the three arrowheads

Figures 6-6 and 6-7. Sample corresponding to the quarry zone in disuse since the end of 1970

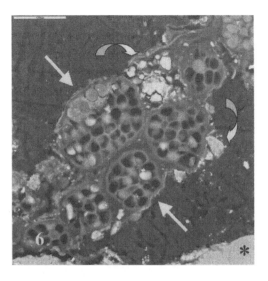

Figure 6-6. Cyanobacteria forming colonies (arrows), possibly *Chroococcus*. Microfragmented minerals adhered to and enveloped by the cyanobacteria are indicated by curved arrows

6. Microbial Biofilms on Carbonate Rocks

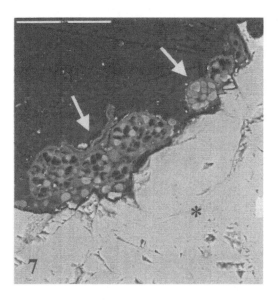

Figure 6-7. The arrows mark cyanobacteria on the lithic substrate (indicated by an asterisk). Note the lack of microorganisms inside the rock

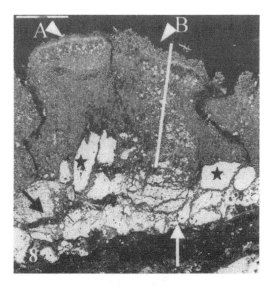

Figure 6-8. Monument stone from Novelda: cross section of the lichen thallus-lithic substrate interface. The thallus is labelled by the arrowheads A and B. Bioclasts, red algae (black arrow), limestone grains (star), and quartz (white arrow) may be observed in the substrate. The white line that cuts across the thallus from bottom to top represents the EDS line-scan (from the substrate to the top of the thallus)

Monument stone from Novelda.

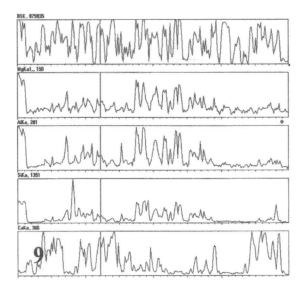

Figure 6-9. Line-scan for Mg, Al, Si and Ca in the area marked by an arrow in Figure 8

Figure 6-10. Enlarged view of the zone marked by the arrowhead A (ascocarp) in Figure 8. Asci are indicated by arrows

6. Microbial Biofilms on Carbonate Rocks

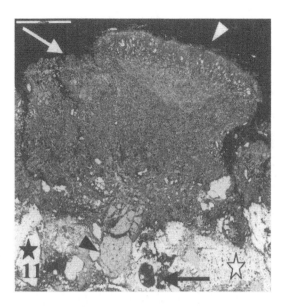

Figure 6-11. Lichen thallus with apothecium showing alteration of the lower part of the substrate and revealing that the microdivided minerals are not enveloped by the hyphae of the apothecium itself. White arrow: thallus; white arrowhead: apothecium; black arrow: Globigerina; black arrowhead: carbonate clast; black star: quartz; white star: red algae

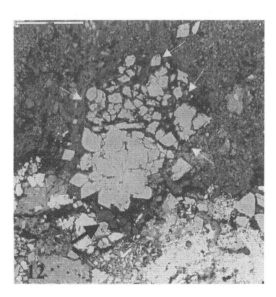

Figure 6-12. Fragment of dolomite rock undergoing obvious microfragmentation on its upper part (indicated by arrows). Note the fissure containing microorganisms (black arrow) under the fragment

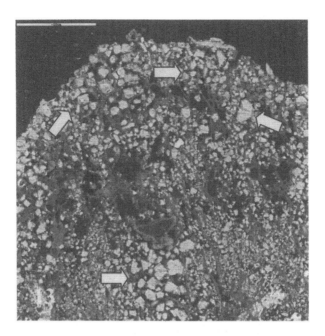

Figure 6-13. Accumulated microfragmented minerals (arrows) in the upper part of the thallus

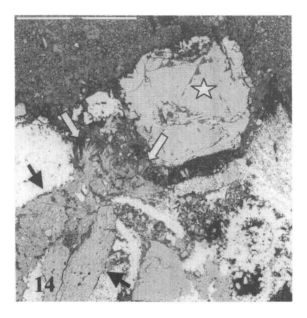

Figure 6-14. Detailed image of detaching substrate minerals (star) provoked by a wedge effect of fungi. The red alga (black arrow) has also suffered alterations (white arrows)

6. Microbial Biofilms on Carbonate Rocks

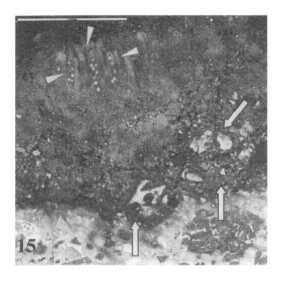

Figure 6-15. Ascocarp with asci and ascospores (white arrowheads), and mechanical disaggregation of the substrate's lower region (white arrows)

Figure 6-16. Soredium (marked by three white arrows) enveloped in microdivided minerals. Rhombi of dolomite (white arrowheads) and glauconite (black arrowheads) can be seen in the substrate

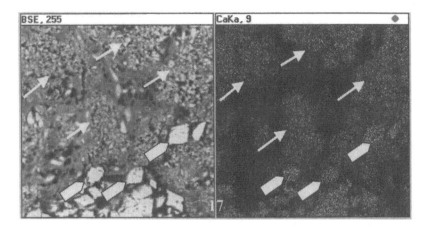

Figure 6-17. Image left: SEM-BSE image showing a detail of the thallus shown in Figure 6-13. Arrows point to calcium oxalate crystals and pentagonal arrows indicate microfragments of the substrate. The figure right is the digital map showing the spatial distribution of calcium in the region of oxalate particles (arrows) and area of microdivided substrate minerals (pentagonal arrows)

Chapter 7

LICHENS ON WYOMING SANDSTONE
Do They Cause Damage?

GIACOMO CHIARI[1] and ROBERT0 COSSIO[2]
[1]*Getty Conservation Institute, 1200 Getty Center Dr, Los Angeles Ca 90049, Email: gchiari@getty.edu;* [2]*Dipartimento di Scienze della Terra,Via Valperga Caluso 35 - 110125 Torino, Italy Email: roberto.cossio@ unito.it*

Abstract: Seven samples of sandstone covered with lichens from a Wyoming canyon that support petroglyphs were studied using a number of different techniques including COLORMOD, a porosimetry measurement based on color mode analysis of images, obtained from thin sections with impregnating resin containing a blue dye. This technique also allows one to measure the porosity gradient. Porosity was less toward the outside, since in the outer layer lichens occluded the pores. When the lichen body was counted as well, the porosity was proved to be the same as in the core of the rock. The sandstone composed largely of quartz, is homogeneous in grain size within a single rock, but differs greatly from one rock to another. ESEM imaging of lichen interaction with the sandstone showed a superficial layer (live lichens and small mineral particles) clearly distinguishable from the bulk (larger grains and pores). Consolidation tests were carried out using ethyl silicate [Wacker OH (with) and Monsanto Silbond (without catalyst)], on samples with the greatest and least porosity. Porosity decreased less for Silbond, since, without a catalyst and an active surface, it hardly polymerized. Therefore, if an ethyl silicate treatment is planned, it is advisable to use catalyzed products. The grain size distribution of the sandstone controls the physical properties: the larger the grains, the greater the porosity, water absorption, fragility and de-cohesion of the rock. Based on observations of thin sections, grain dislodgment due to lichens is difficult to demonstrate because lichen hyphae fill the gaps between grains, which under dry conditions seem to be large enough to accommodate the hyphae without exercising undue pressure on the structure of the rock. This result may help to determine the advisability of removing the lichens from the rock surface.

Keywords: Sandstone; lichens; porosity; image processing; consolidation; conservation

1. INTRODUCTION

This work was carried out as part of an interdisciplinary project, coordinated by Constance Silver, in order to study the deterioration process of sandstone engraved with ancient petroglyphs and covered by lichen encrustations. Other papers (including in this volume) present various aspects of this project: archaeological description, lichen characterization, conservation problems and their possible resolution (Silver 2000, St. Clair 2000; Tratebas Chapter 12 and Silver and Wolbers Chapter 8 in this volume). This paper concentrates on the chemical, physical, mineralogical and petrographic characterization of the stone, and on the interactions of the lichens on the sandstone properties.

The techniques used are:
1. simple microscopic observation of broken samples in reflected light;
2. observation using polarized microscopy of oriented thin sections, with particular care at the junction between the outer border, covered with lichens, and the bulk of the sandstone;
3. powder X-ray diffraction (XRD) both of the undisturbed bulk and of the surface layer, in order to characterize the sandstone and to determine the modifications produced in the rock by lichen activity;
4. environmental scanning electron microscope (ESEM) observation of a freshly cut, untreated section in order to observe, at micrometric level, the lichen-stone interaction;
5. porosity measurements carried out in three different ways, *via* water imbibition, *via* mercury porosimetry (on four samples only) and *via* COLORMOD procedure, which will be described in detail later;
6. liquid chromatography to quantify the soluble salts content (both for anions and cations).

Having characterized the sandstone, a preliminary trial of consolidation was attempted, selecting a silica based consolidant, compatible with the quartz matrix of the sandstone.

2. METHODS AND RESULTS

2.1 Preliminary inspection of the samples by reflected light

Seven rock samples of Early Cretaceous Lakota and Fall River Formation sandstone from the Black Hills were observed by reflected light microscopy, and

7. Lichens on Wyoming Sandstone

photographically documented (both the surface covered with live lichens and the unpolished sections of the stone).

The dimension of the grains is the most important datum that one can obtain from these observations. In fact, there is a great variability in the average grain size between samples collected from different locations in the canyon. In Figure 7-1 the two extreme cases are shown. The other samples were distributed somewhere in between. It is important to notice that the grain size distribution is homogeneous within each sample. This is because all the grains have approximately the same dimension, which causes poor packing, especially if the grain size is large. The empty spaces left among the grains are in fact of the same order of magnitude as the grains themselves. Therefore the total accessible porosity is high, and dependent upon the grain size.

One can also observe that the individual grains are almost all rounded, without sharp edges. This is due to the fact that the quartz sand was already dispersed, by wind or water, and abraded for quite a long time before it became part of the sediment to form the sandstone in question. The pseudo-spherical shape of the grains enhances the poor packing characteristic of the sand.

Figure 7-1. Both images were taken with the same enlarging factor: 5X. a) Broken, unpolished section of Sample #2, clearly showing the quartz matrix, tainted a reddish color by amorphous iron oxides. b) The same for Sample #6.2, which shows a much lighter color. The tip of a toothpick appears in the top right corner as a scale. The large difference in particle size (estimated to be 3:1) between the two samples is evident

Regarding the "strength" of the sandstone, the larger the particles, the smaller the internal adhesion, which is based not only on the forces interacting among the grains but also on the capacity of the grains themselves to interlock. The optimum packing situation is the one in which particles of different size are simultaneously present. Medium size particles fill the voids left by the largest particles; the finer particles, in turn, fill the voids remaining between the medium size particles, and so on. In this way the best packing is obtained and the rock is resistant to compression, to abrasion and to delamination, and presents a low degree of porosity, thus repelling infiltrating water and reducing problems arising from it. Irregular shaped grains and sharp edges also increase adherence, compressive strength, and, in general, physical resistance. As Figure 7-1 shows, the sandstone examined is far from ideally packed. Furthermore, the grain size differs from one location to another, thus presenting conservation problems, which have to be evaluated separately for each location of the petroglyphs.

2.2 Polarized microscopy of oriented thin sections

The use of polarized light did not add materially to the previous microscopic observations other than the determination of the mineral constituent, which was almost pure quartz. In addition to the quartz, a few crystals of phillosilicates (micas) can be detected. In the spaces between quartz crystals there are also very fine deposits of opaque minerals (possibly iron oxides) or dead lichens.

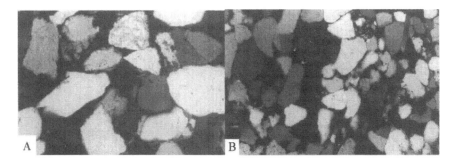

Figure 7-2. Sample #4 and Sample #6.2 seen through crossed polarizers at 10X enlargement. In Sample #6.2 a greater variation in particle distribution is present. As a consequence Sample #6 was very hard and difficult to cut

The size of the grains of Sample #6.2 (Fig. 7-2b) is smaller and they are less homogeneous (better packing), with the consequence that this rock is by far the hardest amongst the samples analyzed. To break and cut it required the use of heavy duty tools and it did not lose any particles during manipulation.

2.3 Powder X-ray diffraction (XRD)

X-ray diffraction analyses of all the samples were carried out in order to establish the mineralogical composition of the sandstone. A SIEMENS D5000 diffractometer was used (Bragg Brentano geometry in reflection mode, collecting data in the range 5 - 50° 2θ), and a HUBER G670 image plate Guinier camera (used in transmission with a capillary mounting, 8 – 100° 2θ). In both cases copper radiation was used.

The main result was that, for all samples, the bulk composition can be considered to be pure quartz, at least at the detection level of X-ray powder diffraction, which is of the order of 1-3 %, depending upon the substance (Fig. 7-3a). In Sample #3, some iron compounds, goethite and bemalite (Fe(OH)$_3$), were found in one part of the rock, but not in others. These iron minerals are probably responsible for the yellow color of the stone. They are poorly crystalline and therefore elusive to the diffraction technique. In Sample #6.2, besides quartz, there are also traces of mica (muscovite) and chlorite (nimite). These two minerals are at trace level and are of no relevance from the conservation point of view; they are detectable only because of the high quality of the pattern, since quartz produces a low background.

In Sample #27.2, besides the dominant quartz, there is a fair amount of goethite (yellow ochre). In this case the ochre must have been deposited with the quartz at the moment of the rock formation, since it uniformly permeates the bulk and it is crystalline (therefore detectable by XRD). In other cases, especially when rings or bands of color are formed on the surface, the iron oxides are transported by water, which dissolved or at least mobilized them, and then deposited them where the water itself evaporated. In this case, the iron compounds are non-crystalline in nature and therefore not visible with this technique.

Taking advantage of the Göblemirrors (a device that allows one to carry out an XRD on whole samples, without the need to reduce them to powder), Sample #55.2 was analyzed directly on the outer stone surface, which was covered with lichens (Fig. 7-3b). Where the lichens are located, the situation is rather complex. Quartz is of course present, being part of the stone substratum. As expected, weddellite, a calcium oxalate, was present since it is a typical by-product of lichens. The other substances present - a feldspar (albite), chlorite, mica (muscovite) - are present as trace minerals in the sandstone. Graphite (carbon) may be due to dry lichen bodies. In fact, the bulk of the same sample showed only quartz, traces of chlorite (nimite) and mica (muscovite). Calcite (c. 3-10%) was also present in the sandstone, while in all other samples carbonates were not found. A second run taking the material from a freshly cut surface confirmed this result. This sample is from the Fall River Formation, while those lacking calcite are Lakota sandstone.

In conclusion, XPD showed that all the samples (with the exception of Sample #55.2) are largely made of quartz. Minor minerals (micas, chlorite, feldspars and iron oxides) are present in an erratic way, mostly not homogeneously distributed; calcite was also present in Sample 55.2.

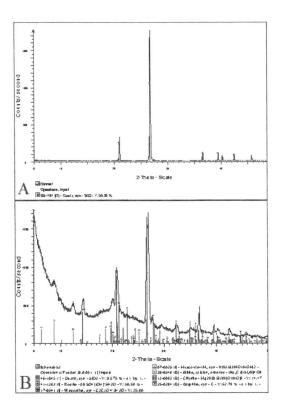

Figure 7-3. XRD patterns of a) the bulk of all the sandstone samples – only quartz is present, and b) the surface covered with lichens, as analyzed by Göblemirrors. The quality of the pattern is obviously much weaker, but a number of minerals (airborne and/or by-products of the lichens) can be detected

2.4 ESEM of a freshly cut, untreated section

Environmental Scanning Electron Microscopy (ESEM) imaging is a very powerful tool, especially when live organisms are present, as in this case. In fact, since the vacuum in the chamber is much reduced, there is no need to dry the organic tissue and gild it as for regular SEM. This allowed us to obtain an interesting series of pictures that were assembled together in order to obtain, at high resolution, a complete cross-section of the sandstone, from the surface to the bulk. These observations allowed us to establish the sandstone-lichen relationship, and possibly the modifications, if any, of the rock caused by the lichens. The ESEM of the Getty Conservation Institute of Los Angeles was employed to analyze a freshly cut section of Sample #55.2, which showed a large colony of lichens on the surface. Figure 7-4, composed of seven pictures, shows the cross section of the sandstone from the

7. Lichens on Wyoming Sandstone

surface, covered with living lichens, to the bulk where no lichen, but only grains of quartz can be observed.

This global picture is important in order to clarify the main issue: are the lichens mainly responsible for the deterioration of the stone or not? In Figure 7-4 it is evident that only the last three photographs portray the undisturbed sandstone toward the interior. Here the dimensions of the grains are fairly uniform and much larger than those shown in the first four photographs. One can estimate that the grains in the sandstone are in the range of 150 x 250 µm, while in the upper part the dimension of 25 µm is seldom achieved.

There is a difference in dimension between the particles in the region close to the surface and those of the bulk of one order of magnitude. The fungal hyphae and the small particles are limited to the four frames toward the surface. The shape of the small particles is characterized by sharp edges, none of which show signs of etching, as would be the case if they were derived from the breakdown of the larger quartz grains constituting the sandstone through chemical attack due to the lichens, but could be due to fracturing by mechanical action. Lichens could also desegregate the grains, by exercising pressure once the hyphae penetrated the pores. If so, at least a few large particles would be present in the superficial layer, but this is not the case. Another important point is that the mineralogical composition of the small particles is different from the bulk, as seen on analyzing the surface by XRD (see Fig. 7-3b). Finally, the dimensions of the small particles are compatible with airborne deposition of dust trapped by the lichen thallus and constituting a composite layer of biological (the living and dead thallus) and mineral (dust carried by the wind) material.

At the junction between the superficial layer and the bulk of the sandstone, it is possible to see parts of hyphae protruding from pores larger than the hyphae themselves. This may indicate that the lichens have enough space to proliferate on the sandstone without exercising pressure and therefore without damaging it. The layer toward the surface, including lichens and dust deposition has a thickness of c. 1200 µm. From the practical point of view of conservation, it remains to be seen if after careful cleaning and removal of this thin layer, the underlying stone is still undisturbed enough to show the petroglyphs or not.

Furthermore, it should be noted that the layer of lichens could actually act as a protection for the stone, isolating it from the direct impact of rain and sun, given the intrinsic weakness of this type of sandstone.

These observations of very minute details are basically in accord with the rest of the data obtained from the various other techniques employed in this study.

Figure 7-4. ESEM imaging of an untreated section of Sample #55.2 showing a layer of live lichens and airborne dust on the top part, and the large quartz grains of the sandstone on the lower part

2.5 Porosity measurements

Porosity of all samples was determined by two methods:

1. By water absorption, in order to determine the capacity of water intake and to measure, via the imbibition coefficient, the volume of pores accessible from the exterior (open porosity). The tests were carried out according to the NORMAL 7/81 specification. The samples were dried and weighed, then immersed in water and left for a specific time, and weighed again. The procedure was repeated until constant weight was reached (complete impregnation). Unfortunately, but necessarily, the specifications could not be followed in respect of the sample volume, since in no instance were cubes of 5 x 5 cm at our disposal. Since the volume of the samples used was small, the first few measurements of water intake versus time are quicker than expected. Nevertheless, the imbibition coefficients, as shown in Figure 7-5, maintain their full significance when used as an internal comparison.

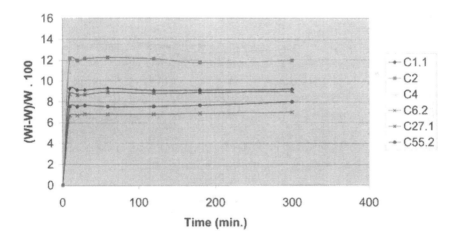

Figure 7-5. Results of the water absorption tests. Sample #3 is not reported since after the first few immersions it completely disaggregated. The water intake correlates nicely with the particle size: rocks with larger particles absorb more water. The porosity measured using COLORMOD gave overall comparable results

Values of the imbibition coefficient in the range 6-14 are very high, indicating that all the samples analyzed have large open porosity and are therefore prone to the intake of large quantities of water from external sources.

The water absorption test may be considered a very useful one since it is very easy to perform (it requires only a precision scale and a limited amount of working time), yet the amount of information that one can obtain from it is substantial and accurate. The empirical observation of resistance to abrasion (manipulation) correlates well with the porosity. Both can be ascribed ultimately to the particle size distribution of the sandstone. This is not surprising since the XRD analysis showed that the rock is almost completely composed of grains of quartz. Therefore there is less "chemical" difference between the various samples than physical difference. This can be found in the way the grains are closely packed, and how they are interlocked with one another. All this depends not only on the pressure exercized on the layers and the time of the formation of the sandstone, but also on the reciprocal dimension of the grains.

2. Measurements of porosity using the new procedure (COLORMOD) recently proposed by the authors, are based on color mode analysis of images (observed under the microscope) on specially prepared thin sections for which the impregnating resin has been colored with a blue dye. Since COLORMOD is based on counting the pixels of different colors, producing as a result the percentage of each selected color range with respect to the total, there is great advantage in enhancing the color contrast. Hence the use of the blue resin for the preparation of the thin sections, which produces blue pores which are clearly distinguished from the rest of the rock. This technique can be used both for macro porosity (as for the sandstone under study, for which photographs taken at the optical microscope are sufficient) and for micro porosity, where pores smaller than a few micrometers require the use of SEM for the image collection. The rock has to be previously impregnated with a marker, for example $AgNO_3$, followed by immersion in a NaCl solution. AgCl precipitates in the micropores and Ag is easily detectable in backscattered mode since it is much heavier than the rest of the atoms present. A further treatment of the image is necessary to transform it from black and white to false colors.

An advantage of this procedure is the possibility to measure the porosity gradient. These data can be obtained by first taking a picture of a cross section of the sample, oriented so that the outside border is visible. The image is then cut in succession into several frames, from the outside toward the inside. If the porosity values measured on the various frames are the same, one can conclude that the rock is homogeneous. This may imply, for example, that there is no particular deterioration process at play toward the surface. On the contrary, if the porosity is larger on the surface, one may conclude that there is deterioration.

7. Lichens on Wyoming Sandstone

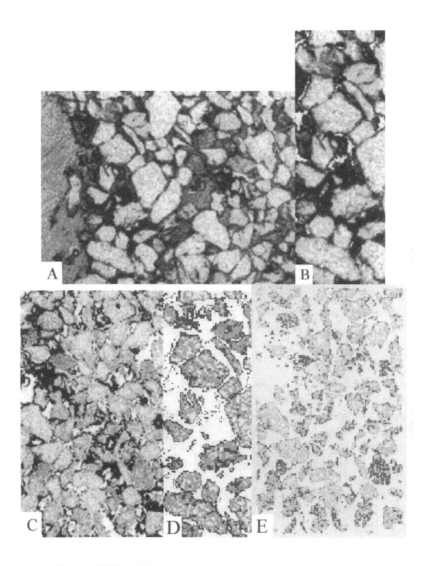

Figure 7-6. Sample #55.2 in thin section (5X) impregnated with blue resin. a) On the left the border covered with lichens. b) Detail near the border after counting the pores (3.6%). c) Detail toward the interior after counting the pores (14.7%). d) and e) are the same as b) and c), in which the dark part (lichens) have been counted as well, obtaining respectively total porosities of 19.4% and 19.7 %, which are statistically indistinguishable from the value of 19.8% obtained for the bulk

In Figure 7-6(a), even a superficial observation shows that the number of pores is far less on the left than on the right side. Conversely, on the left side, there is more of the darker material, interpreted as lichen. This image was therefore divided into two: Figures 7-6(b) and (c) on which COLORMOD was applied subtracting the blue regions which correspond to the resin filling the pores. On the interior (c), the surface area of the pores was 14.7% of the total, compared with 3.6% measured toward the exterior (b). This difference is unexpected, since, if the lichens cause cracks or corrode part of the stone, the porosity should be greater than where they are inactive, namely in the interior. On the other hand, in these images there is a brownish filling between the grains, which is absent in the bulk. COLORMOD was therefore run again taking into account these brownish parts as well, which includes not only the open pores but also the pores filled with the organic material. The cumulative result of both open and filled pores for both images was indistinguishable from the measurement obtained for the bulk, where there are no lichens. This result strongly suggests that the hyphae only fill the pre-existing pores, without enlarging them.

2.6 Determination of soluble salts by liquid chromatography

The presence of soluble salts in the rock, if abundant, could cause problems due to their volume expansion during crystallization. In fact, given the high porosity of the sample, water can have easy access to the interior, and during the evaporation process in warmer seasons can transport and deposit the salt inside the pores near to the surface. Almost all the sample showed a low cohesion of the grains, and such a phenomenon may result in serious desegregation and crumbling of the particles.

Liquid chromatography was therefore applied (using a DIONEX instrument) to quantify the presence of the commonest cations and anions. The results showed a virtual absence of ions, none of which were found in a proportion larger than a few parts per million. One of the potentially serious causes of deterioration on the rock was therefore ruled out.

2.7 Consolidation tests

The consolidant selected to perform preliminary consolidation tests was ethyl silicate, since, given the composition of the rocks (largely quartz), it seemed to be the most promising for compatibility reasons. Wacker OH (with a catalyst) and Monsanto Silbond (without) were applied by full immersion to Samples #2 and #6.2, having the largest and smallest porosity respectively. The rationale for applying a consolidant to these rocks is to increase the cohesion of the grains and at the same time to reduce the porosity. In fact, a reduced water absorption capacity is also an advantage from the point of view of preventing the growth of microorganisms. A less porous stone should remain drier.

7. Lichens on Wyoming Sandstone

Table 7-1. Porosity measurements carried out using the COLORMOD procedure on the samples with larger and smaller porosity, before and after impregnation with two types of ethyl silicate

Sample	#2	#6.2
Untreated	22.5	12.2
Silbond	19.5	9.8
Wacker OH	11.6	4.4

From Table 7-1 it can be seen that the use of Silbond hardly reduces the porosity of either sample, while Wacker OH is effective. This is not surprising, since Silbond is pure TEOS, without any catalyst. Since the sandstone is formed of quartz grains of rather large size, there is little possibility for the TEOS to react with it. Therefore, most of what was absorbed during the treatment evaporated or simply moved toward the outside under the effect of gravity. Wacker OH, on the contrary, reacts equally well without interaction with the stone, since it contains an organic solvent, which evaporates, and a catalyst, which starts the polymerization reaction.

The important conclusion is: if one decides to consolidate these rocks using ethyl silicate, the use of a catalyzed product (such as Wacker OH, Rhone Poulenc, CTS) is recommended if a greater reduction in porosity is desired and if introduction of the organic solvent and catalyst is acceptable (Fig. 7-7).

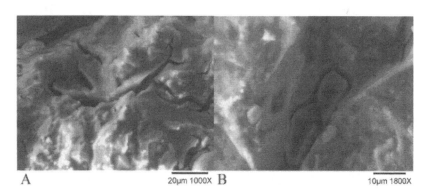

Figure 7-7. Two details of the appearance of the internal surface of the rock after treatment with Wacker OH. The smooth surface is silica gel, which with dehydration and polymerization of the product, cracks. The porosity is strongly but not completely reduced, since water can still penetrate through these cracks

This is a very important datum, which is useful both for comparing the various samples with one another, and for judging the "strength" or durability of a single stone. In fact, especially for a sandstone in which hard and resistant grains are packed together with little cementing matrix, the spaces which are left between the grains are a good indication of the packing. In other words, if the total porosity of our sandstone is high, we may infer that the general properties of the stone are not very good, the grains tend to be easily removed, the cohesion is low, the water penetrability high, the quantity of water absorbed in the same circumstances is high, and the number of microorganisms that can live on it is equally high.

3. CONCLUSIONS

Observations at ESEM show that there is a sharp boundary between the sandstone limit (larger quartz grains) and the superposed layer of lichen and much smaller stone grains, possibly airborne. This may suggest that the superficial layer is developed by the growth of composite material (lichen plus dust) accumulating *over* the stone, rather that *at the expense* of the stone. No evidence of etching was observed. Porosity measurements determined by mode analysis show that the pores are large and homogeneous. The pores are larger in those samples in which the grains are larger, as expected.

Measurements of the porosity gradient show that the total porosity near the surface is the same as in the bulk, provided that the organic material dispersed inside the cavities is counted as pores. If only the open pores are counted, the porosity close to the surface is about six times lower that in the core. This is difficult to justify, since if the lichens disturb the outer layer, cause cracks, dislodge grains or etch the quartz particles, the porosity near the surface should be larger than in the bulk. One can therefore conclude the following:

a) no etching of quartz grains was observed
b) the porosity is the same in the outermost layer and in the inside
c) the grain size is quite different between one rock and another
d) the water absorption coefficient also varies greatly and matches the grain size distribution (large grains = large porosity)
e) the porosity determined by COLORMOD correlates well with the data obtained by water immersion; this technique also shows that the bulk of the sandstone is homogeneous within a rock and that the region near the surface is similar to the bulk if one considers the filling material as pores; these data were confirmed by Mercury porosimetry for a few samples
f) rocks made of smaller grains are much more resistant to abrasion or surface deterioration in general.

All the above is based on microscopic observation, which takes into consideration only phenomena acting at that dimensional scale. Inspection of the petroglyphs shows that where lichens are not present, the condition of the petroglyphs does reflect differences in grain size and porosity of the sandstone. Where lichens are present, the rock surfaces are severely deteriorated despite differences in grain sizes and porosities (A.Tratebas, *pers.comm.*). The results of this study show that the pore spaces in the sandstone are large enough to accommodate lichens and may partially account for the abundance of lichens at the site. Deterioration of the sandstone under lichens would then be due to other factors. Obviously, the decision to keep or remove lichens is influenced by many other considerations, such as aesthetic or presentation reasons, or general site management.

ACKNOWLEDGEMENTS

We are grateful to Constance Silver for proposing this work, and to Larry St. Clair, Alice Tratebas, Bill Ginell and Norman Weiss for many constructive discussions during this project. We particularly thank the Getty Conservation Institute of Los Angeles, especially David Carson, for the ESEM pictures. Carmelo Sibio executed the thin sections, often having to employ novel methods, and Giorgio Carbotta analyzed the soluble salts. Luca Lucarelli of ThermoQuest Italia S.p.A. did the mercury porosimetry measurements. Urs Mueller gave us useful advice during the making of the COLORMOD program. This project was part of a study funded by the US Bureau of Land Management and the US Army Corps of Engineers Fort Worth District. The Progetto Finalizzato Beni Culturali of the CNR financed part of the research reported here.

COLORMOD is available as a freeware program on request to the authors.

REFERENCES

Silver, C.S. (2000) Transcript of team meeting held August 2000 at the Getty Conservation Institute of Los Angeles.
Silver, C.S. and R. Wolbers. (2004) Lichen Encroachment onto Rock Art in Eastern Wyoming: Summary of Conservation Problems and Prospects for Treatment. In: *Biodeterioration of Rock Surfaces.* (L. St. Clair and M. Seaward eds): 115-128. Kluwer Academic Press, Dordrecht.
St. Clair L.L. (2000) Characterization of lichen communities associated with rock art sites in eastern Wyoming. Technical Report submitted to the Bureau of Land Management. 6 pp.
Tratebas, A.M. (2004) Biodeterioration of Prehistoric Rock Art and Issues in Site Preservation. In: *Biodeterioration of Rock Surfaces.* (L. St. Clair and M. Seaward eds): 195-228. Kluwer Academic Press, Dordrecht.

Chapter 8

LICHEN ENCROACHMENT ONTO ROCK ART IN EASTERN WYOMING: CONSERVATION PROBLEMS AND PROSPECTS FOR TREATMENT

CONSTANCE S. SILVER[1] and RICHARD WOLBERS[2]
[1]*Preservar Inc., 310 Riverside Drive, New York, NY 10025, USA;* [2]*Winterthur Museum, Winterthur, and The University of Delaware, DE 19735, USA*

Abstract: Since lichen encroachment onto rock art panels of an important petroglyph site administered by the U. S Bureau of Land Management in eastern Wyoming is extensive, cultural resource managers included an evaluation of the lichen problem as part of a comprehensive management program. During 1999-2001, an interdisciplinary team composed of archaeologists, conservators, conservation scientists and lichenologists studied the problems posed by lichens at this site. This contribution summarizes the results of these studies, including new approaches to the use of non-toxic, enzyme biocides and cleaning methods.

Keywords: rock art conservation; lichens; biodeterioration; sandstone consolidation

1. INTRODUCTION

The study site where more than 100 panels of Paleo-Indian and Archaic rock art are found on sandstone rock faces and outcrops is located in a 12 km canyon at the western edge of the Black Hills in eastern Wyoming. This is one of the longest petroglyph sequences in the world and one of only two known areas of extensive Paleo-Indian rock art in the United States. The site was nominated to the National Register of Historic Places in 2001 in recognition of its outstanding cultural importance.

The petroglyphs of this site were created by chipping through the dark weathering rind of the sandstone to reveal the lighter-toned interior of the rock. The dark surface thus becomes the "background" for the light-toned images of the petroglyphs. To varying degrees, many different lichens occur on several rock art panels throughout the site: some panels have become heavily encroached upon, while others support only a few small colonies. Because some lichens have the capacity to significantly degrade stone over periods of time as brief as a decade (Seaward, Chapter 2, this volume), cultural resource managers determined that the management program for the site must include their study.

Preservar Inc. was engaged from 1999 to 2001 to conduct a comprehensive study of the impact of lichens at the site and investigate methods of treatment that would be applicable to Federally managed rock art sites in general. This project was administered by the U.S. Bureau of Land Management (BLM), which has jurisdiction over the site, and funded by the BLM and the Department of Defense through the U. S Army Corps of Engineers. This contribution, condensed from the full report (Silver, 2004), summarizes the results of the study and makes recommendations for the continuation of conservation research.

2. SUMMARY OF RESEARCH PROBLEMS

Rock art is beautiful and fragile art that also preserves critical data about past cultures. Indeed, in many instances knowledge of prehistoric cultures is derived from surviving rock art. Thus, when this art is destroyed by lichens, its loss impinges upon many fields of study. Lichens are easily killed by any number of biocides; the dead biomass is then removed mechanically, almost always with scrapers, scrubbing brushes and pressure washing. However, at the beginning of this study, it was posited that these standard treatments for lichen control would be inapplicable to rock art sites because:

1. Since lichens can sink hyphae deep into substrata, the mechanical action required to effect removal can easily abrade the surface and dislodge the granular components of the stone. Furthermore, even vigorous mechanical action cannot dislodge hyphae from deep within the substrata.
2. Biocides can provoke staining and have been shown to induce negative physical changes in some stones. In fact, some recent studies have implicated biocides in the creation of indurated crusts that result from the degeneration of lichen thalli (Cameron *et al.,* 1997).
3. Biocides leave behind modern contaminants with the capacity to skew current and possible future scientific methods for dating rock art.
4. Heightened concerns for environmental and occupational safety increasingly militate against the use of biocides.

Consequently, the research program for this site employed a holistic and interdisciplinary approach, so that the overlapping concerns and contributions of archaeologists, conservators, cultural resource managers and scientists could be addressed directly. The following specialists were engaged as a team: Constance S. Silver (Preservar Inc.), project director and conservation specialist; Professor Richard Wolbers (University of Delaware and the Winterthur Museum), conservation scientist and specialist in the cleaning of works of art; Dr. Larry St. Clair (Brigham Young University), lichenologist; Dr. Alice Tratebas (BLM), governmental project supervisor, archeologist and rock art specialist; Professor Giacomo Chiari (Turin University), conservation scientist and crystallographer; Professor Norman Weiss (Columbia University), conservation scientist and specialist in the deterioration and conservation of sandstone.

3. RESEARCH METHODS AND CHRONOLOGY

The study site was initially examined over a two-day period by Silver and Tratebas in 1999. The objectives were to:
a) create an overview of the rock art, including summary descriptions and photographic documentation of panels affected by lichens,
b) develop a general assessment of the nature of the lichen encroachment, and
c) gather samples for scientific analysis and tests of conservation treatments.

Thirty-two panels selected as the focus of the research program were examined by Tratebas, Silver and St. Clair in 2000 and 2001. The objectives of this fieldwork were to:
a) identify each lichen species,
b) document their exact locations on selected rock art panels by recording these positions on color photographs, and
c) study the extent to which micro-environments and possible changes in the environment of the canyon may influence the growth of lichens.

The Getty Conservation Institute (GCI) generously provided conference facilities and use of the environmental scanning electron microscope over the course of four days of meetings in 2000. Team members were able to describe and discuss their respective findings and coordinate on-going research; Dr William Ginnell, Chief Scientist (GCI), also participated.

4. PREVIOUS STUDIES OF LICHEN-INDUCED BIODETERIORATION OF SANDSTONE AND ROCK ART

A comprehensive review of scientific and conservation literature was undertaken. The corpus of literature in both fields is extensive, but ultimately it proved to have little practical application for the conservation problems of, and treatment caveats developed for, rock art.

At the first International Biodeterioration Symposium held in the United Kingdom in 1968 only one paper addressed the biodeterioration of stone by lichens (Lloyd, 1972). The introduction to this paper is quite interesting because it indicates that the deleterious effects of lichens on building materials had already become a cause for concern and that, by 1968, there had been very few published studies (see Piervittori *et al.*, Chapter 14, this volume) that addressed methods for their control:

The first systematic review of the literature on the chemical control of lichen growth established on building materials by Martin and Johnson (1992) lists about 100 articles. All describe the use of some sort of biocide, from relatively mild agents like household bleach to very toxic poisons, followed by mechanical action to remove the dead biomass.

The review of microbial deterioration of building stone by May *et al.* (1993), containing 185 literature sources, cites two very early and important studies, namely Geikie (1880) who measured the rates of weathering of exterior stone by studying tombstones in Edinburgh, and Julien (1884), a chemist on the faculty of Columbia University, who investigated the deterioration of brownstone, a form of sandstone used extensively as a building material in the United States in the 18th and 19th centuries. As will be discussed later, Julien's early studies proved to be relevant for the eastern Wyoming site.

The first published analytical bibliography specifically on lichen deterioration of building stones and some related materials by Piervittori *et al.* (1994) lists 316 papers, one early one by Buchet (1890) of interest for rock art studies because it examines the ability of lichens to colonize a siliceous material, glass, while clearly preferring and utilizing specific pigments (cf. Edwards and Seaward, 1993).

The contribution by Cameron *et al.* (1994) on biological growths on Scottish buildings is useful because it specifically examines the biodeterioration of sandstone, with an emphasis on the deleterious effects of various types of microflora. However, the global remedy for biodeterioration was the use of biocides, followed by mechanical action. Work by Kumar and Kumar (1999) on biodeterioration in tropical environments includes an excellent summary of the mechanisms of lichen deterioration of stone, together with a comprehensive list of biocides with results of tests on their efficacy. However, again the remedy is biocides and mechanical action.

When the focus is directed specifically to the problem of, and remedies for, biodeterioration of rock art, the literature is sparse. UNESCO's groundbreaking study by Brunet *et al.* (1985) was based on conservation studies for the great rock art of the Paleolithic caves of France and Spain. Although lichens were not a problem, other types of damaging microflora were identified. This study is especially noteworthy because it is one of the first to approach the conservation of rock art and the problem of biodeterioration holistically, including alteration of microclimates to inhibit growth of microflora.

The principal search engine for conservation research, *Bibliographic Database of the Conservation Information Network* (BCIN), contains about 500 abstracts specifically on the lichen deterioration of cultural property, the earliest dating from the 1920s. However, out of this large corpus of conservation literature, only about 20 papers since 1970 have specifically addressed the issue of lichen deterioration of rock art. Moreover, all report the use of biocides, followed by mechanical action, as the method for removal and control. Only one paper (Tratebas and Chapman, 1996) examines the dating of rock art by highly sensitive analysis of rock coatings and the potential problem of contamination when biocides are used.

5. LICHEN DETERIORATION OF SANDSTONE AT STUDY SITE

During two visits to the study site, 62 species of lichen were identified on the rock art panels (St. Clair, 2000), all being typical of the local sandstone and for the geographical area (Fig. 8-1); of these, nine were identified as particularly aggressive to the stone, and/or fast-growing including: *Acarospora* spp. (3 species), *Peltula euploca, Sarcogyne regularis, Polysporina simplex, Caloplaca decipiens, Lecanora argopholis* and *Physciella chloantha.*

Microclimates around specific panels were also evaluated. At 16 panels, it was posited that the adjacent growth of the macroflora was creating microclimates that encouraged the growth of lichens; for example, one panel is located adjacent to standing water in a directly shaded area, and it has become heavily encroached upon by the sorediate lichen *Caloplaca decipiens* (Fig. 8-2). Comparisons of photographs taken over the last 20 years indicate that changes in the configuration of the standing water have created a microenvironment that has promoted this lichen's growth. In areas where panels were shaded by vegetation, *Peltula euploca* was identified as a problem. The most aggressive species, defined as such because of their erosive action on the rock surfaces, were *Lecanora argopholis, Acarospora* spp., *Sarcogyne regularis, Polysporina simplex,* and *Caloplaca decipiens.*

Figure 8-1. Lichens on sandstone rock art panels

Figure 8-2. Panel in shaded area above standing pool of water is heavily impacted by *Caloplaca decipiens*

6. CHARACTERIZATION OF SANDSTONE FROM STUDY SITE AND PROSPECTS FOR CONSOLIDATION

Study of the sandstone at the site (Chiari, 2001) had two principal objectives: firstly to characterize the sandstone in order to understand its susceptibility to invasion and deterioration by lichens, and secondly to test consolidants to ascertain if lichen-damaged sandstone can be strengthened by chemical consolidation, and to anticipate any possible negative impacts that might result from its use.

Sandstone samples with lichen attached collected from rock fall adjacent to seven panels located in different areas of the canyon were characterized through a series of laboratory analyses:
1. Preliminary examination by reflected and transmitted light microscopy
2. Characterization of porosity using water-absorption tests
3. Porosimetry examination using the COLORMOD system
4. Ionic chromatography analysis
5. X-ray diffraction (XRD)
6. Environmental scanning electron microscopy (ESEM) examination of the spatial distribution of the lichen on and within the sandstone.

In conjunction with these analyses, tests of consolidation using an ethyl silicate-based consolidant were carried out.

The resulting analytical data identified, for the most part, poorly cemented sandstone composed primarily of quartz grains. Not surprisingly, samples collected at different locations within the canyon varied in grain size and porosity, ranging from a large-grained sample with a high porosity, to a sample with a mixed but smaller mean grain size and lower (but still comparatively high) porosity.

Studies were undertaken to ascertain if the sandstone could be consolidated. Sandstone substrata exposed when lichens are removed are far more vulnerable to weathering, especially freeze-thaw cycles, than the adjacent hard-packed desert varnish surface. Thus, it was posited that they would require a consolidating treatment that would restore some resistance to weathering following removal of lichens.

Samples were treated with a consolidant, *Conservare OH®* (produced by ProSoCo Inc.). Consolidants based on alkali silicates (also referred to as ethyl silicates) have been in existence for more than a century, being, for example, used in Germany to seal egg-shells in order to prolong shelf-life before the invention of refrigeration. Their potential to strengthen degraded architectural stone was recognized relatively early and over the last 25 years alkoxysilanes have been used successfully to consolidate sandstone in many environments (Grissom and Weiss, 1981).

The properties of ethyl silicate consolidants make them highly attractive for use on sandstone. Applied in a liquid state, they have an extremely small molecular structure that permits them to penetrate deeply into deteriorated stone. They collect at contact points between the individual grains of the stone. A catalyst, in conjunction with atmospheric humidity, acts to convert the liquid state to glassy silicon dioxide (SiO_2), which binds the stone's particles together. The resulting deposition is virtually identical to naturally occurring cementing material found in (silica) sandstone, and thus has virtually the same chemical qualities and thermal expansion/contraction properties as the stone. Although ethyl silicates are so compatible with sandstone, there has been scant investigation of their potential use in the conservation of rock art. Some testing of ethyl silicates to consolidate unstable paint at a rock art site in Australia has been attempted, although the outcome has not been reported in the conservation literature (Watchman, 1995).

Following consolidation, the treated samples were examined with the scanning-electron microscope. The resulting images clearly showed the deposition of silica in the pores and between the grains of quartz. However, additional tests will be required to determine the extent to which consolidation increases overall strength and whether it may negatively affect vapor transmission by sealing some pores.

7. FURTHER STUDIES ON SANDSTONE DETERIORATION BY LICHENS

Identification of sandstone composed almost uniformly of quartz prompted questions about the ability of lichens to provoke chemical degradation. To examine this issue, it is essential to return to the work of Julien (1884) who had observed empirically the degradation of quartz in organic environments. However, he could not explain this observation scientifically because it was then believed that only hydrofluoric acid had the capacity to chemically degrade quartz. Since hydrofluoric acid is not found in nature, Julien was at a loss to explain his observations.

During this study, scientists shared Julien's concerns. Indeed, it was even posited that the lichens could not be a major threat to the rock art because they lack the chemicals (assumed to be extremely strong acids) required to corrode the quartz of the sandstone. However, it was obvious to the participating conservators that, where lichens are present on rock art panels, the surfaces are always altered and damaged. The desert varnish had been markedly thinned or removed in ameba-like configurations that are clearly commensurate with lichen colonization. In these areas, there is a marked disaggregation of the grains, abrasion and pitting which leave the vulnerable inner substrata of the sandstone exposed to weathering.

The conservators believed the lichens to be responsible for most of the deterioration of the rock art at the study site and thus came to occupy the position held by Julien more than 120 years earlier: quartz clearly degrades in an organic environment, but the mechanisms of degradation remain elusive. However, with additional research by the conservators, mechanisms that have the potential to degrade the quartz-based sandstone of the site were located in the scientific literature. Clearly, issues of communication between scientists and conservators were some of the unexpected topics that was raised in the course of the interdisciplinary study.

First, and most obviously, lichen thalli have the ability to penetrate into even the smallest pores of sandstone (see Chiari and Cossio Chapter 7, Fig.7-4, this volume). Once established in the pores, they are able to swell and contract, exerting pressure against the grains. As grains are wedged out, water can enter and continue to exert internal pressure through freeze-thaw cycles. However, these mechanisms do not explain the observed chemical erosion and corrosion of the sandstone's surfaces in conjunction with the lichens.

Julien's scientific mystery, the chemical degradation of quartz in organic environments, apparently remained unsolved for 100 years until several scientists in different fields began to re-examine this phenomenon (Silvermann, 1979). In some cases, the re-examination was made possible by advances in analytical techniques. However, it had been known for some time that there is a class of enzymes referred to as chemi-lithotropics, produced by bacteria and working on various inorganic substrata, employing all the characteristic range of activities found on organic substrata. Bond-breaking lyases or oxido-reductases, are enzymes that can break chemical bonds and oxidize or reduce either organic or inorganic materials. Although Hallbauer and Jahns (1977), with the aid of the scanning electron microscope, reported lichens attacking quartzite rock surfaces in "real time" and several other excellent studies in the 1980s and 1990s focused on lichen deterioration of sandstone in Antarctica (Weed and Norton, 1991), these studies do not provide cogent explanations for the mechanisms at work.

The work by Banfield *et al.* (1999) provides comprehensive answers to many questions raised about the biodeterioration of silica-based rocks. These authors pose the question: what can we learn from the lichen-mineral microcosm? Perhaps the most important answer they give is that lichen colonies in fact provide a safe harbor for microbes which break down stone as follows:

> A wide variety of microorganisms colonize mineral surfaces, among the most familiar of which are lichens. Although classically described as symbiotic associations between photosynthetic microorganisms and fungi, lichens are actually extremely complex microbial communities. A mass of fungal hyphae, or thallus, composes the majority of any lichen. Photo-synthetic microorganisms lie just beneath the upper surface. Although these are typically green algae, other photosynthetic microbes such as diatoms and cyanobacteria occur. This upper zone is a region of carbon transfer, in the form of sugars, between the photosynthesis deeper within the lichen thallus, but other prokaryotes reside among the fungal hyphae. Little is known regarding the biodiversity of this non-

photosynthetic assemblage, and the role these organisms play in nutrient transfer from substratum to fungus is a fertile area for research.

The authors point out that acid production is the essential mechanism through which microbes affect the degradation of stone:

> Solutions of organic acids in concentrations comparable to or slightly higher than ground water show increase in dissolution rates of less than one order of magnitude. These relatively small effects may hide much larger responses in natural systems in which local microenvironments may be characterized by very high acid concentrations because of cell proximity.

> It is possible to demonstrate directly that microbes can cause low pH microenvironments at mineral surfaces . . . As acidity increases, below pH = 5, the rates of silicate mineral dissolution increase by a factor of a^n_{H+}. Lowering pH to 3-4 corresponds to a 10- to 1000-fold increase in dissolution rate. . . In addition to inorganic acid production, microbes also can catalyze mineral weathering rates by production of organic ligands. Ligands can complex with ions on the mineral surface and can weaken metal-oxygen bonds. Alternatively, ligands indirectly affect reactions by forming complexes with ions in solution, thereby decreasing solution saturation state. Experimental studies using relatively dilute solutions of compounds such as oxalic acid, citric acid, pyruvate, α-ketoglutarate, acetate, propionate, lactate, etc. have shown rate enhancement for silicate dissolution of up to one order of magnitude. The effect is somewhat similar to that of acid production because organic ligands affect silicate mineral dissolution stoichiometry by complexing with, and increasing the solubility of, less soluble major ions such as Al and Fe.

The most detailed and conclusive study of the effects of organic compounds on quartz was by Bennett *et al.* (1988) who pointed out that high concentrations of dissolved silica are often present in organically rich environments, such as soil pore liquid, as observed by Julien (1884). The study also described how the organic complextion of silica had been hypothesized for more than 100 years to explain an observed natural phenomenon, dissolved silica in organically rich environments. Although precise explanations still remain somewhat elusive, the authors present comprehensive and convincing scientific evidence for this phenomenon:

> The relative ability of the various organic acid anions to accelerate quartz dissolution provides further evidence of the nature of the interaction from the unique structure of each organic acid molecule. In our study, only the multi-protic acids (oxalate and citrate) or the multi-functional acids (salicylic acid) accelerated quartz dissolution. Acetic acid, a mono-functional mono-protic acid did not accelerate the dissolution of quartz, though the results of the dissolution

experiments, which showed a decreasing dissolution rate with increasing acetate concentration, suggest that acetate may absorb onto the surface of the quartz grains. This adsorbed layer of organic molecules would have the charged carboxylic sites pointed toward the quartz surface, leaving a hydrophobic outer surface.

At the conclusion of their experiments, the authors suggest that:

> ... simple organic acids at concentrations encountered in organic rich soils and weathering zones indeed complex silica aqueous systems at neutral pH. The formation of a silica-organic acid complex lowers the activity of free monomeric silica in solution, allowing continued dissolution of quartz until equilibrium is re-established. Complextion at the hydrated quartz surface may also decrease the activation energy of the rate limiting step of dissolution in water. The combined effect is to increase the apparent solubility of quartz in aqueous system as well as increasing the rate of dissolution. Silica/organic acid complextion also would aid in the accelerated dissolution of alumino-silicate minerals under pH conditions that are unfavorable for aluminum complextion.

8. A NEW APPROACH TO THE PROBLEMS INVOLVED IN LICHEN REMOVAL

Research was undertaken by the authors to develop methods for the removal of lichens on and within the sandstone. Avoidance of the use of biocides and limitation of the extent of mechanical action to remove the dead biomass were research goals. An additional goal was to develop a method that would not affect the rock chemistry or provoke alterations and contaminations that could compromise current and future methods for the dating of the rock art.

Enzymes have been used in conservation for many years to remove small quantities of organic materials, such as bloodstains and glue residues from fragile objects and works of art. It was posited that enzymes might also work effectively to kill lichens and facilitate their removal from substrata. Enzymes presented many potentially attractive qualities because, if successful, they would function both as non-toxic biocides and low-impact cleaning agents by:

a) essentially liquefying the lichen biomass, thus permitting its removal with little mechanical stress applied to the stone,
b) dissolving the sub-surface elements of the lichens, facilitating all or much of their removal, and
c) avoiding contamination of the stone with chemicals.

Lichen-covered sandstone from the site was used by the authors to test the efficacy of enzymes as a biocide and cleaning agent. The experiments were conducted in the Conservation Laboratory of the Winterthur Museum. The enzymes were mixed into a gel medium using lysing enzyme (Sigma Catalogue no. L.2260) from *Trichoderma hazarium* and lyophilized powder containing cellulose, protease, and chininase (prepared for Sigma Chemical).

The gel was applied to the lichens with a brush and left in contact for 20 minutes. The lichens were effectively liquefied and could easily be removed by rinsing with deionized water. Only minimal mechanical action, some light brushing with a soft brush, was required to remove residues The treated rocks were cut into cross-sections and studied at 125X magnification before and after treatment in normal light and fluorescent light after staining with fluorescent dyes. Considerable amounts of lichen material had been removed from the sandstone.

In theory, this treatment should not contaminate the sandstone because it employs organic compounds similar to those found in living organisms such as lichens. However, the complex scientific analyses required to ascertain whether enzymatic cleaning is truly free of contaminants were beyond the scope of this project and must await future research efforts.

9. CONCLUSIONS AND RECOMMENDATIONS FOR FURTHER RESEARCH

Several conclusions were reached in the course of this interdisciplinary study of the eastern Wyoming site. First, visual examination of the sandstone indicated that the lichens do pose a substantial threat to the rock art. The exact mechanisms of deterioration require further study, but the lichens clearly have the capacity to cause considerable disruption to the packing of the quartz grains through pressure exerted by fungal hyphae insinuated into the pores of the sandstone.

Generally overlooked research since the late 1970s on the dissolution of quartz in organic environments provided important data that explain how quartz is profoundly degraded by organic acids. These data may provide avenues for future research that may further explain how lichens and other microorganisms chemically degrade sandstone

A comprehensive review of conservation literature was less fruitful. Many papers describe the deterioration of rock art by a variety of agents, including lichens. However, there were few articles that discussed successful conservation treatments. Only one article discussed the potential to compromise present and future dating of rock art by chemical contamination from biocides used during treatment. While there are many outstanding papers that detail successful site preservation and management, the general tenor of the conservation literature is that most rock art, like rock itself, is inexorably linked to geological time and is thus fated to deteriorate.

Conservation treatments researched during this study suggest that procedures can be developed for long-term control of some agents of deterioration. Use of enzymes appears to provide a more benign biocide and cleaning system for the control of lichens. Consolidation tests using ethyl silicate-based consolidants show that silica was successfully deposited within the fabric of the site's sandstone and could help restore strength and resistance to areas degraded by lichens.

These treatments will require further studies to determine if there are any negative side effects associated with their use. Many recent advances in technology may well offer additional avenues for conservation treatments of rock art that will be both benign and effective. Observations at this study site indicate that conservation of rock art is an issue that should now be re-evaluated and re-examined.

ACKNOWLEDGEMENTS

The authors wish to express their gratitude to the U.S. Bureau of Land Management, the U.S. Army Corps of Engineers (Fort Worth District) for sponsoring this research and the Getty Conservation Institute for use of their facilities. We are also grateful to Dr. Alice Tratebas (Bureau of Land Management, Newcastle, Wyoming) for her support throughout the various components of this project and to Professor M.R.D.Seaward and Dr. L.L. St. Clair for their considerable help in the preparation of this paper.

REFERENCES

Banfield, J.F., Barker, W.W., Welch, S.A. and Taunton, A. (1999) Biological impact on mineral dissolution: application of the lichen model to understanding mineral weathering in the rhizosphere. *Proceedings of the National Academy of Science* 96: 3404-3411.

Bennett, P.C., Melcer, M.E. Siegel, D.I. and Hassett, J.P. (1988) The dissolution of quartz in dilute aqueous solutions of organic acids at 25 degrees centigrade. *Geochimica et Cosmochimica Acta* 52: 1521-1530.

Brunet, J., Vidal, P. and Vouve, J. (1985) *Conservation de l'Art Rupestre*. UNESCO, Paris.

Buchet, G. (1890) Les lichens attaquent le verre et, dans les vitraux, semblent préférer certaines couleurs. *Céreal Séance Société Biologique*, Sér.2: 13.

Cameron, S., Urquehart, D., Wakefield, R. and Young, M. (1997) *Biological Growths on Sandstone Buildings*. Masonry Conservation Research Group, Robert Gordon University, Edinburgh.

Chiari, G. (2001) *Report on the Study of the Sandstone from Whoopup Canyon (48WE33)*. Unpublished report on file, Bureau of Land Management, Newcastle Field Office, Newcastle, Wyoming.

Edwards, H.G.M. and Seaward, M.R.D. (1993) Raman microscopy of lichen-substratum interfaces. *Journal of the Hattori Botanical Laboratory* 74: 303-316.

Geikie, A. (1880) Rock weathering as illustrated in Edinburgh churchyards. *Proceedings of the Royal Society of Edinburgh*, 518-532.

Grissom, C.A. and Weiss, N.R. (1981) Alkoxysilanes in the conservation of art and architecture: 1861-1981. *AATA Supplement* 18: 149-204.

Hallbauer, D.K. and Jahns, H.M. (1977) Attack of lichens on quartzitic rock surfaces. *The Lichenologist* 9: 119-122.

Julien, A.A. (1884) The durability of building stones in New York City and vicinity. In: *US 10th Census 1880, Vol.X. Special Report on Petroleum Coke Building Stone*: 364-384. New York, New York.

Kumar, R. and Kumar, A.V. (1999) *Biodeterioration of Stone in Tropical Environments*. The Getty Conservation Institute, Los Angeles.

Lloyd, A.O. (1972) An approach to the testing of lichen inhibitors. In: *Biodeterioration of Materials*, Vol. 2. (A.H.Walters and E.H.Hueck van der Plas, eds.): 185-191. Wiley, New York.

Martin, A.K. and Johnson, G.C. (1992) Chemical control of lichen growths established on building materials: a compilation of published literature. *Biodeterioration Abstracts* 6: 101-117.

May, E., Lewis, F.J., Pereira, S., Tayler, S., Seaward, M.R.D. and Allsopp, D. (1993) Microbial deterioration of building stone – a review. *Biodeterioration Abstracts* 7: 109-123.

Piervittori, R., Salvadori, O. and Laccisaglia, A. (1994) Literature on lichens and biodeterioration of stonework. I. *The Lichenologist* 26: 171-192.

Silver, C.S. (2004) *The Rock Art of Site 48WE33: research on lichen-induced deterioration and prospects for conservation treatment*. Unpublished report, Bureau of Land Management, Newcastle, Wyoming.

Silvermann, M.P. (1979) Biological and organic chemical decomposition of silicates. *Studies in Environmental Science 3, Biogeochemical Cycling of Mineral-forming Elements* (P.A.Trudinger and D.J.Swaine, eds.): 445-465. Elsevier, Oxford.

St. Clair, L.L. (2000) Characterization of lichen communities associated with rock art sites in eastern Wyoming. Technical Report submitted to the Bureau of Land Management, Newcastle Field Office, Newcastle, Wyoming.

Tratebas, A.M. and Chapman, F. (1996) Ethical and conservation issues in removing lichens from petroglyphs. *Rock Art Research* 13: 129-133.

Watchman, A. (1995) *Paint Stabilization by Using an Artificial Silica Gel Consolidant*. Unpublished report, Australian Institute of Aboriginal and Torres Strait Islander Studies, Canberra.

Weed, R., and Norton, S.A. (1991) Siliceous crusts, quartz rinds and biotic weathering of sandstones in the cold desert of Antarctica. In: *Diversity of Environmental Biochemistry* (J. Berthelin, ed.): 327-339. Elsevier, Amsterdam.

Chapter 9

LICHEN BIODETERIORATION AT INSCRIPTION ROCK, EL MORRO NATIONAL MONUMENT, RAMAH, NEW MEXICO, USA

KATHRYN B. KNIGHT, LARRY L. ST. CLAIR and JOHN S. GARDNER
Department of Integrative Biology, Brigham Young University, Provo 84602, U.S.A.

Abstract: El Morro National Monument is located in Cibola County, New Mexico, about 200 km west of Albuquerque, New Mexico. The main attraction at the Monument is Inscription Rock, a large sandstone formation which has become a natural repository for more than 700 years of comments and notations recorded by travelers through the region. Recently, encroachment of lichens onto several inscription panels has been documented. Because lichens are known to decompose rock surfaces physically and chemically, lichen communities on several panels have been studied. Species and relative abundance data have been compiled. Distribution of lichen species on each panel has been photographically documented in order to monitor future growth/encroachment trends, exfoliation patterns, and changes in species composition. Samples of common lichen species on the Inscription Rock panels were collected from a large, separate boulder field north of the Rock. These samples have been analyzed using thin-layer chromatography and SEM and light microscopy. PIXE analysis was also performed on *Usnea hirta* samples from five locations in the Monument to determine if air pollution is influencing substrate degradation patterns. Since establishment of the Monument, protection from grazing and wood cutting/gathering has resulted in changes in the microenvironment in and around several of the panels. SEM and light microscopy analysis have shown that mycobiont hyphae regularly penetrate up to 20mm into rock surfaces. TLC analysis has shown that virtually all the major lichen species on the panels produce at least 4-5 secondary chemicals including several organic acids. Elemental analysis data from *Usnea hirta* samples confirm that all potential pollutant elements are well within background levels. Preemptive recommendations include: 1) continue to monitor growth plots; 2) set up biocide studies; and 3) selectively remove the woody vegetation adjacent to some panels.

Keywords: lichens; petroglyphs; rock art; conservation

1. INTRODUCTION

Lichens are known to actively decompose rock surfaces. In natural settings this phenomenon contributes to a process called pedogenesis, or soil formation. The process is both physical and chemical in nature. Physical degradation occurs as mycobiont hyphae invade and destabilize rock surfaces. In addition to physical degradation, organic acids produced by saxicolous lichens enhance chemical decomposition processes.

Lichen-mediated biodeterioration of rock surfaces has become a serious concern at El Morro National Monument in west central New Mexico. The Monument is a historic and prehistoric landmark with Inscription Rock as its central feature. This formation has several prehistoric petroglyphs as well as many historically significant inscriptions. In some cases lichens are growing on or over some of the petroglyphs and inscriptions. Because many lichens actively degrade rock surfaces, the National Park Service commissioned this study to evaluate the nature and extent of the lichen biodeterioration problem at Inscription Rock.

2. PROJECT OBJECTIVES

The objective of this study was to determine if lichens are actively involved in the biodeterioration of petroglyphs and/or inscriptions at El Morro National Monument. The following six sub-objectives were used to specifically define the nature and extent of the biodeterioration problem at El Morro:
1. Summarize the known effects of lichens on sandstone in arid habitats.
2. Identify lichen species in specific focal areas and provide information about any threatened, endangered or endemic species.
3. Determine the extent of mycobiont encroachment into sandstone substrates using scanning electron and light microscopy.
4. Evaluate potential air pollution-related impact on local substrates by determining concentrations of airborne pollutant elements in sensitive indicator species using PIXE analysis.
5. Characterize secondary chemistry of lichens from focal areas using thin-layer chromatography.
6. Photographically document distribution and growth patterns of lichens in specific focal areas. Where possible, taxa from these images have been identified to species level.

3. BACKGROUND HISTORY AND GENERAL HABITAT DESCRIPTION OF EL MORRO NATIONAL MONUMENT

El Morro National Monument, located in Cibola County, New Mexico, c. 200 km west of Albuquerque, was established in 1906 and includes 1278 acres. The main attraction at the Monument is Inscription Rock, a natural repository for more than 700 years of comments and notations by travelers throughout the region, and more than 2000 petroglyphs and inscriptions are preserved on its lower reaches.

The climate at El Morro National Monument is hot and dry in the summer with afternoon thunderstorms common in July and August. In the winter, temperatures commonly drop below freezing. The area receives an average of 15 to 16 inches of precipitation annually, most of which falls as rain in July and August or as snow in from December to February. Elevation at the base of Inscription Rock is about 7200 feet, increasing by about 200 feet at the top of the mesa.

Common vascular plants in the Monument include *Pinus edulis, P. ponderosa, Juniperus scopulorum, J. monosperma, J. deppeana, Bouteloua gracilis, Yucca baccata, Y. glauca, Rhus trilobata, Mahonia repens, Opuntia spp., Symphoricarpos utahensis, Atriplex canescens, Eurotia lanata, Artemisia tridentata, Chrysothamnus nauseosus, C. viscidiflorus, Gutierrezia sorothrae, Quercus gambelii, Amelanchier mormonica,* and *Ribes inebrians.*

The geology of Inscription Rock includes Jurassic-age Zuni Sandstone capped by Cretaceous-age Dakota Sandstone. Surface water runoff, capillary movement of groundwater, and vertical infiltration all contribute to deterioration of rock surfaces. Lichens are most abundant in shaded locations, seep lines and spillways and appear to be playing an active role in the deterioration of rock surfaces.

4. EFFECTS OF MICROFLORA ON ROCK SUBSTRATA

Lichens, algae, cyanobacteria, and bryophytes commonly occur on soil, bark, and rock substrates in arid habitats. The commonest biological component of exposed rock surfaces in the western United States is lichens. Lichens are symbiotic systems consisting of a fungus (usually an ascomycete) and a eukaryotic alga and/or a cyanobacterium. Generally, lichens in arid habitats are slow-growing. However, growth rates vary depending on growth form, microhabitat, and species. Crustose lichens, the most common growth form in arid locations, tend to have the slowest growth rate. In contrast, foliose and fruticose lichens have much faster growth rates (St. Clair, 1999).

Weathering of rock surfaces (pedogenesis) is a complex, natural activity involving mechanical, chemical, and biological processes (Syers and Iskandar, 1973). Mechanical weathering typically precedes and enhances chemical weathering by increasing substrate surface area and are known to play an important role in pedogenesis (Syers and Iskandar, 1973). Frequently, lichens are the first living things to occupy newly exposed rock surfaces. The degree to which they contribute to pedogenesis is dictated by a number of factors, including the chemical nature and hardness of the rock; local environmental conditions (e.g. temperature and precipitation patterns); and the specific structural and chemical attributes of the lichen flora.

Lichens contribute to mechanical weathering of rocks in four ways:
1. penetration of mycobiont hyphae (up to 15-20 mm) and rhizines into naturally occurring crevices and cracks in rock surfaces;
2. expansion and contraction of lichen thalli due to changes in temperature and hydration status (Nash 1996);
3. swelling action of organic salts produced by lichens; and
4. fracturing and incorporation of mineral fragments by lichen thalli (Chen et al., 1999).

Water is essential for many of the chemical reactions associated with the breakdown of rock substrates. Because lichens are able to absorb water in either the liquid or vapor phase, chemical and physical weathering processes are expedited on lichen-covered rock surfaces. The mixing of respiratory CO_2 with water in lichen tissues results in the formation of carbonic acid, which also enhances the solubility of rock surfaces by lowering the pH of the substrate microenvironment adjacent to lichen thalli (Chen et al., 1999).

Many lichens produce secondary chemicals, including various weak organic acids, which actively chelate substrate cations, and thus modify the chemical and physical structure of mineral substrates (Jones 1988); for example, oxalic acid produced by many lichen species forms chemical complexes with substrate cations (Chen et al., 1999). Specifically, oxalic acid reacts with calcium carbonate substrates, forming the insoluble compound calcium oxalate. Because calcium oxalate is insoluble, it accumulates on the surface and within lichen thalli, or at the lichen-rock interface (Edwards et al., 1998). Calcium oxalate also accumulates on substrate surfaces, leaving significant and often unsightly white deposits. This can be a problem for delicate and intricate stone monuments because white calcium oxalate deposits often obscure the detail and historical significance of the structure (Seaward & Edwards, 1995). Oxalic acid also solubilizes magnesium silicates (Jones, 1988). Other mineral substrate elements also form oxalates, which can be equally destructive; for example, breakdown of sandstone surfaces beneath many lichens can be explained by dissolution of the sandstone cement, a process which is enhanced by the formation of Si, Fe and Al oxalates (Johnston and Vestal, 1993).

9. Lichen Biodeterioration at Inscription Rock

Lichens appear to be affecting several of the inscriptions at El Morro National Monument. The purpose of this research project is to assist the National Park Service to better understand the impact of lichens on sandstone substrates in general and more specifically the various notations on Inscription Rock. National Park Service personnel will use the information from this project to improve monitoring protocol and to develop more effective treatment methodologies.

5. METHODS

5.1 Collecting, curation, identification and deposition of lichen specimens

In order to accurately document lichen species in the focal areas, specimens were collected from adjacent, less sensitive areas, including the rocky area north of Inscription Rock. To fully characterize lichen biodiversity in the Monument, specimens were also collected from all bark and soil. Because distribution of lichens is directly influenced by substrata, moisture, and sunlight, all available substrata and habitats at each reference site were carefully examined and lichen specimens from them were collected. Small samples of each species were either removed directly from the substratum, or, depending on the species, with a small amount of the substratum (bark, wood, rock, or soil).

Species were identified using standard lichen keys and taxonomic treatises. Chemical spot tests, and where necessary, thin-layer chromatography techniques were used to finalize species identifications.

All specimens were placed in carefully labeled paper sacks and returned to the BYU Herbarium of Nonvascular Cryptogams, where they were curated, identified, placed in permanent herbarium envelopes, and labeled with current epithets and authors' names as well as detailed information about the collection site, habitat, and substrate. Herbarium (BRY C-) and collection numbers were also assigned. Herbarium specimens are on long-term loan to the BYU Herbarium of Nonvascular Cryptogams in Provo, Utah. Numbering, labeling and accountability for specimens will follow National Park Service protocol. Collection data will be entered into ANCS.

5.2 Elemental analyses of sensitive indicator species

In order to document potential air pollution effects on Inscription Rock, pollution-sensitive indicator species were collected from several locations in the Monument. Following careful consideration of species, substrata, growth forms, documented/suspected pollution sensitivities, and abundance, one pollution-sensitive indicator species was collected and returned to BYU where elemental analyses were performed using PIXE technology. All elemental analysis samples were collected using a ceramic knife to avoid metal contamination.

At each site sufficient material of one sensitive indicator species was collected for elemental analyses (3-6 g dry wt). Samples were placed in Nasco sterile plastic bags (to avoid contamination) and transported back to the BYU Herbarium of Nonvascular Cryptogams. Excess material is permanently stored in Nasco sterile plastic bags in the Archival Collection of Elemental Analysis Samples at the BYU Herbarium of Nonvascular Cryptogams; this material is available for additional testing upon request.

At the Herbarium, surface debris and substratal material were removed from all elemental analysis samples using Teflon-coated forceps. Clean, 1 g samples of one sensitive indicator species from each site were then delivered to the Elemental Analysis Laboratory at Brigham Young University.

Samples were prepared for PIXE analysis using the methods of Duflou et al. (1987). Samples were placed in Teflon containers with a Teflon-coated steel ball, cooled to liquid nitrogen temperature, powdered by brittle fracture techniques using a Braun Mikro-Dismemberator II, and then dried in an Imperial IV Microprocessor oven for 14 hours at 80° C. Sub-samples weighing 150 mg were then placed in Teflon containers and spiked with 1 ml of a 360 ppm Yttrium solution. Samples were oven-dried again for 14 h at 80° C and then homogenized using the Mikro-Dismemberator. Approx. 1 mg of powdered lichen tissue was carefully weighed out onto a thin polycarbonate film in an area of 0.5 cm^2. A 1.5% solution of polystyrene in toluene was used to secure the sample to the film.

Samples were analyzed using a 2 MV Van de Graaff accelerator with a 2.28 MeV proton beam, which passed through a 1.1 mg -1/cm^2 pyrolytic graphite diffuser foil. The proton beam was collimated to irradiate an area of 0.38 cm^2 on the sample. Typically, 10-100 nA proton beam currents were used. X-rays were detected using a Tracor X-ray Spectrometer (model TX-3/48-206) with a 10 mm^2 by 3 mm thick Si(Li) detector positioned at 90° to the proton beam. Samples were analyzed twice using different X-ray absorbers between the samples and the detector. One was a 49 mg/cm^2 Mylar absorber with a 0.27mm^2 pinhole (2.8% of detector area) backed with a 8.5 mg/cm^2 beryllium foil. A 98 mg/cm^2 Mylar absorber was also used.

To ensure adequate quality control, samples of NIST SRM 1571 orchard leaves and other standards were prepared and analyzed using the same protocol.

5.3 Microscopy

5.3.1 Scanning Electron Microscopy

Five lichen-substratum samples were examined to determine the depth of fungal penetration into sandstone substrata. The thickest samples examined were 20 mm in cross section. Cross-sections were cut using a diamond saw. Cut specimens were then treated with a 15% solution of hydrofluoric acid for 10-20 minutes. The acid treatment relaxed the substrate structure so that the development and degree of penetration of the fungal hyphae into the rock were more readily discernable. Specimens were then mounted on aluminum stubs with super-glue and sputter-coated with a thin layer of gold, using a Polaron E5300 Freeze Drier with a sputter-coater attachment. A JEOL JSM840A Scanning Electron Microscope was then used to examine sample cross sections. Degree of hyphal encroachment into rock samples was documented photographically.

5.3.2 Light Microscopy

Untreated thallus-substrate sections were also observed using a Zeiss LSM10 compound light microscope and a Zeiss dissecting microscope. Cross-sections were prepared using a diamond saw and specimens of them were photographed using a Pentax (35mm) Photomicrographic System.

5.4 Thin-layer chromatography

Secondary chemistry for 14 lichen specimens (based on the species lists from the focal areas) was determined using thin-layer chromatography (Culberson, 1972). In some cases, High performance thin-layer chromatography was also used to specifically characterize some groups of chemicals not adequately defined by standard TLC methods (Arup et al., 1993).

6. RESULTS

6.1 Field observations and notes from focus areas

Detailed notes and photographs were taken at 11 sites, nine of which are at Inscription Rock; the remaining two are located at the "ancient inscription site" near the superintendent's house. Table 9-1 contains a species list from each focal area together with the total number of species and the average number of species per focal area.

Site No. 1: along Inscription Rock, around corner from marker G. Lichens: *Acarospora* sp. 1 (cracked, brown, pruinose areoles); *Lecidea* sp. (thallus eroded, apothecia black). Photographs: Roll 2 #33.

Site No. 2: along Inscription Rock, to the left of marker J; water flow zone; partially shaded by junipers. Species: *Collema* sp. (crustose); *Peltula* sp. (most abundant species at this site); *Lecanora* cf. *valesiaca* (placodioid); *Xanthoparmelia* sp.; *Lepraria* sp.; *Physcia* sp. (fine lobes, small). Photographs: Roll 2 #32.

Site No. 3: along Inscription Rock, to left of marker P, water flow zone, some shading from nearby trees. Lichens: *Acarospora* sp. 2 (brown, smooth areoles, with one apothecium per areole); *Collema* sp. (crustose); *Acarospora* sp. 3 (white areoles); *Lecanora* sp. 1 (apothecial margins white, disk black; thallus eroded); *Lecidea* sp. 1 (thallus white, apothecia black, rough); *Lecidea* sp. 2 (thallus absent, small, black apothecia abundant). Note: 15 feet up water flow area there are several large thalli of *Dermatocarpon* sp. Photographs: Roll 1 # 30 – *Lecanora* sp. 1; Roll 1 #31 – *Lecidea* sp. 1; Roll 1 32 – *Lecidea* sp. 2.

Site No. 4: along Inscription Rock, to left of marker X. Lichens: *Lecidea* sp. 2, *Lecidea* sp. 3 (thallus white-gray; apothecia black, convex, pruinose); *Lecidea* sp. 1; *Collema* sp. Photographs: Roll 1 #33 – general shot of panel; Roll 1 # 34 & 35 – *Lecidea* sp. 3. Notes: heavy encroachment on panels by *Lecidea* sp. 1 and *Lecidea* sp. 2; there are some bryophytes (moss) from ground level up 1-2 feet on rock but below panel; this area is heavily shaded by juniper and pinyon pine.

Site No. 5: along Inscription Rock, immediately right of marker X. Lichens: *Lecidea* sp.1; *Lecidea* sp. 2; *Acarospora* sp. 1; *Caloplaca fraudans*, *Caloplaca* sp. 1. Photographs: Roll 1 # 36 – general shot of panel; Roll 1 #37 – specific shot of portion of panel; Roll 2 #1 – *Acarospora* sp. 1. Note: partial shading at this site by woody vegetation.

Site No. 6: along Inscription Rock, immediately right of marker BB. Lichens: *Caloplaca decipiens* (sorediate), common; *Lecidea* sp. 4 (thallus white to gray, well developed; apothecia black, abundant, flat to convex); *Caloplaca* sp. 1; *Caloplaca fraudans*; *Lecidea* sp. 1; *Lecidea* sp. 2; *Aspicilia* sp. 1 (thallus gray; apothecia black, flexuous; abundant); *Candelariella* sp. (mostly apothecia; large patches); *Staurothele* sp. (common, round patches, up to 1 cm in diameter); *Lecanora* sp. 1. Photographs: Roll 2 #2 – general shot of panel; Roll 2 #3 – *Lecidea* sp. 4; Roll 2 #4 – *Caloplaca decipiens*; Roll 2 #5 – *Caloplaca fraudans*; Roll 2 #6 – *Aspicilia* sp. 1; Roll 2 #7 – *Staurothele* sp.; Roll 2 #8 – *Candelariella* sp.; Roll 2 #9 – *Caloplaca* sp. 1; Roll 2 #10 – general shot of bryophytes (moss); Roll 2 #11 – specific shot showing bryophytes. Note:

abundant bryophyte development from ground level up to 1-3 feet on rock, below main panel.

Site No. 7: along Inscription Rock, immediately right of marker GG (between GG and HH). Lichens (lightly lichenized lower portion of panel): *Caloplaca fraudans*; *Caloplaca* cf *sideritis*; *Lecidea* sp. 1; *Caloplaca* sp. 1; *Peltula* sp. (heavily lichenized portion below panel to ground level): *Candelariella* sp.; *Caloplaca* sp. 1; *Caloplaca fraudans*; *Caloplaca decipiens*; *Physcia/Phaeophyscia* sp.; *Melanelia* sp., *Caloplaca* cf *sideritis*; *Lecidea* sp. 4; *Lecidella* sp.; *Aspicilia* sp. 1; *Rhizocarpon disporum*; *Collema* sp. (crustose); *Lecanora cenisia*, *Toninia* sp. (squamules small); *Rinodina* sp.; *Acarospora* cf *fuscata*; *Dermatocarpon* sp. Photographs: Roll 2 #12 – general shot of panel; Roll 2 #13 – *Caloplaca sideritis*; Roll 2 #14 – *Dermatocarpon* sp.; Roll 2 #15 – *Lecanora cenisia*; Roll 2 #16 – *Lecidella* sp.; Roll 2 #17 – *Melanelia* sp.

Site No. 8: along Inscription Rock, between markers HH and II. Lichens: *Physcia* sp. (abundant); *Acarospora* sp. 3 (brown, somewhat cracked areoles); *Caloplaca fraudans*; *Melanelia* sp.; *Candelariella* sp.; *Physcia caesia*; *Lecidea* sp. 1; *Lecidea* sp. 4; *Collema* sp. (crustose). Photographs: Roll 2 #18 – *Physcia* sp.; Roll 2 #19 general shot of panel. Note: water flow area.

Site No. 9: along Inscription Rock, immediately left of marker JJ. Lichens: *Acarospora fuscata*; *Collema* sp. (crustose); *Candelariella* sp.; *Lecidea* sp. 1; *Lecidea* sp. 4; *Melanelia* sp.; *Lecidea* sp. 2; *Aspicilia* sp. 1; *Lecidella* sp.; *Caloplaca* sp. 1; *Acarospora* sp. 4 (areoles black, pruinose; 1 apothecium per areole). Photographs: Roll 2 #20 – general shot of panel; Roll 2 #21 – *Acarospora* sp. 4.

Site No. 10: "ancient inscription site", near superintendent's house, left panel – sundial image. Lichens: *Physcia* sp.; *Melanelia* sp.; *Lecanora novomexicana*; *Lecanora* sp. 2 (thallus areolate with abundant apothecia; disks black); *Candelariella* sp.; *Rhizocarpon disporum*; *Collema* sp. (crustose); *Lecanora* sp. 1; *Xanthoparmelia* sp. Photographs: Roll 2 #22 – general photograph of left panel (sundial image). Note: this panel is dominated by *Physcia* sp.

Site No. 11: "ancient inscription site", near superintendent's house, right panel – Kokopele (chipped out) figures. Lichens (specifically from chipped out areas): *Candelariella* sp.; *Caloplaca* sp. 2 (thallus crustose, black; apothecia orange, waxy with prominent dark margins); *Toninia* sp. (squamules dark gray, small); *Acarospora* sp. 4; *Lecidea* sp. 4; *Aspicilia desertorum*; *Lecidea* sp. 2; *Aspicilia* sp. 1; *Rhizocarpon disporum*. Photographs: Roll 2 #23 – general photograph of right panel (Kokopele, chipped out, figures).

Table 9-1. Listing of lichen species by focal area

Species	Site 1	Site 2	Site 3	Site 4	Site 5	Site 6	Site 7	Site 8	Site 9	Site 10	Site 11	Totals
Acarospora fuscata							X		X			2
Acarospora sp. 1	X				X							2
Acarospora sp. 2			X									1
Acarospora sp. 3			X					X				2
Acarospora sp. 4									X		X	2
Aspicilia desertorum											X	1
Aspicilia sp. 1						X	X		X		X	4
Caloplaca decipiens						X	X					2
Caloplaca fraudans					X	X	X	X				4
Caloplaca sideritis							X					1
Caloplaca sp. 1					X	X	X		X			4
Caloplaca sp. 2										X		1
Candelariella sp.						X	X	X	X	X	X	6
Collema sp.		X	X	X			X	X	X	X		7
Dermatocarpon sp.			X				X					2
Lecanora cenisia							X					1
Lecanora novomexicana										X		1
Lecanora valesiaca		X										1
Lecanora sp. 1			X			X				X		3
Lecanora sp. 2										X		1
Lecidea sp. 1	X		X	X	X	X	X	X	X			8
Lecidea sp. 2			X	X	X	X			X		X	6
Lecidea sp. 3				X								1
Lecidea sp. 4						X	X	X	X		X	5
Lecidella sp.							X		X			2
Lepraria sp.		X										1
Melanelia sp.						X	X	X	X			4
Peltula sp.		X					X					2
Physcia caesia								X				1
Physcia sp.		X						X		X		3
Physcia/Phaeophyscia sp.							X					1
Rhizocarpon disporum							X			X	X	3
Rinodina sp.							X					1
Staurothele sp.					X							1
Toninia sp.							X				X	2
Xanthoparmelia sp.		X								X		2
Total Species	2	6	7	4	5	10	19	9	11	9	9	43
Ave no. species/site												8.27

9. Lichen Biodeterioration at Inscription Rock 139

6.2 Growth rate photographic record

Below is a list of the growth study plots established at selected sites in the Monument. These plots will be photographically monitored over time to document development and encroachment of lichen species in the focal areas. Figures 9-1 through 9-14 show the growth plots from May 2000 and May 2001.

Site No. 1 (May 2000): growth study plot (around corner from marker G, photograph: (Roll 2 #33); re-photographed May 2001 (Roll 1 #1)

Site No. 2 (May 2000): growth study plot (left of marker J, right side of water flow zone, 4.5 feet from ground level, photograph: (Roll 2 #32); re-photographed May 2001 (Roll 1 #2 and #3)

Site No. 3 (May 2000): growth study plot (left of marker P, right side of water flow zone, 4 feet from ground level, photograph: (Roll 2 #31); re-photographed May 2001 (Roll 1 #5).

Site No. 4 (May 2000): growth study plot (left of marker X, photograph: (Roll 2 #30); re-photographed May 2001 (Roll 1 #6)

Site No. 5 (May 2000): growth study plot (right of marker X, photograph: (Roll 2 #29); re-photographed May 2001 (Roll 1 #7)

Site No. 6 (May 2000): growth study plot (right of marker BB, immediately right of large carved inscription, photograph: (Roll 2 #27); re-photographed May 2001 (Roll 1 #8); immediately right of photograph Roll 2 #27 is photograph Roll 2 #28 (May 2000); re-photographed May 2001 (Roll 1 #9)

Site No. 7 (May 2000): growth study plot (right of marker GG, between marker GG and HH, photograph: (Roll 2 #26); re-photographed May 2001 (Roll 1 #10)

Site No. 8 (May 2000): growth study plot (between markers HH and II, water flow zone, photograph: (Roll 2 #25); re-photographed May 2001 (Roll 1 #12)

Site No. 9 (May 2000): growth study plot (left of marker JJ, photograph: (Roll 2 #24); re-photographed May 2001 (Roll 1 #13)

Site No. 10 (May 2000): growth study plot (left panel, at ancient inscription site, 5 feet above ground (from rock ledge), right of sundial image, photograph: (Roll 2 #34); re-photographed May 2001 (Roll 1 #17); 1.5 feet right of sundial figure, 6 inches down (from previous photograph, Roll 2 #34) photograph (Roll 2 #35); re-photographed May 2001 (Roll 1 #18)

Site No. 11 (May 2000): growth study plot (right panel, at ancient inscription site, snake figure, photograph: (Roll 2 #36); re-photographed May 2001 (Roll 1 #19); upper middle of right panel, bird figure, photograph Roll 2 #37; re-photographed May 2001 (Roll 1 #20)

Figure 9-1. Site No. 1, Growth Study Plot A-2000, B-2001

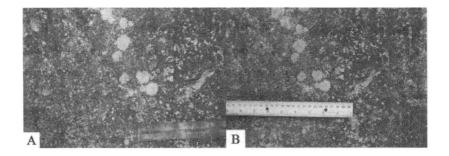

Figure 9-2. Site No. 2, Growth Study Plot A-2000, B-2001

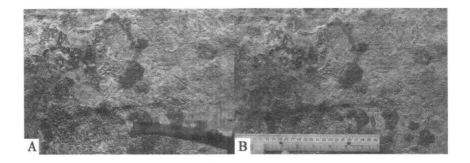

Figure 9-3. Site No. 3, Growth Study Plot A-2000, B-2001

9. Lichen Biodeterioration at Inscription Rock

Figure 9-4. Site No. 4, Growth Study Plot A-2000, B-2001

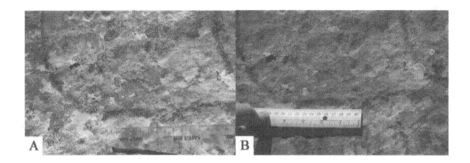

Figure 9-5. Site No. 5, Growth Study Plot A-2000, B-2001

Figure 9-6. Site No. 6, Growth Study Plot A-2000, B-2001

142 *Biodeterioration of Rock Surfaces*

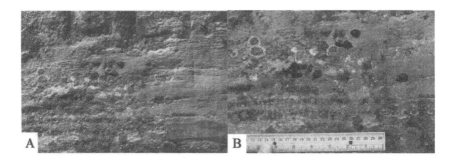

Figure 9-7. Site No. 6, Growth Study Plot A-2000, B-2001

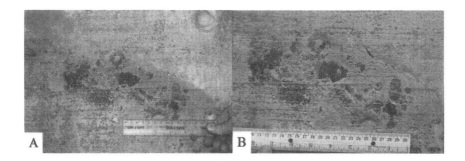

Figure 9-8. Site No. 7, Growth Study Plot A-2000, B-2001

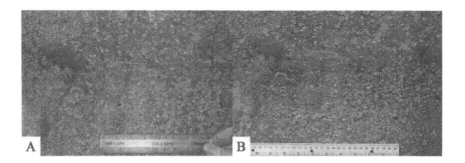

Figure 9-9. Site No. 8, Growth Study Plot A-2000, B-2001

9. Lichen Biodeterioration at Inscription Rock

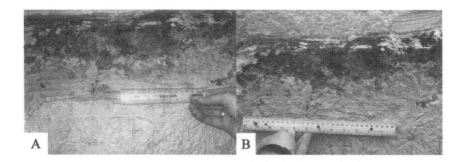

Figure 9-10. Site No. 9, Growth Study Plot A-2000, B-2001

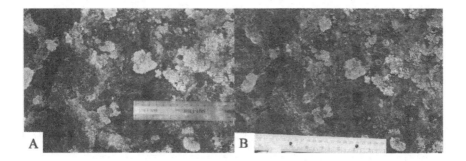

Figure 9-11. Site No. 10, Growth Study Plot A-2000, B-2001

Figure 9-12. Site No. 10, Growth Study Plot A-2000, B-2001

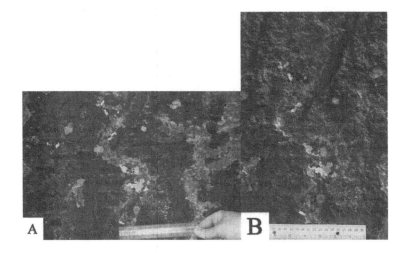

Figure 9-13. Site No. 11, Growth Study Plot A-2000, B-2001

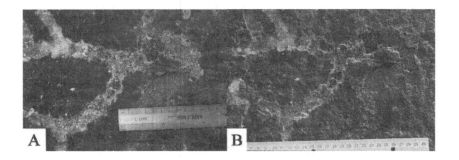

Figure 9-14. Site No. 11, Growth Study Plot A-2000, B-2001

6.3 Thin-layer chromatography samples

In order to evaluate biochemical degradation of panels by lichens, samples of 14 common lichen species were collected and analyzed to characterize their secondary chemistry. Below is a list of samples collected for TLC analysis:

EM.TLC.1   *Aspicilia* sp.
EM.TLC.2   *Caloplaca fraudans*
EM.TLC.3   *Candelariella* sp.
EM.TLC.4   *Collema* sp. (crustose)
EM.TLC.5   *Lecanora* sp. 1
EM.TLC.6   *Lecanora* sp. 2
EM.TLC.7   *Lecidea* sp. 1

9. Lichen Biodeterioration at Inscription Rock

EM.TLC.8   *Lecidea* sp. 2
EM.TLC.9   *Lecidea* sp. 4
EM.TLC.10   *Melanelia* sp.
EM.TLC.11   *Physcia* sp. (from ancient site, site No. 10)
EM.TLC.12   *Physcia* sp. (from site No. 8)
EM.TLC.13   *Staurothele* sp.
EM.TLC.14   *Xanthoparmelia* sp.

The above samples were analyzed using standard TLC techniques. Table 9-2 contains a list of the secondary chemicals detected for each species. This table also indicates the frequency of each chemical across all species and the total number of chemicals per species. A total of 35 different secondary chemicals are reported for all species, at least 9 of these being organic acids, with divaricatic acid being the most frequent.

Table 9-2. Results of thin-layer chromatography analysis of selected species

Species	anziaic acid	argopsin	barbatate	benzoic acid	calploicin	calycin	confluentic acid	dione	diploicin
Aspicilia sp.			X						
Caloplaca fraudans									
Candelariella sp.					X	X			X
Collema sp.									X
Lecanora sp. 1	X								X
Lecanora sp. 2	X								X
Lecidia sp. 1									X
Lecidia sp. 2								X	X
Lecidia sp. 3				X					X
Melanelia sp.									
Physcia sp. (ancient)									
Physcia sp. (site 8)									
Staurothele sp.		X					X		
Xanthoparmelia sp.									X
Totals	2	1	1	1	1	1	1	1	8

Table 9-2.1. (Continued) Results of thin-layer chromatography analysis of selected species

Species	divaricatic acid	epiphorellic acid	eriodermate	evernic acid	glomellic acid	hiascic acid	homosekikiac acid
Aspicilia sp.	X						
Caloplaca fraudans	X						
Candelariella sp.	X				X		X
Collema sp.							
Lecanora sp. 1							
Lecanora sp. 2	X				X		
Lecidia sp. 1			X				
Lecidia sp. 2				X			
Lecidia sp. 3	X						
Melanelia sp.	X					X	
Physcia sp. (ancient)		X					
Physcia sp. (site 8)	X						
Staurothele sp.	X						
Xanthoparmelia sp.							
Totals	8	1	1	1	2	1	1

Table 9-2.2. (Continued) Results of thin-layer chromatography analysis of selected species

Species	lecanorate	leprolomin	lichexanthone	lobaridone	paludosic acid	pannarin	parietin
Aspicilia sp.	X					X	
Caloplaca fraudans	X			X			X
Candelariella sp.					X		
Collema sp.							
Lecanora sp. 1		X					
Lecanora sp. 2							
Lecidia sp. 1				X		X	
Lecidia sp. 2							
Lecidia sp. 3							
Melanelia sp.			X				
Physcia sp. (ancient)	X		X		X		
Physcia sp. (site 8)			X			X	
Staurothele sp.			X				
Xanthoparmelia sp.			X				
Totals	3	1	5	2	2	3	1

9. Lichen Biodeterioration at Inscription Rock

Table 9-2.3. (Continued) Results of thin-layer chromatography analysis of selected species

Species	planaic acid	protolichesterinic acid	psoromic acid	scrobiculin	solorinic acid	sphaeric acid	squamatic acid
Aspicilia sp.					X		
Caloplaca fraudans					X		X
Candelariella sp.		X	X			X	
Collema sp.							
Lecanora sp. 1				X			
Lecanora sp. 2							
Lecidia sp. 1							
Lecidia sp. 2							
Lecidia sp. 3		X					
Melanelia sp.	X	X				X	
Physcia sp. (ancient)						X	
Physcia sp. (site 8)						X	
Staurothele sp.							
Xanthoparmelia sp.						X	
Totals	1	3	1	1	2	5	1

Table 9-2.4. (Continued) Results of thin-layer chromatography analysis of selected species

Species	Stictic acid	thiomelin	usnic acid	vicanicin	xanthone	Totals
Aspicilia sp.					X	6
Caloplaca fraudans	X					7
Candelariella sp.					X	11
Collema sp.		X		X	X	4
Lecanora sp. 1		X				5
Lecanora sp. 2						4
Lecidia sp. 1			X			5
Lecidia sp. 2	X	X		X		6
Lecidia sp. 3	X					5
Melanelia sp.						6
Physcia sp. (ancient)						5
Physcia sp. (site 8)						4
Staurothele sp.						4
Xanthoparmelia sp.	X		X			5
Totals	4	3	2	2	3	

6.4 Microscopy samples

In order to evaluate hyphal encroachment into rock surfaces, samples of 15 common lichen species were collected. Eight samples were analyzed using SEM and light microscopy. Below is a list of the samples collected for microscopic examination:

SEM.EM.1* *Candelariella rosulans* – fracture line
SEM.EM.2* *Acarospora* sp.
SEM.EM.3* *Acarospora* sp.
SEM.EM.4a* *Xanthoparmelia plittii* (light green)
SEM.EM.4b* *Melanelia* sp. (brown foliose)
SEM.EM.5* *Staurothele* sp.
SEM.EM.6* *Physcia caesia* (light gray)
SEM.EM.7* *Lecidea* sp.
SEM.EM.8 *Lecidea* cf. *tessellata*
SEM.EM.9 *Staurothele* sp.
SEM.EM.10 *Lecidea* sp.
SEM.EM.11 *Lecidea* sp.
SEM.EM.12 *Caloplaca* sp.
SEM.EM.13 *Aspicilia desertorum*
SEM.EM.14 *Lecidella* sp.
SEM.EM.15 *Xanthoparmelia coloradoensis*

* Samples examined using SEM and/or light microscopy.

Five of the above samples (SEM.EM.1, SEM.EM.2, SEM.EM.4b, SEM.EM.6 & SEM.EM.7) were examined using standard SEM techniques. Three additional samples (SEM.EM.3, SEM.EM.4a & SEM.EM.5) were also examined using standard light microscopy techniques. Images from the scanning electron microscope and light microscope show that hyphae are well developed throughout the sample and often extend beyond the bottom of the rock.

Table 9-3 contains information on the degree of hyphal development and penetration in the eight samples examined. Figures 9-15 through 9-22 are light microscopy images and Figures 9-23 through 9-27 are SEM images. Figures 9-15, 9-16, 9-19 and 9-20 show the hyphae extending across a fracture line. In Figure 9-17, the arrow is pointing to a rock particle. Figures 9-21 and 9-22 show the hyphae extending beyond the edge of our sample; this is easily observed since the samples are in water allowing the hyphae to hydrate and float. Figures 9-26 and 9-27 are detailed montages of cross-sections of two entire samples from top to bottom.

9. Lichen Biodeterioration at Inscription Rock

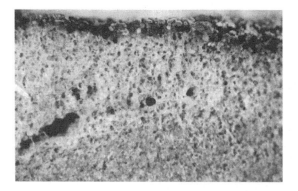

Figure 9-15. Candelariella rosulans 1.6X

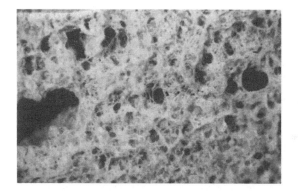

Figure 9-16. Candelariella rosulans 4X

Figure 9-17. Candelariella rosulans 4X

150   *Biodeterioration of Rock Surfaces*

Figure 9-18. Candelariella rosulans 5X

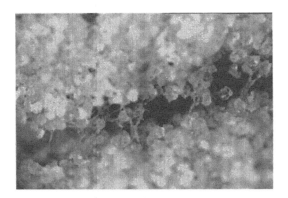

Figure 9-19. Candelariella rosulans 4X

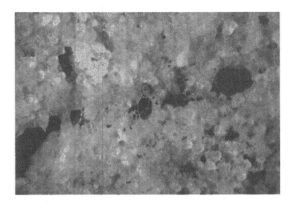

Figure 9-20. Candelariella rosulans 4X

9. Lichen Biodeterioration at Inscription Rock 151

Figure 9-21. Physcia caesia 1.6X

Figure 9-22. Physcia caesia 1.6X

Figure 9-23. Melanelia sp. 50X

152 *Biodeterioration of Rock Surfaces*

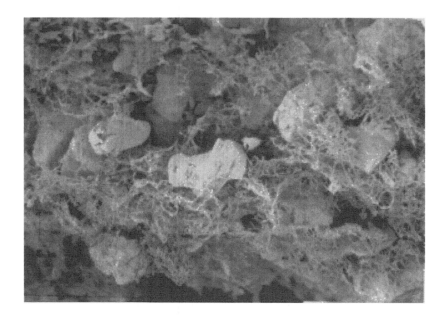

Figure 9-24. Acarospora sp. 130X

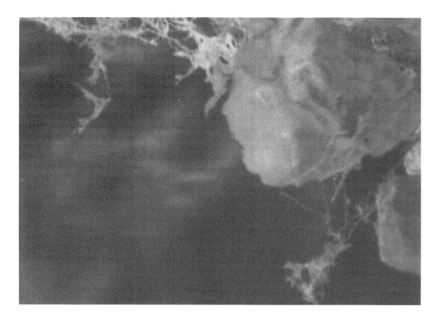

Figure 9-25. Acaropsora sp. 400X

9. Lichen Biodeterioration at Inscription Rock

Figure 9-26. Montage of endolithic hyphae from *Melanelia* sp., A-top, 35X

154 *Biodeterioration of Rock Surfaces*

Figure 9-26. Montage of endolithic hyphae from *Melanelia* sp., B-middle, 35X

9. *Lichen Biodeterioration at Inscription Rock* 155

Figure 9-26. Montage of endolithic hyphae from *Melanelia* sp., C-bottom, 35X

156   *Biodeterioration of Rock Surfaces*

Figure 9-27. Montage of endolithic hyphae from *Acarospora* sp., A-top, 35X

9. *Lichen Biodeterioration at Inscription Rock* 157

Figure 9-27. Montage of endolithic hyphae from *Acarospora* sp., B-middle, 35X

Figure 9-27. Montage of endolithic hyphae from *Acarospora* sp., C-bottom, 35X

9. *Lichen Biodeterioration at Inscription Rock* 159

Table 9-3. Degree of hyphal development and encroachment into samples of substratum

Species	Depth of sample (mm)	depth of penetration (mm)	degree of development
Candelariella rosulans (1)	18	18	Abundant
Acarospora sp. (2)	20	20 +	Abundant
Acarospora sp. (3)	19	19 +	Abundant
Xanthoparmelia plittii (4a)	16	16	Abundant
Melanelia sp. (4b)	16	16 +	Abundant
Staurothele sp. (5)	20	20 +	Abundant
Physcia caesia (6)	17	17 +	Common
Lecidea sp. (7)	20	20	Abundant

6.5 Elemental analysis

In order to document the air quality status of the Monument, samples of the pollution-sensitive indicator species *Usnea hirta* were collected from bark at five locations and analyzed for 21 potential pollutant elements using PIXE technology (Table 9-4). All these elements are within background concentrations. Cu/Zn and Fe/Ti ratios are also within background levels. Replicates of the specimens collected for elemental analysis listed below have also been archived in the Herbarium of Non-vascular Cryptogams at Brigham Young University.

Table 9-5 lists the lichen species collected from all available substrata within the Monument and Table 9-6 lists the pollutant-sensitive indicator species. The lichen flora is diverse (61 species in 32 genera) and well developed, and contains no threatened, endangered or endemic species.

Table 9-4. Elemental analysis data for Usnea sp.

Collection Site	Elements (ppm except where indicated)											
	P	S%	CL	K%	Ca%	Ti	Ba	V	Cr	Mn	Fe	Fe/Ti
along switchbacks, north side of Inscription Rock, Sample #779	1280	0.11	360	0.4	0.52	99	32	5	4.7	52	600	6.06
west of trail, where trail reaches top of mesa, Sample #780	600	0.13	290	0.3	0.76	70	35	5	1.6	34	510	7.29
north side of Inscription Rock, near rock field, Sample #781	1840	0.135	440	0.45	0.45	83	39	6	3.9	59	600	7.23
descent portion of trail below mesa, west of visitors center, Sample #782	730	0.169	340	0.29	1.25	116	37	4.9	1.58	34	680	5.86
along state road no. 53, west of monument visitors center, Sample #783	1300	0.092	230	0.37	0.72	43	37	5.2	1.48	43	320	7.44

Table 9-4.1 (Continued) Elemental analysis data for Usnea sp.

Collection Site	Elements (ppm except where indicated)										
	Co	Ni	Cu	Zn	Pb	As	Se	Br	Rb	Sr	Cu/Zn
along switchbacks, north side of Inscription Rock, Sample #779	3.7	1.62	5.2	23	5.6	1.5	1.23	6.6	3.2	17.7	0.23
west of trail, where trail reaches top of mesa, Sample #780	4	0.68	5	23	5.4	1.49	1.62	5.6	3.3	19	0.217
north side of Inscription Rock, near rock field, Sample #781	3.6	1.62	5.7	23	4.1	1.34	1.36	7.3	3.4	18.3	0.25
descent portion of trail below mesa, west of visitors center, Sample #782	4.6	0.88	5.9	22	4.9	1.14	1.09	8.9	3.3	21	0.27
along state road no. 53, west of monument visitors center, Sample #783	3.8	1.18	4.7	22	4.8	1.98	1.75	3.3	3.9	19.3	0.21

Table 9.5 Lichens of El Morro National Monument

Taxa	BRYC No.	Collection No.	Substratum	Growth form
Acarospora strigata	38079	11035	Rock	Crustose
Acarospora fuscata	38091	11047	Rock	Crustose
Acarospora badiofusca	38092	11048	Rock	Crustose
Amandinea punctata	38038	10994	Bark	Crustose
Aspicilia cinerea	38065	11021	Rock	Crustose
Caloplaca chrysophthalma	38024	10980	Bark	Crustose
Caloplaca holocarpa	38036	10992	Bark	Crustose (scant)
Caloplaca arizonica	38026	10982	Bark	Crustose
Caloplaca epithallina	38093	11049	Rock	Crustose
Caloplaca. fraudans	38062	11018	Rock	Crustose
Caloplaca sideritis	38078	11034	Rock	Crustose
Caloplaca trachyphylla	38061	11017	Rock	Crustose
Candelariella rosulans	38097	11053	Rock	Crustose
Catapyrenium zahlbruckneri	38066	11022	Rock	Crustose
Cladonia fimbriata	38055	11011	Soil/Moss	Squamulose
Cladonia pocillum	38056	11012	Soil/Moss	Squamulose
Collema nigrescens	38042	10998	Bark	Foliose
Dermatocarpon miniatum	38112	11068	Rock	Foliose
Diploschistes muscorum	38057	11013	Soil/Moss	Crustose
Flavopunctelia soredica	38022	10978	Bark	Foliose
Lasallia papulosa	38111	11067	Rock	Foliose
Lecanora meridionalis	38039	10995	Bark	Crustose

Table 9.5.1 (Continued) Lichens of El Morro National Monument

Taxa	BRYC No.	Collection No.	Substratum	Growth form
Lecanora impudens	38035	10991	Bark	Crustose
Lecanora mughicola	38049	11005	Lignum	Crustose
Lecanora muralis	38059	11015	Rock	Crustose
Lecanora novomexicana	38064	11020	Rock	Crustose
Lecidea tessellata	38101	11057	Rock	Crustose (scant)
Lecidea atrobrunnea	38082	11038	Rock	Crustose
Lecidella stigmatea	38084	11040	Rock	Crustose
Lecidella viridans	38063	11019	Rock	Crustose
Leptogium furfuraceum	38043	10999	Bark	Foliose
Megaspora verrucosa	38044	11000	Bark	Crustose
Melanelia elegantula	38077	11033	Rock	Foliose
Melanelia exaperatula	38069	11025	Rock	Foliose
Melanelia subolivacea	38030	10986	Bark	Foliose
Peltigera horizontalis	38058	11014	Soil/Moss	Foliose
Phaeophyscia kairamoi	38023	10979	Bark	Foliose
Physcia aipola	38040	10996	Bark	Foliose
Physcia albinea	38072	11028	Rock	Foliose
Physcia biziana	38032A	10988A	Bark	Foliose
Physcia caesia	38060	11016	Rock	Foliose
Physcia dimidiata	38032B	10988B	Bark	Foliose
Physcia dubia	38100	11056	Rock	Foliose
Physcia stellaris	38027	10983	Bark	Foliose
Physconia elegantula	38028	10984	Bark	Foliose
Placidium squamulosum	38115	11071	Soil	Squamulose
Polysporina simplex	38099	11055	Rock	Crustose
Rhizocarpon disporum	38096	11052	Rock	Crustose
Rhizoplaca chrysoleuca	38068	11024	Rock	Foliose
Rhizoplaca melanophthalma	38067	11023	Rock	Foliose
Sarcogyne regularis	38083	11039	Rock	Crustose (scant)
Staurothele areolata	38088	11044	Rock	Crustose
Staurothele drummondii	38081	11037	Rock	Crustose
Tuckermannopsis fendleri	38050	11006	Bark	Foliose
Usnea hirta	38031	10987	Bark	Fruticose
Xanthoria fallax	38021	10977	Bark	Foliose
Xanthoria polycarpa	38046	11002	Bark	Foliose
Xanthoparmelia chlorochroa	38053	11009	Soil (vagrant)	Foliose
Xanthoparmelia coloradoensis	38113	11069	Rock	Foliose
Xanthoparmelia plittii	38110	11066	Rock	Foliose
Xanthoparmelia wyomingica	38054	11010	Soil (vagrant)	Foliose

Table 9-6. List of pollution sensitive indicator species. S = sensitive, I = intermediately sensitive, S-I = sensitive to intermediately sensitive, I-T = intermediately sensitive to tolerant. 14 species are sensitive to one or more air pollutants

Species	SO_2	Ozone	NO_X/PAN	F
Caloplaca holocarpa	I			
Cladonia fimbriata	S-I			
Collema nigrescens	S			
Melanelia exaperatula	I			
Melanelia subolivacea	I-T	I-T		
Physcia aipolia	I			S
Physcia caesia	I			
Physcia dubia				S-I
Physcia stellaris	I			
Rhizocarpon chrysoleuca	S			
Rhizocarpon melanophthalma	S		S	
Usnea hirta	S-I			
Xanthoria fallax	S-I		S	
Xanthoria polycarpa	S-I			

7. DISCUSSION, CONCLUSIONS AND RECOMMENDATIONS

The fact that all eight lichen species examined using light and scanning electron microscopy showed significant encroachment (15-21 mm) into rock samples suggests a significant potential for physical degradation of inscriptions and petroglyphs at El Morro National Monument. Natural wetting and drying cycles of rock surfaces causes swelling and contracting of hyphae and thalli, resulting in physical disturbance and general breakdown of rock surfaces. In addition, production of organic acids by all lichen species sampled contributes to the chemical degradation of the substratum, thus enhancing deterioration of rock surfaces. Finally, growth study plots (after one year) show very slow growth rates for all species examined.

Shading due to establishment of woody vascular plant vegetation around and near Inscription Rock appears to promote the development and growth of lichens in several of the focal areas. Shading tends to alter the "normal" microhabitat parameters of the site, specifically slowing water evaporation and reducing mean summer temperature.

Concentrations of airborne pollutant elements in thalli of the sensitive indicator species Usnea hirta are well within background levels (Table 9-4). A total of 14 air pollution sensitive lichen species were identified from the Monument. Within a specific location, any number of pollution-sensitive species over 10 suggests low pollution impact (Table 9-6). Therefore, air quality-related issues do not appear to be contributing to the deterioration of rock surfaces in the Monument.

The following recommendations for future monitoring of lichen biodeterioration processes in El Morro National Monument have been communicated to the National Park Service: 1) continue (for at least the next five years) documentation of lichen growth in the focal areas on Inscription Rock using the established photographic plots; 2) set up enzyme digestion plots in the rocky area north of Inscription Rock to test the feasibility and effectiveness of this technology in controlling lichen growth on sensitive substrata; and 3) selective removal of woody plant vegetation from some of the focal areas, while photographically documenting changes in lichen communities in response to changing microhabitat parameters.

REFERENCES

Arup, U., Ekman, S., Lindbolm, L. and Mattson, J.E. (1993) High performance thin layer chromatography (HPLTC), an improved technique for screening lichen substances. *The Lichenologist* 25: 61-71.

Chen, J., Blume, H. and Beyer, L. (1999) Weathering of rocks induced by lichen colonization - a review. *Catena* 39: 121-146.

Culberson, C.F. (1972) Improved conditions and new data for the identification of lichen products by a standardized thin-layer chromatographic method. *Journal of Chromatography* 72:113-124.

Duflou, H., Maenhaut, W. and DeReuck, J. (1987) Application of PIXE analysis to the study of regional distribution of trace elements in normal human brain tissue. *Biological Trace Element Research* 13: 1.

Edwards, H. G. M., Holder, J.M. and Wynn-Williams, D.D. (1998) Comparative FT-Raman spectrometry of *Xanthoria* lichen-substratum systems from Temperate and Antarctic habitats. *Soil Biology and Biochemistry* 30: 1947-1953.

Johnston, C. G., and Vestal, J.R. (1993) Biogeochemistry of oxalate in the Antarctic cryptoendolithic lichen-dominated community. *Microbial Ecology* 25: 305-319.

Jones, D. (1988) Lichens and pedogenesis. In: *Handbook of Lichenology, Vol. III*. (M. Galun ed): 109-124. CRC Press, Boca Raton.

Nash, T. H., ed. (1996) *Lichen Biology*. Cambridge University Press, Cambridge.

Seaward, M. R. D. and Edwards, H.G. M. (1995) Lichen-substratum interface studies, with particular reference to Raman microscopic analysis. *Cryptogamic Botany* 5: 282-287.

St. Clair, L.L. (1999) *A Color Guidebook to Common Rocky Mountain Lichens*. M. L. Bean Life Science Museum of Brigham Young University, Provo.

Syers, J. K. and Iskandar, I.K. (1973) Pedogenetic significance of lichens. In: *The Lichens*. (V. Ahmadjian & M. E. Hale eds): 225-248. Academic Press, London.

Chapter 10

LICHENS OF DIFFERENT MORTARS AT ARCHAEOLOGICAL SITES IN SOUTHERN SPAIN

An Overview

X. ARIÑO and C. SAIZ-JIMENEZ
Instituto de Recursos Naturales y Agrobiologia, CSIC, Apartado 1052, 41080 Sevilla, Spain

Abstract: Lichens on mortars at archaeological sites in southern Spain are reviewed. Excavation of archaeological remains and their subsequent colonization provided an opportunity to investigate the lichen vegetation on this artificial building material and to examine lichen/mortar interactions. Studies were carried out in three different Roman settlements at Baelo Claudia, an industrial town located near the Gibraltar Strait, Italica, a residential town near Hispalis (present day Seville), and the Necropolis of Carmona, 30 km from Seville. Several topics, such as factors affecting lichen ecology, lichen-mortar interactions, chromatic changes on monuments and efficacy of biocide applications, are discussed.

Keywords: Algophase; biocide; biodeterioration; chromatic changes; lichens; lithobionts; mortar

1. INTRODUCTION

By definition, mortar is a mixture of sand grains joined together by a binder (lime, cement, etc.) of plastic consistency, traditionally used as a building material. Ancient mortars are made mainly of lime and sand; on some occasions other materials were added, such as clay, in order to improve their properties. Lime mortars are characterized by high porosity, low mechanical strength and a relatively high deformation capacity, together with excellent permeability to water (Puertas *et al.*, 1994). Mortars are used in stone and brick constructions, and for coating stone surfaces.

The use of mortar to cover stonework was common among Roman builders, not only for ornamental purposes but also for protecting the stone from environmental agents; there are many examples of this construction technique in the settlements and towns of Roman Hispania (Spain). Excavation and exposure of these archaeological sites is usually associated with rapid biological colonization, especially by pioneer plant communities of cyanobacteria, algae and lichens, a natural process creating a living mosaic of colors and textures.

Deterioration of mortars has been studied in recent years (Ariño and Saiz-Jimenez, 1996a, 1996b, 1997). In terms of bioreceptivity, the ability of any material to support biological growth (Guillitte, 1995), mortar proved to be the most bioreceptive material in laboratory tests. In fact, mortar is a substratum easily colonized by lichens due to its composition and structure (Saiz-Jimenez, 2001; Saiz-Jimenez and Ariño, 1996a, 2001).

Mortars of archaeological sites in southern Spain have provided an opportunity to investigate the lichen vegetation on this artificial building material and to examine lichen/mortar interactions. In this overview we present studies carried out in three different Roman settlements at Baelo Claudia, an industrial town devoted to the fishing and processing of tuna, located near the Gibraltar Strait, Italica, a residential town near Hispalis (present day Seville), and the Necropolis of Carmona, 30 km from Seville.

2. FACTORS AFFECTING LICHEN DISTRIBUTION

As far as lichen distribution is concerned, it has been suggested that while both the composition and the structure of the substratum are important in determining the nature of communities of lithobionts, especially lichens, it is the microclimatic factors which determine the abundance and distribution of these organisms.

2.1 Solar irradiation

The mortars of the Roman town of Baelo Claudia, subjected to different microclimatic conditions, provide an interesting example which can be used to study the effect of environmental factors on lichen colonization. The distribution of the lichens is closely related to the microclimatic conditions on the walls of the three temples located in the Forum. These are determined by the orientation of each wall, as the level of exposure to sunlight seems to be a primary factor, influencing both the likelihood of lichen colonization and the thallus type (Ariño and Saiz-Jimenez, 1996a). Twenty-two relevés (Table 10-1) were taken on the west-facing wall of the temple of Jupiter subjected to different light intensities, from the shadowed side (relevé number 1) to the more exposed site (relevé number 22). Slight changes in solar irradiation (differences in temperature and relative humidity were negligible)

10. Lichens of Different Mortars in Southern Spain

modify the community composition and also affect the type of thalli (crustose endolithic, crustose epilithic or leprose) and their distribution.

Table 10-1. Relevés from the west wall of the Temple of Jupiter (Baelo Claudia)

	1	2	3	4	5	6	7	8	9	10	11	12	13	14	15	16	17	18	19	20	21	22
Verrucaria viridula	+	+	+	+	+	+	+	+	+	+	+	+	+	+	+	+	+	+	+	+	+	
Verrucaria hochstetteri	+	+	+	+	+	+		+	+	+	+			+	+	+	+	+	+	+	+	+
Caloplaca citrina	+	+	+	+		+	+	+		+		+	+	+	+		+					+
Verrucaria nigrescens	+			+	+						+					+	+					
Collema cristatum	+	+	+		+									+	+	+	+	+	+	+		
Verrucaria macrostoma			+	+						+	+	+			+	+				+		
Botryolepraria lesdainii	+	+	+	+	+		+	+	+	+	+	+	+	+	+	+	+	+	+			
Catapyrenium squamulosum		+						+											+			
Toninia aromatica		+																				
Verrucaria muralis					+															+		
Caloplaca flavescens			+			+			+					+				+	+	+	+	
Lecania turicensis				+					+			+	+	+	+	+	+	+	+	+		
Collema auriculatum												+	+	+	+		+	+				
Pseudosagedia linearis												+	+	+	+	+	+	+				
Opegrapha calcarea												+	+									
Clauzadea immersa																+						
Caloplaca velana														+	+				+	+	+	
Caloplaca lactea										+				+	+	+	+	+	+	+		
Caloplaca erythrocarpa																					+	
Lecanora albescens																				+		
Sarcogyne regularis																				+		+

On the sheltered side, epilithic colonization was abundant, with lichen covering c. 50 % of the surface at the beginning of the transect. *Verrucaria viridula* was the dominant species and the non-corticated *Botryolepraria lesdainni* was also frequent in fissures and holes. On the opposite side, which was subjected to longer periods of sunlight, epilithic growth was very restricted, and although lichens were abundant, the majority of the thalli were deeply immersed in the mortar. In this situation, the percentage cover achieved almost 90 % of the surface, and the community was dominated by *Verrucaria hochstetteri*. Other heliophilic species, such as *Caloplaca lactea, Sarcogyne regularis* and *Lecania turicensis*, were also frequent. There was a noticeable increase in the number of lichen taxa on this wall, and in the dominance of species with bright colors, yellow, orange, and white, an adaptative strategy to a higher light intensity. On the other hand, cyanophilous lichens seem to be less dependent on the environmental conditions of the walls, and more so on the presence of fissures and run-off water tracks.

Comparison of relevés showed there to be one main factor affecting the distribution of the species. As temperature and humidity are almost constant, the variation can only be explained by differences in solar irradiation. Several seasonal measurements of irradiation at the wall of the temple showed that the exposed side has a daily period of 2-3 hours with an irradiation between 2500-3000 µmol m^{-1}s^{-1}, whilst the sheltered side has under 1 hour at 1500-2000 µmol m^{-1}s^{-1}. Hence, irradiation strongly influences both lichen colonization and the composition of lichen assemblages.

2.2 Composition of substrata

To study the influence of substrata on lichen colonization, mortars from the three archaeological sites (Baelo Claudia, Italica and Carmona) were compared. The mortar of Baelo Claudia is composed mainly of calcite and contains only traces of quartz and dolomite. The amount of quartz increases in the samples from Italica, whereas in the mortar of Carmona calcite and quartz are in approximately the same proportion, together with feldspar and gypsum. The hardness varies according to the proportions of these components, Baelo Claudia being soft, Italica medium and Carmona hard.

Mortars at the sites studied were extensively colonized by a variety of lichen taxa (Table 10-2). A frequently found group of rather nitrophytic taxa, abundant in exposed habitats, includes *Aspicilia contorta, Caloplaca citrina, C. velana, Catapyrenium squamulosum, Lecania turicencis, Lecanora albescens, Verrucaria macrostoma* and *V. nigrescens*. The growth and distribution of species is strongly influenced by the prevailing microclimatic parameters, which were directly related to exposure of the sampled surfaces. Thus, mortar subjected to direct solar irradiation was colonized by xerophytic species, strongly pigmented and with crustose thalli (mainly *Caloplaca* and *Verrucaria* species). In contrast, in the sheltered environments, such as at the entrances of the hypogeal tombs of the Necropolis of Carmona, large areas of the walls are coated with the leprariod thalli of *Botryolepraria lesdainii* and *Leproplaca xantholyta*.

Table 10-2. Relevés on mortars from three archaeological sites: Baelo Claudia (F, I1, I2, J1, J2), Carmona (EL1, EL2, EL3, G1, MC1, MCII1, MCII2, MCIIS, MCIIN, MCIIO, T118, ANF) and Italica (IT1, IT2, IT3, IT4, IT5, IT6)

	F	I1	I2	J1	J2	EL1	EL2	EL3	MCII1	MCII2	G1	MCIIN	MCIIS	MCIIO	MCI	T118	ANF	IT1	IT2	IT3	IT4	IT5	IT6
Aspicilia contorta ssp. *hoffmanniana*						+		+	+		+							+	+	+	+	+	+
Botryolepraria lesdainii			+	+	+						+	+			+	+	+						
Caloplaca aurantia						+		+	+			+	+	+				+	+	+		+	
Caloplaca citrina		+		+	+	+		+	+			+	+					+	+	+		+	
Caloplaca crenulatella																		+	+			+	
Caloplaca erythrocarpa	+																				+		+
Caloplaca flavescens	+		+								+												
Caloplaca lactea	+	+		+				+			+	+						+	+	+	+	+	+
Caloplaca teicholyta							+						+					+	+	+		+	+
Caloplaca variabilis						+	+											+	+	+	+	+	+
Caloplaca velana	+	+		+		+	+											+	+	+		+	+
Catapyrenium squamulosum				+	+													+	+		+		
Clauzadea immersa				+																			
Collema auriforme	+		+	+	+																		
Collema crispum var. *metzleri*											+							+	+	+			+
Collema tenax			+	+	+																		
Dirina massiliensis															+								
Endocarpon pusillum							+				+												
Fulgensia fulgida						+	+																
Fulgensia subbracteata											+												
Lecania turicensis	+	+		+				+			+		+					+	+		+	+	
Lecanora albescens	+							+										+	+	+		+	+
Lecanora dispersa																		+				+	
Lecanora pruinosa										+			+	+									
Lecidea cuprea															+								
Lepraria nivalis								+	+	+					+								
Leproplaca xantholyta								+	+	+					+	+							
Opegrapha calcarea				+																			
Polyblastia albida														+									
Pseudosagedia linearis				+											+								
Rinodina bischoffii							+																
Sarcogyne regularis	+	+																					
Squamarina concrescens										+	+				+								
Squamarina oleosa											+												
Staurothele hymenogonia						+										+							

Table 10-2.1 (Continued) Relevés on mortars from three archaeological sites: Baelo Claudia (F, I1, I2, J1, J2), Carmona (EL1, EL2, EL3, G1, MC1, MCII1, MCII2, MCIIS, MCIIN, MCIIO, T118, ANF) and Italica (IT1,IT2,IT3, IT4, IT5, IT6)

	F	I1	I2	J1	J2	EL1	EL2	EL3	MCII1	MCII2	G1	MCIIN	MCIIS	MCIIO	MC1	T118	ANF	IT1	IT2	IT3	IT4	IT5	IT6
Thelidium incavatum															+								
Thelidium olivaceum																+							
Toninia aromatica	+								+	+	+												
Toninia sedifolia							+																
Verrucaria calciseda												+		+									
Verrucaria hochstetteri	+		+		+				+		+												
Verrucaria macrostoma	+		+	+	+	+	+	+	+		+	+	+					+	+	+	+	+	+
Verrucaria muralis			+																				
Verrucaria nigrescens			+	+						+						+	+	+		+	+		+
Verrucaria viridula	+	+			+	+		+	+		+	+	+	+									

Twenty-three relevés undertaken from the three archaeological sites were analysed by clustering methods (Figure 10-1). Two main groups can be easily detected:

1. composed of relevés from dry surfaces with high solar radiation (F, I1,EL1, MCII1, MCIIS, MCIIO), with one subgroup composed of seven relevés from horizontal surfaces with high insolation (8,18,19,20, 21,22,23).

2. composed of relevés from sheltered and humid areas without direct insolation (I2, J2, J1, ANF, MCII2, MCIIN, G1, MC1, T118).

One of the relevés (EL2) cannot be assigned to any group; it was taken from an east-exposed wall, with the mortar layer heavily deteriorated and colonized by some terricolous species.

10. Lichens of Different Mortars in Southern Spain

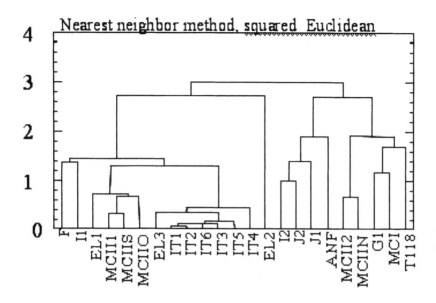

Figure 10-1. Cluster ordination of mortar relevés from three archaeological sites: F, I1, I2, J1, J2 (Baelo Claudia). EL1, EL2, EL3, MCII1, MCII2, MCIIN, MCIIS, MCIIO, MCI, ANF, T118, GI (Carmona). IT1, IT2, IT3, IT4, IT5, IT6 (Italica)

These analyses show that microclimate is the primary factor affecting the composition of a lichen community. Nevertheless, differences in substratum also play a role, particularly in respect of hardness. Lichens with an endolithic thallus (e.g. *Caloplaca lactea, Sarcogyne regularis, Clauzadea immersa, Verrucaria hochstetteri*) were frequent and abundant on south- and east-facing walls with high insolation of the temples of Baelo Claudia. This type of thallus was lacking or rare on the walls of the Necropolis (having the same exposure but a different mortar), the community here being composed of epilithic lichens. At Italica, although the site is not comparable with the two previous cases (vertical walls), the mortar of the pavements and mosaics was also colonized mainly by epilithic lichens, and, due to the accumulation of sand, dust and airborne particles, some species typical of soils (*Catapyrenium squamulosum, Collema crispum*) were particularly abundant, together with cyanobacteria and mosses. Thalli grew initially on the mortar, followed by subsequent invasion of the limestone *tesserae*, suggesting that mortar is the first material to be colonized (Saiz-Jimenez and Ariño, 2001).

Results show that slight differences in microclimate are more important than differences in mortar composition in determining the colonizing lichens. However, in groups of relevés in similar microclimatic conditions, the type of mortar is the factor determining the differences between relevés.

3. LICHEN-MORTAR INTERACTION: INFLUENCE ON BIODETERIORATION

Study of the interface between the lichen thallus and the substratum provides evidence that although the communities which develop are similar on the three types of mortar, there are differences in the depths of penetration of the hyphae and the subsequent effect on the deterioration of the surface layer. The three mortars showed considerable alteration at the interface with lichen thalli. Clusters of hyphae forming bundles of variable thickness penetrated in different directions through the calcite cement, either by dissolution of the material or via the natural fissures and pores.

3.1 Case study 1: mortar of Baelo Claudia (Figure 10-2)

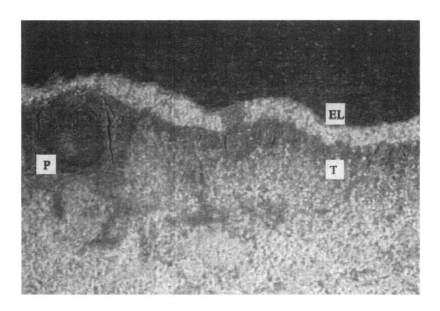

Figure 10-2. Mortar at Baelo Claudia: the surface few millimeters are perforated by the endolithic thallus of *Verrucaria* (T) with perithecium (P), and showing the external layer of calcium oxalate (EL)

Transverse sections from the mortar interface with *Caloplaca lactea* and *Verrucaria viridula* showed regular penetration of hyphae up to a depth of 1.5-2.0 mm. The contact zone between lichen and substratum was extensively pitted due to hyphal bundles affecting not only the surface, but also the internal layers of mortar. As the majority of the thalli developed endolithically, the mortar was considerably altered in the zone immediately beneath the surface, particularly around the fruiting bodies (in this case, perithecia). A mineral layer 50-100 μm thick, mainly composed of calcium oxalate, developed beneath the thallus (Ariño *et al.*, 1997).

Oxalic acid excreted by the mycobiont plays an active role in the weathering of carbonate substrata, leading to the formation of crystals of calcium oxalate (Jones *et al.*, 1980). The calcium oxalate crystals are more insoluble than calcium carbonate and usually accumulate at the interface between thallus and mortar, although some oxalate is deposited on the surface of the thalli, producing a pruinose aspect. This calcium oxalate produced *de novo* in the superficial layers can be incorporated into the mortar through processes of dissolution and precipitation of calcium carbonate, which acts as a cementing agent. This process initiates the formation of a superficial white or orange patina composed mainly of calcium carbonate and calcium oxalate.

3.2 Case study 2: mortar of Carmona (Figure 10-3)

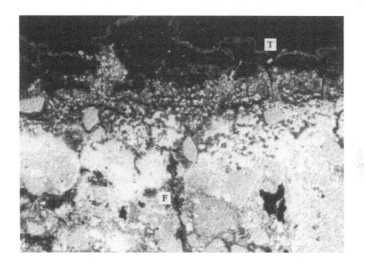

Figure 10-3. Mortar at Carmona: beneath a thallus of *Caloplaca flavescens* (T), the mortar is disintegrated and bundles of hyphae have mainly penetrated the fissures (F)

The study of the interface between this mortar and the thalli of certain lichens, such as *Caloplaca flavescens*, *Dirina massiliensis* and *Verrucaria macrostoma*, showed that the hyphae scarcely penetrated into the substratum, and bundles of hyphae dissolving the carbonate matrix were only observed near the surface (0.3-0.5 mm depth). Deeper penetration only occurred in fissures. A large accumulation of mineral particles, not evident at the macroscopic level, was found just beneath the placodioid thallus of *Caloplaca flavescens*, possibly due to disaggregation of the mortar which had undergone considerable alteration.

3.3 Case study 3: mortar of Italica (Figure 10-4)

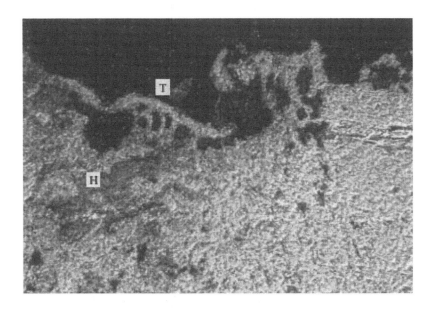

Figure 10-4. Mortar of Italica: the layer near to the surface is clearly deteriorated by the thallus (T) and via penetration of hyphal bundles (H)

Here, the surface of the mortar was strongly altered due to the dissolution of carbonate cements, which produced mortar disintegration and loss of quartz particles. Pits were similar to those described for the mortar of Baelo Claudia. Hyphal penetration was limited to areas near the surface and to some fissures.

The mineralogical composition and proportion of aggregates play an important role in the interaction between mortar and the lichen thallus. In the mortar of Carmona there is no evident pitting and the stronger alteration is in the external layer, reflected in the disintegration produced by the loss of the matrix (Ariño *et al.*, 1997). Zones beneath the surface are scarcely altered as thalli are epilithic and only occasionally penetrate into the substratum. In contrast, the mortar of Baelo Claudia, composed almost exclusively of calcite, shows stronger chemical alterations and deeper penetration of the hyphae by dissolution of the carbonate matrix, with the presence of both macroscopic and microscopic pitting. The mortar of Italica is intermediate in its behavior. The presence of hyphal bundles in the mortar could be related to the attachment to the porous substratum and to the mechanisms of dispersion and permanence of the thallus, acting as a regenerative mechanism after rupture and detachment from the superficial layer. The presence of such structures in the substratum must be taken into account if it is decided to eliminate lichens from a surface, as the thallus can rapidly regenerate.

4. LICHEN COMMUNITIES AND CHROMATIC CHANGES ON MONUMENTS

Another interesting case study was observed in the Roman amphitheatre at Italica. Due to the absence of natural stone in the surrounding area, it was built from a mixture of siliceous and calcareous pebbles cemented with mortar. Analysis of 37 relevés randomly undertaken revealed several factors affecting lichen vegetation in the amphitheatre (Nimis *et al.* 1998). In this case, the pH of the substratum is highly influential in supporting a mixture of silicolous and calcicolous species; other important factors affecting lichen distribution are eutrophication caused by birds, which induce the growth of ornithocoprophilous species, and differences in solar irradiation.

As a result of this combination of ecological factors, differences in the color of the resulting lichens produced chromatic changes to the amphitheatre. Three color groups were defined: orange-yellow, prevailing along the outer rim of the amphitheatre, white (including light grey and light green), which dominates the south-facing half, and dark, dominating the north-facing half.

Chromatic effects of lichen colonization were also evidenced in Baelo Claudia in terms of the various substrata, including mortars. As well as those lichens colonizing the sandstone pavement of the Forum (Ariño *et al.*, 1995), *Caloplaca* species were much in evidence, being well adapted to alkaline substrata and sunny exposed situations. These species are predominantly crustose, but two are placodioid; other taxa present include the foliose *Xanthoria calcicola* and the squamulose *Collema crispum* var. *metzleri*. Dominance of crustose thalli can be considered typical on rocks in arid environments, their morphology being well adapted to preventing physical alteration and avoiding excessive water loss.

Most thalli are intensely colored with yellow, red, brown or black pigments; their role in protecting the photobiont from an excess of insolation is obvious. The lichen community on the well-preserved flagstones is dominated by light pigmented (yellow or orange) or pruinose (due to accumulation of calcium oxalate crystals on their surface) species. Otherwise, the community developed on deteriorated flagstones (with higher proportion of clay) is dominated by dark pigmented species (brown or black). Table 10-3 shows the percentage cover by such lichens. We can easily compare the two groups and statistical analysis seems unnecessary. As a result, the pavement of the Forum has two types of colored flagstones, the well preserved are yellow-orange, and the deteriorated are dark grey-brown.

Table 10-3. Forum of Baelo Claudia: percentage colonization by lichens on seven flagstones (1 to 4 – well-preserved flagstones; 5 to 7 deteriorated flagstones)

	1	2	3	4	5	6	7
Caloplaca velana	27,82	27,23	35,61	31,29	0,66		
Caloplaca lactea	0,49			0,57			
Caloplaca variabilis	0,19	0,42	1,58	0,34	1,18		
Caloplaca flavescens		0,13	0,38				
Aspicilia contorta ssp. hoffmanniana		0,36	0,38	0,09	0,45		
Caloplaca citrina			0,05	0,14	0,55	0,07	0,36
Xanthoria calcicola			1,01				
Lecanora albescens			0,17	0,93	1,78	0,06	
Lecania turicensis			0,18	2,35	20,16	0,37	
Verrucaria macrostoma					7,23	2,05	3,60
Collema crispum var. metzleri					0,18	7,83	10,56
Caloplaca teicholyta					1,05	0,13	5,07
Caloplaca erythrocarpa					0,15		
Lecanora dispersa					0,01		
Total covering (%)	28,50	28,14	39,36	34,70	33,40	10,51	19,59

The color of the lichen thalli is important for understanding the interaction between lichens and substrata, as it is well known that one of the main causes of biodeterioration by lichens are the cycles of drying and wetting, which is strongly temperature dependent (Seaward, 1988). In the Necropolis of Carmona, the temperature of bare substratum was compared with the interfaces of substrata supporting *Diploschistes gypsaceous* (white) and *Toninia sedifolia* (black). Results showed that the bare rock suffered from higher temperature fluctuations than lichen-covered substrata. Bare rock and the interface of *D. gypsaceous* reached similar temperatures, but the interface of *T. sedifolia* was on average 5-6°C higher (Ariño and Saiz-Jimenez, 1997).

5. PROTECTING MORTAR FROM LICHEN COLONIZATION

As exposure of archaeological remains results in a gradual deterioration caused by environmental and/or biological agents, some consolidation techniques have been carried out; for example, at Baelo Claudia, some heavily biodeteriorated stuccoed capitals were consolidated with a polyester resin to preserve their surfaces. Ten years later, these treated surfaces were significantly colonized by the lichens *Caloplaca lactea, C. citrina, Sarcogyne regularis, Verrucaria muralis* and *Lecania turicensis*. Observation of samples by electron scanning microscopy showed that the lichens were growing between the mortar and resin, forming a dense network of hyphae and algae, with fruiting bodies emerging through the resin layer, producing extensive perforation. A combination of physical and chemical action is necessary to penetrate the resin layer (Ariño and Saiz-Jimenez, 1996b).

Another case study concerns the Guadalupe Monastery (Caceres), constructed of bricks and mortar which, before restoration, was heavily colonized by phototropic communities. After restoration and cleaning, the monument was treated in 1991 with Hydrophase (silicone) and Algophase (biocide). This monument provided an opportunity to investigate the effectiveness of separate applications of silicone and biocide, the performance of the surface treatment, the efficacy of the biocide, species composition and microbial strategies, if any, re-colonizing the mortars (Ariño *et al.*, 2002).

In the majority of the samples, there was no lichen re-colonization, and in some restricted areas thalli were small in comparison with the original colonization. In other areas, where significant lichen growth was detected, environmental conditions were not especially favorable for preventing the growth of phototropic organisms, as constant rising damp washed out or eliminated the effect of the biocide. Deposition of organic matter and salts caused by bird droppings prevented the inhibition of lichens which grew directly on the droppings, particularly on roof tiles. It was apparent that under normal conditions, biocide/silicone can inhibit re-colonization of phototropic organisms for a period ranging from five to eight years, as demonstrated on the Scurano Portico at Florence, Italy (Gomez-Bolea *et al.*, 1999), but not under conditions of continuous rising damp or under the influence of certain environmental agents, such as strong nitrification.

6. A NEW OUTLOOK ON LICHENS

The capability of lichens to act as biomonitors of environmental parameters is well known. Their presence is consequent upon a complex of factors affecting a monument in the long-term. For this reason, the close relationship between lichen communities and microclimatic factors suggests that it may be possible to use them for microclimatic monitoring of monuments. A protective role of lichens in an aggressive environment has even been noticed, the biodeterioration *versus* bioprotection being an unstable equilibrium at the mercy of environmental conditions, the type of substratum and the different colonizing species (Ariño *et al.*, 1995).

The presence of lichens on stone, natural or man-made, is due to the natural relationship between the lithosphere and the biosphere. It should probably be accepted in the artificial struggle to keep the surface of the stones free from colonization by the continuous application of biocides. It is very often ourselves who induce colonization by bringing about changes in microclimatic conditions, such as exposure, temperature, and nitrification, and therefore small changes in these conditions could solve the problem. Only when the colonization is proved to be pathological, through in-depth studies of its causes and consequences, should cleaning techniques be employed.

ACKNOWLEDGEMENT

This work was supported by the Consejeria de Cultura, Junta de Andalucia and the European Commission, project EVK4-CT1999-20001.

REFERENCES

Ariño, X. and Saiz-Jimenez, C. (1996a) Colonization and deterioration processes in Roman mortars by cyanobacteria, algae and lichens. *Aerobiologia* 12: 9-18.

Ariño, X. and Saiz-Jimenez, C. (1996b) Lichen deterioration of consolidants used in the conservation of stone monuments. *The Lichenologist* 28: 391-394.

Ariño, X. and Saiz-Jimenez, C. (1997) Deterioration of the Elephant tomb (Necropolis of Carmona, Seville, Spain). *International Biodeterioration and Biodegradation* 40: 233-239.

Ariño, X., Ortega-Calvo, J.J., Gomez-Bolea, A. and Saiz-Jimenez, C. (1995) Lichen colonization of the Roman pavement at Baelo Claudia (Cadiz, Spain): biodeterioration vs. bioprotection. *The Science of the Total Environment* 167: 353-363

Ariño, X., Gomez-Bolea, A. and Saiz-Jimenez, C. (1997) Lichens on ancient mortars. *International Biodeterioration and Biodegradation* 40: 217-224.

Ariño, X., Canals, A., Gomez-Bolea, A. and Saiz-Jimenez, C. (2002) Assessment of the performance of a water-repellent/biocide treatment after 8 years. In: *Protection and Conservation of the Cultural Heritage of the Mediterranean Cities* (E. Galan and F. Zezza, eds.): 121-125. Balkema, Lisse.

Gomez-Bolea, A., Ariño, X., Balzarotti, R. and Saiz-Jimenez, C. (1999) Surface treatment of stones: consequences on lichen colonization. In: Of *Microbes and Art. The Role of Microbial Communities in the Degradation and Protection of Cultural Heritage* (P. Tiano and G. Mastromei, eds.): 233-237. Consiglio Nazionale delle Ricerche, Florence.

Guillite, O. (1995) Bioreceptivity: a new concept for building ecology studies. *The Science of the Total Environment* 167: 215-220.

Jones, D., Wilson, M.J. and Tait, J.M. (1980) Weathering of a basalt by *Pertusaria corallina*. *The Lichenologist* 12: 277-289

Nimis, P.L., Seaward, M.R.D., Ariño, X. and Barreno, E. (1998) Lichen induced chromatic changes on monuments: a case study on the Roman amphitheater of Italica (S. Spain). *Plant Biosystems* 132: 53-61.

Puertas, F., Blanco-Varela, M.T., Palomo, A., Ortega-Calvo, J.J., Ariño, X. and Saiz-Jimenez, C. (1994) Decay of Roman and repair mortars in mosaics from Italica, Spain. *The Science of the Total Environment* 153: 123-131.

Saiz-Jimenez, C. (2001) The biodeterioration of building materials. In: *A Practical Manual on Microbiologically Influenced Corrosion*, vol. 2. (J. G. Stoecker ed.): 4.1-4.20. NACE Press, Houston.

Saiz-Jimenez, C. and Ariño, X. (1995) Biological colonization and deterioration of mortars by phototrophic organisms. *Materiales de Construcción* 45: 5-16.

Saiz-Jimenez, C. and Ariño, X. (2001) Microbial corrosion of cultural heritage stoneworks. In: *A Practical Manual on Microbiologically Influenced Corrosion*, vol. 2. (J. G. Stoecker ed.): 11.25-11.33. NACE Press, Houston.

Seaward, M.R.D. (1988) Lichen damage to ancient monuments: a case study. *The Lichenologist* 10: 291-295.

Chapter 11

OBSERVATIONS ON LICHENS GROWING ON ARTIFACTS IN THE INDIAN SUBCONTINENT

S. SAXENA[1], D.K.UPRETI[1], AJAY SINGH[1] and K.P. SINGH[2]
[1]*Lichenology Laboratory, Plant Biodiversity and Conservation Biology DivisionNational Botanical Research Institute, Lucknow-1, India* [2]*Botanical Survey of India, Central Circle, 10 Chetham Lines, Allahabad, India*

Abstract: A detailed account of lichens growing on different artifacts in the Indian subcontinent is provided. Some interesting distributional and ecological patterns of lichens growing on historical monuments and buildings are also presented.

Keywords: Lichens; man-made substrata; India; historical monuments

1. INTRODUCTION

Lichens are hardy autotrophic symbiotic entities with frugal life sustaining requirements: besides a little moisture and minimum nutrition, all they need is a foothold of a fairly permanent nature for their establishment and growth. Provided they have the first two factors, they can survive the most adverse living conditions; as for the third factor, they are found thriving on a wide variety of substrata. How can man-made objects defy invasion by such determined survivors?

The association of lichens with artifacts has attracted the attention of many researchers, the explorative work undertaken as a consequence being not only floristic in nature but also concerned with the effects lichen growth has on them.

A detailed account of lichens on man-made substrata in general was presented by Brightman and Seaward (1978); their review includes a survey of the lichens associated with various building materials, but not their alleged role in biodeterioration and the need for artifact conservation. A number of researchers, viz. Bech-Anderson (1986), Bravery (1981), Chaffer (1972), Del Monte (1991), Del Monte and Sabbioni (1983), Del Monte *et al.* (1987), Gehrmann *et al.* (1988), Hale (1975), Hyvert (1978), Jones and Wilson (1985), Jones *et al.* (1980), Keen (1976),

Krumbein (1987), Lloyd (1973), Saiz-Jimenez (1981), Salvadori and Zitelli (1981), Schatz (1963), Schatz *et al.* (1954), Seshadri and Subramanian (1949), and Swindale and Jackson (1956), with an eye on assessment of damage caused to the substratum and its conservation, have enumerated many lichen taxa considered as culprits in this regard. Studies of these researchers are mainly concerned with European monuments, more particularly those relating to the Roman period spread over Italy and Spain.

In the context of South Asia, reference can be made to the studies of Chatterjee *et al.* (1995), Gairola (1968), Gayathri (1980), Lal (1970), Riederer (1981), Singh and Dhawan (1991), Singh and Sinha (1993), Singh and Upreti (1984), Upreti (1988), and Upreti and Büdel (1990), nearly all of which are concerned with monuments; however, other artifacts are colonized by lichens. The above references and the following review provide a total scenario of lichen colonization on a variety of man-made objects so far observed in this region of the world. In the light of the amazingly wide climatic diversity of the Indian subcontinent it is not difficult to realize that there is plenty of scope for long-term scientific exploration in this field, so the following communication can only be regarded as a preliminary study on the subject.

2. SUBSTRATA

2.1 Plastic netting

There are a few reports of lichen growth on plastic material, but none from the geographical region in question. Brightman and Seaward (1978) in their coverage of lichen growth on wide ranging man-made objects recorded three species from plastic and fibreglass. Sipman (1994) noted the growth of 12 foliicolous lichens on plastic tape used to mark forest trails in French Guiana, and Lücking (1998) documented 63 such species on a plastic sign-board (29.5 x 19.5 cm) in the rain forest at La Selva Biological Station, Costa Rica.

From the South Asian region, specifically Nepal, Upreti and Dixit (2002) found 11 lichen species growing on plastic nets commonly used to fence nursery plots at the Royal Botanic Garden, Kathmandu. Because of their texture, the moisture and dust-retaining ability, the woven plastic threads would appear to facilitate lichen growth over the net. Lichens enumerated from this substratum are:

Heterodermia diademata (Taylor) Awasthi
H. firmula (Nyl.) Trevisan
H. incana (Stirton) Awasthi
Lecanora flavidofusca Müll. Arg.

Micarea sp.
Parmotrema nilgherrense (Nyl.) Hale
P. tinctorum (Nyl.) Hale
Pertusaria indica Srivastava & Awasthi
Phaeophyscia endococcina (Körber) Moberg
P. hispidula (Ach.) Essl.
Xanthoria candelaria (L.) Arn.

Generally speaking, a large percentage of lichen species is substrate-specific, but none of the above appear to exhibit obligatism in terms of plastic material.

2.2 Metal objects

Early observations of Schaerer (1850) and Arnold (1875) demonstrated how some species of *Aspicilia, Caloplaca, Candelariella, Pertusaria, Physcia, Verrucaria* and *Xanthoria* were capable of living on, and occasionally thriving on, iron-rich substrata.

There have also been some similar observations in India from temperate Himalaya where excessive moisture exposes iron objects to rusting, the roughened surfaces of which provide a convenient niche for the establishment of lichen propagules. Corrugated iron sheets (commonly called tin sheets) used for roofing purposes, as well as a wide range of other man-made metalliferous materials, with time, support thalli of *Caloplaca* sp. and *Xanthoria parietina*.

2.3 Roof tiles

Tiles made of two entirely different materials are used for roofing purpose in India: one is man-made to the required shape and size from burnt clay and the other comprises thin stone sheets manually flaked off from loosely stratified slate stone rocks; the latter, natural objects dressed to size, are in use in the mountainous Himalayan region where this variety of rock is plentiful. The calcareous slate tiles in the temperate parts of this region provide a favorable substratum for the growth of *Caloplaca irrubescens*.

Burnt clay tiles, commonly used in the rest of the country, such as at Tirunevelli, Tamil Nadu, support *Peltula euploca* and *P. patellata*. These species, under constant exposure to direct sunlight for almost the whole day, develop a prominent upper cortex which acts as a photo-protective screen to the photobiont and helps to retain absorbed moisture in the thallus for a longer period.

2.4 Wood

The timber of *Cedrus* and *Pinus* is commonly used in the Himalayan region for the construction of electricity poles, fence poles, wooden planks for roofing and small bridges over streams. These wooden objects on exposure to the vagary of nature become the victims of lichen invasion. In general, these substrata support *Buellia tincta*, *Tephromela fimbriatula* and *Caloplaca* sp., but foliose species of *Physcia*, *Dirinaria* and *Phaeophyscia* are commonly found, particularly on electricity poles.

3. HISTORICAL MONUMENTS AND BUILDINGS

Indian cultural heritage comprises a variety of artifacts, including buildings. A wide range of building material has been used for their construction, the predominant component being stone, including granite, sandstone, limestone, marble, khondalite and mica schist. Other materials were brick and mortar, timber and metals. Lichen taxa collected from Indian monuments are enumerated in Table 11-1.

Table 11-1. Distribution of lichens on Indian monuments and buildings

Species	Collection site				
	Karnataka	Tamilnadu	Orissa	Uttar Pradesh	Madhya Pradesh
Bacidia sp. no.1			D		
B. sp. no.2			Ra		
Buellia posthabiti	Cha				
B. disjecta					Ab
B. spp.	Ti				
Caloplaca sp. type 1			Su,		
C. sp. type 2			S,A,Ko		
C. sp. type 3	Sh		Su		
C. sp. type 4	Chan,Ch,P, Ti		Ko		
C. holochracea	P, Y		Su		
C. poliotera					Bu
Candelaria concolor	Ch,Cha, Chan,T				
Cladonia ramulosa	Sh				
Coccocarpia palmicola	Sh				
Diploschistes euganeus	Y				

11. Lichens Growing on Artifacts in the Indian Subcontinent

Table 11-1.1 (Continued) Distribution of lichens on Indian monuments and buildings

Species	Karnataka	Tamilnadu	Orissa	Uttar Pradesh	Madhya Pradesh
Dirinaria confluens	C				
D. consimilis	Y				
D. papillulifera			A,Bh		
Endocarpon nanum	Ch, K		A	Ca	
E. pusillum	Cha			Ca	
E. rosettum				Ca	
Heterodermia hypocaesia	Sh				
H. incana	Sh				
H. leucomela ssp. *boryi*	Sh				
H. microphylla	Sh				
Lecanora sp. type 1	Y				
L. sp. type 2	Y				
L. sp. type 3	P				
L. sp. type 4	Cha				
L. sp. type 5	Y				
L. sp.					Ab
Lepraria sp.	Cha		R, S, A		
Leptogium indicum	Sh				
L. pichneum	Ch				
Lichina sp.					Bu
Parmelinella wallachiana	Sh				
Peltula euploca	H, Ti				Ab
P. patellata	K, P			N	
P. obscurans		G			Bu
P. zahlbruckneri					Bu
Phylliscum indicum	Cha, Chan				Bu
P. tenue	K, Ti				
P. macrosporum	P				I
Physcia phaea	Sh				
P. spp.					I
Pyxine cocoes var. *cocoes*	Ch, Ti		A, Bh, Su		
P. petricola var. *petricola*	P, Cha, H		B		
P. petricola var. *pallida*	K		S, B, R		
Roccella montagnei			Su		

A = Amreshwar temple, Chaurasi, Dist. Puri; Ab = Abandoned Khajuraho temple, near Parvati temple, Dist. Chattarpur; Bh = Bhairingeshwar temple, Dhauli, Dist. Puri; B = Chamundi temple, Chamundi Hill, Dist. Mysore; Bu = Bukhar Pahar, Mahoba, Dist. Chattarpur; Ca = Carlton Hotel, Dist. Lucknow; Ch = Chennakeshava temple, Belur, Dist. Hassan; Cha = Chennakeshava temple, Hullekere, Dist. Hassan; Chan = Chennakeshava temple, Turuvekere, Dist. Tumkur; D = Dilkusha Fort, Dilkusha gravewall, Dist. Lucknow; G = Gopalsamudrum village, Dist. Tirunelveli; H = Hoyseleshwara temple, Halibed, Dist. Hassan; I = Imambara, Dist. Lucknow; K = Keshava temple, Somanathpur, Dist. Mysore; Ko = Kotitirth temple, Bhubneswar, Dist. Puri; N = NBRI boundary wall, Dist. Lucknow; P = Panchlingeshwara temple, Somanthpur, Dist. Mysore; R = Rameshwar temple, Bhubneswar, Dist. Puri; Ra = Raja Sahib's Kila, Dist. Mahmudabad; S = Shatruganeshwar temple, Bhubneswar, Dist. Puri; Sh = Sivappa's fort complex, Kavaledurga, Dist. Shimoga; Shi = Shiva temple, Tikhatganj, Dist. Malihabad; T = Suntemple, Konark Dist. Mysore; Ti = Tippu's Jail, Srirangapatna, Dist. Mysore; Y = Yognarsimhaswami temple, Baggavalli, Dist. Chickmangalur

A mere floristic enumeration does not fully describe the lichen scenario of these monuments. Some interesting distributional and ecological patterns were noted and briefly presented herewith.

As a rule, the nature and composition of vegetation, including lichens, are determined by the prevailing climatic conditions. This view may appear very simplistic, since a number of other factors play their part in breaking down vegetational uniformity over a given area. Lichen communities on monuments amply display this aspect of ecology.

The building materials of Indian monuments show great diversity, some being constructed of bricks and mortar, and others from a variety of stone materials enumerated earlier. Each of these materials, with their own subtypes, display a wide spectrum of physical (texture, porosity, etc.) and chemical (mineral composition, pH, etc.) properties. Different types of stones used in the construction of a single monument, coupled with the pointing material, creates innumerable growth conditions over their surface for lichens.

The architecture of the monuments is another factor instrumental in determining the pattern of lichen growth, the different microclimates created providing ecological niches. Moreover, each building has exterior faces exposed to the complex of weathering agents; generally the orientation of these faces is either horizontal or vertical.

The horizontal face of a building comprises mainly roof and terraces. These are exposed to direct sun for the greater part of the day and provide uniformly xeric conditions for plant growth. The same conditions are displayed in the case of tiles, as discussed earlier. The roofs of most Indian monuments are made of stone slabs, either with calcareous pointing at their joints or completely covered with a thick

11. Lichens Growing on Artifacts in the Indian Subcontinent

layer of lime plaster. This face of the monument is a preferred substratum for squamulose lichens (Fig. 11-1), such as species of *Peltula* and *Endocarpon*, and also for some crustose forms, such as *Phylliscum indicum, P. tenue* and *P. macrosporum*. All these lichens are devoid of lichen acids. Thus, chemical weathering of the substratum by them can safely be ruled out, but that cannot be said with any measure of certainty about physical or mechanical damage caused by them.

Figure 11-1. Squamulose lichen genera *Petula* and *Endocarpon* on calcareous pointing

The ecological situation concerning the vertical faces of building comprising mainly walls is more complicated. Walls facing different directions that receive varying degrees of sunlight, rain showers, wind velocity, etc. present differing microclimatic conditions that affect their overall vegetational composition and frequency.

Different heights of the same wall also appear to enjoy different levels of light (and heat) and moisture. The base of the wall may be protected from sun to a greater extent than its upper reaches. Shadows cast by surrounding vegetation also play a role in this respect, leading to lower evaporation rates of moisture. Moreover, the water content there is more or less constantly replenished from the ground by capillary action in the case of porous building material. The lichen flora of this part of the wall of Karanataka and Orissa monuments is dominated by granular sterile thalli of *Lepraria* and *Lecanora* species (Fig. 11-2). Though there is as yet no direct evidence of biodeterioration of the substratum caused by these lichens, the usnic

188   *Biodeterioration of Rock Surfaces*

acid, zeorin and divaricatic acid detected in their thalli have the potential to cause damage to it by chelation action.

Figure 11-2. Sterile thalli of *Lepraria* and *Lecanora*

The upper reaches of walls, with their greater exposure to sun, show a high evaporation rate of moisture and are hence drier than lower parts. Foliose lichen species are in greater profusion here, with taxa of *Physcia, Pyxine, Dirinaria, Heterodermia* and *Candelaria* commonly occurring (Fig. 11-3). There are also occasional splashes of other foliose lichens, as well as a frequent occurrence of many crustose forms.

11. Lichens Growing on Artifacts in the Indian Subcontinent

Figure 11-3. Members of the *Physciaceae* on exposed monuments

Of all the foliose species, *Pyxine petricola* and *P. cocoes* are universally present on the monuments of Karnataka and Orissa, sometimes alone, sometimes mixed with others. The varietal distribution of these species shows an interesting pattern. *P. cocoes* var. *cocoes* is equally frequent on the monuments of both states, whereas var. *prominula* is confined to southern Indian states. Similarly *P. petricola* var. *petricola* is more frequent on monuments of southern India, while var. *pallida* has dominance on those of Orissa (Fig. 11-4). A yellow foliose lichen *Candelaria concolor* forms a frequent association with these species of *Pyxine*.

Species such as Coccocarpia palmicola, Parmelinella wallichiana, Leptogium indicum, L. pichneum, Heterodermia hypocaesia, H. incana, H. leucomela var. boryi and H. microphylla were found on a complex of monuments situated in a wooded, relatively more moist area in southern India.

The corticolous species *Dirinaria papillulifera*, *D. confluens* and *D. consimilis* were found growing on stone monuments, the first species from Orissa and the other two from southern India. The lichen acids of all three species are potential chelators.

Figure 11-4. Pyxine petricola var. *pallida* on Orissa monument

Places subjected to atmospheric pollution have their own ecology. This factor coupled with artificially created dry conditions in urban areas cause almost total destruction of the local lichen flora. Other lichens found growing on calcareous lime plaster of monuments at Lucknow, comprising species of *Endocarpon* and *Phylliscum*, were other exceptions. They are well able to withstand dryness, perhaps because of their inherent resistance to this factor, but against acidic pollutants the basic (alkaline) nature of the substratum provides chemical protection to these lichens. The xerophytic genera *Endocarpon*, *Peltula* and *Phylliscum* collected from different monuments seem to prefer man-made calcareous material, such as lime or cement layers for their growth and are not confined to any particular face (horizontal or vertical) of the building.

Besides ecological considerations of monument lichens, another interesting observation on the biodeterioration of the substratum is worth mentioning. It is well known that crustose lichens, being in closer contact with the substratum, are more efficient biodeteriorating agents than foliose lichens. An example of this came to light during studies of Indian monuments in the form of a patch of *Lecanora* sp.

growing on a calcareous substratum. The thallus was so rich in crystals of calcium oxalate that a large number of them protruded out of it. For formation of these crystals, calcium obviously came from the substratum, the loss of which clearly leads to weakening of rock material.

In the 1990s, the authors visited Meghalaya state and examined rock pillars (monoliths) erected by educated and culturally advanced tribal people in memory of their departed kin (Fig. 11-5). In general, the rocks chosen for monoliths are longish in shape and made to stand erect after fixing their bases tightly in holes of suitable

Figure 11-5. Monoliths in Meghalaya

dimensions dugout in the ground. These memorials if sufficiently old (in terms of centuries) and related to some important person may qualify for consideration as protected monuments. These memorials consisting of purely local rock in its original condition have no specific lichen flora of their own, being the same as that of other *in situ* rocks nearby.

REFERENCES

Arnold, F.C.G. (1875) Die lichenen des Fränkischen Juras. *Flora, Jena* 58: 524-528.
Bech-Anderson, J. (1986) Biodeterioration of natural and artificial stone caused by algae, lichens, mosses and higher plants. Proceedings of the VIth International Biodeterioration Symposium: 126-131. Washington.
Bravery, A.F. (1981) Preservation in the construction industry. In: *Principles and Practice of Disinfection, Preservation and Sterilization* (A.D. Russell. W.B. Hugo and G.A.J. Ayliffe, ed): 379-402. Blackwell Scientific, Oxford.

Brightman, F.H. and Seaward, M.R.D. (1978) Lichens of man-made substrates. In: *Lichen Ecology* (M.R.D. Seaward, ed): 253-293. Academic Press, London.

Chaffer, R.J. (1972) *The Weathering of Natural Stones*. DSIR Building Research Special Report No. 18, London.

Chatterjee, S., Singh, A. and Sinha, G.P. (1995) Lichens from some monuments in Karnataka and Orrisa, India. *Geophytology* 25: 81-88.

Del Monte, M. (1991) Trajan's Column: lichens don't live here anymore. *Endeavour* 15: 86-93.

Del Monte, M. and Sabbioni, C. (1983) Weddellite on limestone in the Venice environment. *Environmental Science and Technology* 17: 518-522.

Del Monte, M., Sabbioni, C. and Zappia, G. (1987) A study of patina called 'scialbatura' on imperial Roman marbles. *Studies in Conservation* 32: 114-121.

Gairola, T.R. (1968) Examples of the preservation of monuments in India. In: *The Conservation of the Cultural Property*: 139-152. UNESCO. Rome.

Gayathri, P. (1980) Effects of lichens on granite statues. *Birla Archeological Cultural Research Institute Research Bulletin* 2: 41-52.

Gehrman, C.K., Peterson, K. and Krumbein, W.E. (1988) Silicole and calcicole lichens on the Jewish tombstones – interactions with environment and biocorrosion. In: *IVth International Congress on Deterioration and Conservation of Stone,* supplement: 33-38. Nicholas Copernicus University, Torun.

Hale, M.E. (1975) Control of biological growth on Mayan archeological ruins in Guatemala and Honduras. *National Geographic Society Research Reports, 1975 Projects*. 305-321.

Hyvert, G. (1978) Weathering and restoration of Borobudur Temple, Indonesia. In: *Decay and Preservation of Stone* (E.M. Winkler, ed): 95-100. Engineering Geology Case Histories No. 11. Geological Society of America, Boulder.

Jones, D. and Wilson, M.J. (1985) Chemical activity of lichens on mineral surfaces – a review. *International Biodeterioration* 21: 99-104.

Jones, D., Wilson, M.J. and Tait, J.M. (1980) Weathering of basalt by *Pertusaria corallina*. *The Lichenologist* 12: 277-289.

Keen, R. (1976) *Controlling Algae and other Growths on Concrete*. Advisory Note 45.020. Cement and Concrete Association, Slough.

Krumbein, W.E. (1987) Microbial interaction with mineral materials. In: *Biodeterioration 7* (D.R. Houghton, R.N. Smith and H.O.W. Eggins, eds.): 78-100. Elsevier Applied Science, Amsterdam.

Lal, B.B. (1970) Indian rock paintings and their preservation. In: *Aboriginal Antiquities in Australia: their nature and preservation* (F.D. McCarthy, ed): 139-146. Australian Aboriginal Studies no. 22, Canberra.

Lloyd, A.O. (1973) Lichen attack on marble at Torcello-Venice. In: *Atti Congresso Petrolio e Ambriente Congresso*: 221-224. Artioli, Modena.

Lücking, R. (1998) 'Plasticolous' lichens in a tropical rain forest at La Selva Biological Station, Costa Rica. *The Lichenologist* 30: 287-290.

Riederer, J. (1981) The restoration of archeological monuments in Sri Lanka. In: *The Conservation of Stone-II* (R. Rossi-Manaresi, ed): 737-757. Bologna.

Riederer, J. (1984) The restoration of archeological monuments in the tropical climate. In: *ICOM Committee for Conservation. 7th Triennial Meeting*. Copenhagen.

Saiz-Jimenez, C. (1981) Weathering of the building materials of the Giralda (Seville, Spain) by lichens. In: *ICOM Committee for Conservation, 6th Triennial Meeting, Working Group: Stone*. Ottawa.

Salvadori, O. and Zitelli, A. (1981) Monohydrate and dihydrate calcium oxalate in living lichen incrustations biodeteriorating marble columns of the Basilica of Santa Maria Assunta on the Island of Torcello (Venice). In: *Conservation of Stone*. Part A: *Deterioration* (R. Rossi-Manaresi, ed): 379-390. Manaresi. Bologna.

Schaerer, L.E. (1850) *Enumeratio lichenum Europaeorum*. Staemplf, Bern.

Schatz, A. (1963) Soil microorganisms and soil chelation. The pedogenic action of lichens and lichens acids. *Agricultural and Food Chemistry* 11: 112-118.

Schatz, A., Cheronis, N.D., Schatz, V. and Trelawny, G.S. (1954) Chelation (sequestration) as a biological weathering factor in pedogenesis. *Proceedings of the Pennsylvania Academy of Science* 4: 233-239.

Seshadri, T.R. and Subramanian, S.S. (1949) A lichen (*Parmelia tinctorum*) on a Java monument. *Journal of Scientific and Industrial Research* 8: 170-171.

Singh, A. and Dhawan, S. (1991) Interesting observation on stone weathering of an Indian monument by lichen. *Geophytology* 21: 119-123.

Singh, A. and Sinha, G.P. (1993) Corrosion of natural and monument stone with special reference to lichen activity. In: *Recent Advances in Biodeterioration and Biodegradation*. Volume 1. (K.L. Grag, N. Garg and K. G. Mukerji. eds): 355-377. Naya Prokash, Calcutta.

Singh, A. and Upreti, D.K. (1984) The lichen genus *Endocarpon* from India. *Candollea* 39: 39-548.

Sipman, H.J.M. (1994) Foliicolous lichens on plastic tape. *The Lichenologist* 26: 311-312.

Swindale, L.D. and Jackson, M.L. (1956) Genetic process in some residual podzolised soils of New Zealand. *Transactions of the 6th International Congress on Soil Science* 4: 233-239.

Upreti, D.K. (1988) A new species of lichen genus *Phylliscum* from India. *Current Science* 57: 906-907.

Upreti, D.K. and Büdel, B. (1990) The lichen genera *Heppia* and *Peltula* in India. *Journal of the Hattori Botanical Laboratory* 68: 274-284.

Upreti, D.K. and Dixit, A. (2002) Lichens on plastic net. *British Lichen Society Bulletin* 90: 66-67.

Chapter 12

BIODETERIORATION OF PREHISTORIC ROCK ART AND ISSUES IN SITE PRESERVATION

ALICE M. TRATEBAS
*Bureau of Land Management, 1101 Washington Blvd.,Newcastle, WY 82701, USA,
Email:Alice_Tratebas@blm.gov*

Abstract: Lichens, algae and mosses growing on prehistoric rock art pose a major worldwide threat for its preservation and conservation. Most rock art is surficial, as paintings, or shallowly pecked or incised into the rock surface. Breakdown of the rock surface by lichens and other microflora can easily erase these images. Lichens and other damaging flora inhabit a wide variety of environments and have destroyed entire panels of rock art. Methods of lichen control developed for buildings and large historic monuments may be inappropriate for more delicate rock art. Rock art researchers must consider conservation, research and ethical issues for site preservation where biodeterioration is a factor.

1. INTRODUCTION

Research and publications on biodeterioration specific to rock art are minimal, in contrast to a voluminous literature on biodeterioration of historic buildings and monuments and natural rock. Rock art has completely different issues from historic structures. Rock art has fine-scale, delicate details that are easily destroyed. Rock coatings on rock art contain valuable scientific data for dating the images and reconstructing paleoenvironments. Paintings and painted petroglyphs contain scientific data, including paint formulas and organic materials useful for dating and paleoenvironmental reconstruction. The context of use for most prehistoric rock art depends upon an unaltered natural environment, as opposed to the culturally constructed environment of historic buildings and monuments. Microflora treatments on rock art, consequently, may negatively alter the aesthetics and settings of sites.

Rock art recorders or conservators have often taken action to halt biodeterioration without thinking through all the issues. Where mosses, lichens, and algae densely cover rock art, researchers in many localities around the world have mechanically removed microflora and/or washed the rock art. Publications often mention cleaning prior to recordation, although the practice is so widespread that for many sites we may not have a published record of cleaning treatment. An early method (no longer used) for recording the Valcamonica petroglyphs in the Italian Alps was to remove the microflora mechanically and wash the surface, then brush on a coating of white Pelican paint, followed by sponging black paint onto the rock face to highlight the glyphs for tracing (Anati 1964: 24-25, 1977: 9-22; Taylor et al. 1977: 309-310). Purely by chance, phenol in the white paint acted as a biocide to prevent quick regrowth of microflora (Taylor et al. 1977: 315, 318). Casein in the paint was thought to harden the rock exterior and serve as a sort of consolidant. However, when I visited the Valcamonica sites in 1995, some of the surfaces that had been cleaned and painted looked more friable than the rock surfaces nearby, which still had natural rock coatings, and glyph edges looked more eroded compared with the published photographs taken several decades earlier.

Between 1984 and 1993, the Torrey Lake (Wyoming, USA) petroglyphs were treated with orthophenylphenol in denatured alcohol and Clorox (sodium hypochlorite) in solutions ranging from 1:4 to full strength (Childers 1994). Unfortunately, no off-site tests of the biocides were conducted prior to using them on the petroglyphs. Before and after scientific studies to determine the effectiveness or the residual effects of the biocides (either positive or negative) were not conducted, although results were monitored visually and documented on before and after photographs. Prior to treatment, no consultations were conducted with interested parties, including Native Americans from the neighboring Indian reservation, who use the site, and rock art researchers and interested scientists. The treatment left fresh exposures of unweathered, friable, eroded and spalled sandstone that have never been scientifically evaluated for preservation problems (Fig. 12-1). The treated areas are surrounded by healthy lichens, which can easily recolonize freshly exposed rough surfaces (Fig. 12-2).

12. Biodeterioration of Prehistoric Rock Art

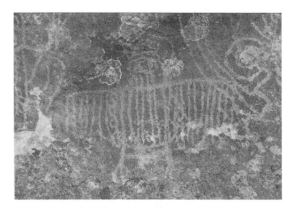

Figure 12-1. Biocide treatments typically leave eroded surfaces and remnants of lichen thalli (Photograph by Alice M. Tratebas)

Figure 12-2. Petroglyphs uncovered by biocide treatment are surrounded by healthy lichen (Photograph by Alice M. Tratebas)

Cases like these show the need for careful consideration of all the issues in biodeterioration of rock art. Because the article on the Torrey site treatment gives more details than many publications, it is easier to criticize, but may not differ much from other treatments that have been applied to rock art. Prior to my critique of the latter case and outlining a program for the evaluation of the issues before making decisions concerning lichen treatment (Tratebas and Chapman 1996), the only previous cautionary publication on the key issues in lichen treatment was by Florian (1978). The major issues she addressed were the possible use of lichenometry to date the rock art, the unanswered question of whether lichen removal prevents or increases rock surface deterioration, the fact that the effectiveness of treatment methods is unknown, and the potential deterioration that many control methods may cause. This chapter expands on the issues addressed in these two previous publications. Although the primary focus is lichen mediated biodeterioration, the chapter also covers damage by other types of microflora.

2. BACKGROUND

Previous work on biodeterioration of rock art has not been systematic, nor have adequate scientific studies and controlled experiments been conducted to fully delineate all the issues and assess the effects of treatments. The small scattering of publications has been worldwide. Rebricova and Ageeva (1995) recognized the environmental factors that favored microfloral growth and pointed out the need to restore the original microenvironment to inhibit biodeterioration. They experimented with several biocides (Table 12-1).

Table 12-1. Some Biocides Tested or Used on Rock Art

Reference	Location	Substrate	Rock Art Type	Biocide
Anati 1977; Taylor et al. 1977	Italian Alps	sandstone, schist	petroglyphs	phenol in Pelican paint
Childers 1994	Wyoming USA	sandstone	petroglyphs	(1) orthophenylphenol in denatured alcohol; (2) "Clorox" (sodium hypochlorite)
Clarke 1976	Western Australia	granite	petroglyphs	(1) pentachlorophenol in methylated spirit; (2) dichlorophen, (3) salicylanilide, (4) ammonium hydroxide, (5) non-ionic detergent; followed by zinc or magnesium fluorosilicate

Table 12-2.1 (Continued) Some Biocides Tested or Used on Rock Art

Reference	Location	Substrate	Rock Art Type	Biocide
Hygen 1999	Norway	granite	petroglyphs	10% quaternary ammonium
Lal 1970	India	sandstone or quartzite	paintings	ammonia; followed by consolidation with polyvinyl acetate in toluene with dibutyl phthalate added
Lefèvre 1974; Brunet and Vidal 1980	France	limestone	paintings	dilute formol and microenvironmental controls
Rebricova and Ageeva 1995	Siberia	marble; sandstone	petroglyphs	(1) quaternary ammonium salt, (2) hydrogen peroxide, (3) tinorgano-compound
Young and Wainwright 1995	Ontario Canada	marble	laboratory tests on algae samples from petroglyph site	tested (1) chelates of copper, (2) monuron, simazine, and atrazine, (3) ammonium chloride monohydrate, (4) dichlorophen, (5) quaternary ammonium chloride and bromide, (6) ethanol, (7) silver nitrate; applied microenvironmental controls

to treat algae and lichen on white marble at a Lake Baikal site and lichen on sandstone at the Shishkino site on the Lena River. Algae were growing several millimeters under marble surfaces for protection from excessive insolation. On sandstone, lichen favored water run off areas and surfaces with a positive slope. To evaluate the treatment success, they examined the microflora cells, a more scientific assessment than most conservators have carried out for rock art projects. The researchers are monitoring the treatments over a long time and reported their results after a period of six years.

In Australia, Clarke (1976) experimented with several biocides (Table 12.1), as well as washing with non-ionic detergent, to treat lichen biodeterioration of petroglyphs on granite. When monitored after 10 years, the treated areas were still lichen free, but in more humid Australian climates, recolonization has occurred within 6 to 12 months (Lambert 1989: 26-27). To treat prehistoric paintings, Boustead (1970) recommended a 5% ammonia solution to remove algae and lichen, followed by magnesium or zinc fluorosilicate to inhibit regrowth, and a 2% solution of sodium pentachlorphenate for fungi, but does not report on the results of any trials of these treatments.

Norwegian archaeologists have developed a regular program of cleaning soil and vegetation from petroglyphs to reduce the acidity of surface water, which seems to encourage the growth of some lichen species, and then treating lichen thalli with 10% quaternary ammonium compounds (Hygen 1999). Because observations indicated that the colonizing phase appeared to be most damaging to the substrate, the recommended re-treatment interval was reduced from 6 or 7 years to every 3 years.

The effects of algicides, on algae collected from the Peterborough site in Canada were tested in the laboratory by Young and Wainwright (1995). Algae growing on and below the surface of the marble had pitted the calcite grains (Wainwright 1985). Part of the rationale for constructing a building over the Peterborough site was control of the algae. If the building succeeded in slowly killing the algae by removal of moisture, biocides would not be needed (Wainwright 1996).

To control lichen and algae on carved monuments in Mesoamerica, Hale (1975) used Clorox and borax. Post treatment, he tested surfaces for the presence of ions to check for chemical residues. Hale recommended alteration of the microenvironment to inhibit regrowth (by cutting nearby trees), rather than consolidation because adequate data on consolidants are lacking and available evidence suggests that they will ultimately damage the stone. He published more detailed observations of the treatment results than is typical of most rock art literature as well as providing information about the location of the project records in order to maximize the effectiveness of future researchers.

In the Mahadeo Hills of India, Lal (1970) carried out extensive treatments of paintings in rockshelters. The treatment involved cleaning away soot, dust, and other surface accretions, as well as using ammonia to kill cryptogamic growth and applying a post-treatment consolidant. The treatment required delicacy and care on and near the paintings.

After manually removing moss from a small test area in a California lava cave and investigating the chemical components of a commercial biocide (which contained toxic dioxin), Silver (1989: 5) decided not to treat the moss obscuring paintings for two reasons: 1) the possibility that the biocide might alter the cave ecosystem and 2) the possibility that the moss might actually be protecting the paintings.

Despite the recognition that detailed scientific studies of rock art biodeterioration and treatments are lacking (Florian 1978; Taylor et al. 1977), and the cautions that methods suitable for rock art would have more severe constraints than treatments on buildings and monuments (Taylor et al. 1977: 315, Wainwright 1986: 51), most treatments carried out directly on rock art have lacked adequate consideration of treatment issues and adequate post-treatment scientific assessments of the results. The sparse published literature on rock art treatments also suggests that rock art recorders and investigators may have conducted many more treatments that have never been published or even documented. Some treatments are documented only in unpublished government reports that are difficult to obtain. Present investigators may be unaware of past treatments on a site and their residue of chemical alterations and consequent preservation problems. For example, an investigator treated petroglyphs at Castle Gardens (Wyoming, USA) with consolidant in the 1970s, but the Federal agency that manages the site received no records of which panels on the large site were treated, consequently, cannot monitor the treatment for long-term effects or take it into consideration in preservation planning.

Aside from specific site treatments, the published literature on rock art conservation in general is sparse. Rosenfeld (1988) has published a highly useful general manual that contains a summary of biodeterioration and treatments. Although the Kumar and Kumar (1999) literature summary on biodeterioration focuses more on buildings and monuments, it contains an extensive discussion and bibliography useful also for rock art. These few studies, plus the extensive literature for buildings and monuments show that there are many conservation issues in biodeterioration of rock art.

3. CONSERVATION ISSUES

The first and most basic issue is to determine that the microflora are in fact damaging the rock art and, if so, to identify the nature of the damage. What species are present, what kind of damage does each cause, and how fast growing or aggressive to the surface are the species? Some rock art investigators have perhaps been hasty in their efforts to remove lichen thalli. In most cases, lichen are very slow growing and the researcher has the time to test, experiment, consult, and weigh the issues and possibilities.

The substantial research on lichen weathering of rocks leaves no doubt that lichen deteriorates rock art by both mechanical and chemical actions, including rhizine penetration, expansion and contraction of the thallus, production of oxalic and other lichen acids, and chelation of metal ions (Syers and Iskander 1973, Jones 1988, Seaward 1997). Bacteria, fungi, and algae, as well as higher organisms, also participate in rock weathering (Dorn 1998: 43, 48, 49, 57, 59-61). Fungi and algae may colonize surfaces faster than lichen, but their effect on the substrate in many cases may be less severe (Dorn 1998: 49, Hale 1975). Although a few researchers have argued that lichens do little damage to the substratum, others have convincingly demonstrated the damage using appropriate scientific methods. For example, Jones et al. (1981) showed corroded and pitted grains and relic minerals that had metal ions extracted. Elemental energy dispersive spectroscopy has shown calcium concentrated at the rock surface immediately under lichen, while silicon was sparser in the same area compared with deeper in the substratum (Silver 2002: Figs. 31-33). A recent study has documented differences in rock chemistry beneath lichen thalli and adjacent lichen-free areas (Dandridge 2001, Dandridge and Meen 2003). This research shows that the weathering zone that normally provides some protection from deterioration is absent under lichen thalli.

The issue for rock art is not so much whether the lichen will deteriorate the rock art as (1) whether under natural conditions the lichen is expanding or receding, (2) if expanding, how fast, and (3) how damaging the specific lichen is to the surface. An endolithic species may be causing more damage than a foliose species. A species that is quickly expanding by asexual structures under favorable conditions may require more urgent assessment and action to preserve the rock art. Changes in the environment or local microenvironment may be keeping the lichen in stasis or causing it to recede. An investigator may miss the trend if assessing the potential for biodeterioration from a single site visit. For example, an ancient petroglyph panel in eastern Wyoming (site 48WE33) showed rapid spread of the sorediate lichen *Caloplaca decipiens* following several peak precipitation years that resulted in an expanded permanent pool of water at the base of the panel (St. Clair 2000). Several years later, however, flash floods scoured the rock face and left the lichen in a condition similar to its state before the expansion. In rare cases, a lichenologist has participated in biodeterioration work on rock art (e.g. Rebricova and Ageeva 1995; also Hale 1975 on monuments). In other cases, though, the work has been conducted without the input of a lichenologist to fully identify and assess the specific site situation, including viability of the species involved and knowledge of their growth patterns and preferred habitat.

A second issue for rock art preservation is whether removal of lichens would cause the rock art to deteriorate more rapidly. If lichens have undermined the surface, the rhizines and thalli may now be holding the surface together and removal of the lichen will result in immediate surface loss. Lichens may also stabilize surfaces damaged by other weathering processes, such as frost action (Walderhaug and Walderhaug 1998: 126). Most lichen treatments on rock art have not sampled the biomass removed to determine how much rock surface was also removed. Experiments by Silver and Wolbers (on exfoliated rock samples collected near rock art panels) have recovered rock particles from the biomass removed (Silver 2002). Lichens typically hold loose rock materials in place and removal of lichen thalli may increase the rate of erosion (Dorn 1998:59). A scanning electron microscope cross-section of rock under lichen shows a layer of fine particles (which may be fractured substrate grains, airborne dust, or precipitated lichen products) that would erode if the lichen were removed (Chiari, this volume, chapter 7, Fig. 7-4). Lichens may remove cement and minerals from the substrate and replace them with substances that are weathered easily, chemically and mechanically, if the lichens are removed (Dandridge and Meen 2003). Complex factors of climate, nature of the substrate, and lichen species influence whether lichen are protecting or stabilizing the substrate (Dorn 1998: 59, 101). Answering this question would require investigations by specialists in lichenology and rock weathering.

Another aspect of this issue is the exposure of unweathered rock, which lacks the protection of a weathering rind (Dandridge and Meen 2003), a rock coating such as varnish, or a case hardening agent, such as silica, iron film, or oxalate (Dorn 1998:

85-107). Lichen removal at the Torrey site shows fresh spalls and unweathered rock now exposed (Fig. 1; Childers 1994: Figure 6)). For natural conditions to replace the weathering rind, rock coating, or case hardening takes considerable time. Based on sampling historic engravings and faced rock, Dorn (1994:15) has estimated that it takes about 100 years for the initial varnish to begin forming. Studies of varnish accretion rates based on examining radiometrically dated geomorphic surfaces show variable rates that average a few micrometers per millennium (Liu and Broecker 2000). In some climates, though, silica coatings may form comparatively quickly given the necessary conditions (Dorn 1998: 92, Rosenfeld 1988: 25). Besides weathering more rapidly, the freshly exposed rough surface would be favorable for the growth or re-growth of microorganisms. Lichens and other microflora frequently grow in water run off locations and the surfaces exposed by their removal may erode more quickly due to run off previously absorbed by the microflora. Childers (1994: 104) observed water flowing across glyphs following removal of lichen thalli.

A third aspect of accelerated deterioration if lichens and other microflora are removed is unanticipated consequences of altering the ecosystem. Moisture retained by lichen or other flora might be needed to prevent painted rock art from drying and flaking off the surface. Moisture retention might contribute to formation of silica, calcite, oxalate, or varnish coatings that are preserving the rock art (Watchman 1993b, Dorn 1998, Kaluarachchi *et al.* 1995, Rosenfeld 1988: 21-27). Drying of the surface might create conditions for salt efflorescence that would cause the rock art to spall (Silver 1989). Because of the complexity of micro-ecosystems, researchers should first remove the microflora under near-identical conditions at a location without rock art and then monitor the results over time before removing microflora from rock art sites. In addition, studies should be conducted to determine how much rock surface would be removed with the microflora and what the condition of the newly exposed substrate would be. An adequate method of scientifically assessing the results of removal should be devised, rather than just taking pre- and post-treatment photographs and making a subjective visual assessment of post-treatment condition.

Another issue in removal of lichen and other microflora is the expected regrowth rate. Is the environment such that lichens will recolonize the site quickly? In rare cases, the local environment may have changed enough that initiation of regrowth is unlikely. If surfaces favorable for lichen growth are still present, such as cracks or spalls, pockets of dust or soil for some species, or calcareous rocks for other species, lichens may recolonize rapidly. If regrowth is likely, either periodic application of biocides would be necessary or biocides with a residual effect (lingering or bonded chemical) might be applied. In such cases, the investigator should assess whether long-term repeated treatments or residual chemicals in the rock are feasible or advisable.

3.1 Biocide and mechanical removal issues

Biodeterioration treatments include both mechanical and chemical methods. Although mechanically picking or brushing off organisms may not leave a chemical residue on the rock surface (except when detergent is used or the water is not de-ionized), such treatments inevitably result in mechanical damage to the rock surface. Lichens cling so tightly to the substratum that they will usually take part of the rock with them when removed mechanically. For algae, Wainwright (1985: 29) found that washing the black accretion on the Peterborough marble left a brown stain. Other investigators have observed that washing smears algae, rather than removing it. Water contributes to the swelling and shrinking of lichen thalli that produces mechanical damage. Water can destabilize a poorly cemented sedimentary rock. It may also provide a moist environment favorable for microorganism re-growth. Observations on mechanical removal and washing have shown that the result is temporary (Clarke 1976, Kumar and Kumar 1999). If frequent treatments are necessary, the investigator should question the value of any treatment. Repeated treatments, each with new mechanical damage to the rock surface could undermine the long-term preservation of the site. No careful, scientifically controlled experiments of mechanical removal or washing have been conducted to adequately assess the effect on various substrates and the length of time the treatment is effective in different environments. Published references documenting mechanical removal likely represent only a minor percentage of the sites that have been "cleaned" prior to recordation, with no assessment of the effects or documentation in the rock art conservation literature.

Among the vast number of chemical biocides that have been used on buildings and monuments (Taylor *et al.* 1977, Kumar and Kumar 1999, Martin and Johnson 1992), a smaller number have been used on rock art (Table 12-1). The list of rock art treatments tabulated is more representative than definitive, because publications of rock art conservation efforts are often in widely varied, obscure, and difficult to locate resources or in unpublished government reports. As with mechanical "cleaning" of rock art, unreported biocide applications are probably more extensive. Motivations for using biocides have been avoidance of mechanical damage (Hale 1975), longer effectiveness (Kumar and Kumar 1999), greater ease of application (less manual labor), and difficulty of actually removing lichens mechanically (Wainwright 1986: 51).

Biocides present both general problems and problems specific to the chemical used. All biocides that have been used on cultural sites and features chemically contaminate the rock surface. Such contamination prevents scientists from dating rock art using more recently developed techniques (see Research Issues below). It also prevents analysis of other data, such as prehistoric paint formulas. Adequate investigations of the long-term effects of biocides on various rock types are lacking. Biocides may chemically alter or break down some rock types by etching or

12. Biodeterioration of Prehistoric Rock Art

dissolution of minerals and salt crystallization induced exfoliation (Kumar and Kumar 1999: 38, Taylor *et al.* 1977: 317). Biocides may also react chemically with or deteriorate rock coatings, such as calcite, that may be preserving paintings or petroglyphs and, of course, can chemically and mechanically damage prehistoric paintings.

Clarke (1976) speculates that zinc fluorosilicate may have a residual biocide effect due to chemically bonding with some of the rock minerals. Zinc and magnesium fluorosilicate form a hard crust, especially on calcareous rocks, which discolors the surface and causes it to blister and exfoliate (Rosenfeld 1988: 64, Taylor *et al.* 1977: 316). Florian (1978: 97) suggests that the fluoride hardens some mineral components and the silica coats the surface, resulting in the observed exfoliation. The sodium salt in phenolic biocides reacts with iron in the substrate to stain light colored stones (Kumar and Kumar 1999: 38, Taylor *et al.* 1977: 316). Ammonium hydroxide leaches iron oxide and discolors the rock surface (Clarke 1976, Andersson 1986). Clarke advises against using this chemical on clay bonded rocks or paintings. Copper compounds cause corrosion and staining of the substrate (Martin and Johnson 1992: 115, Florian 1978: 97). In addition, some lichen species tolerate copper and other heavy minerals, reducing the effectiveness of copper as a biocide for some suites of lichens (Martin and Johnson 1992: 115). Silver nitrate tested in the laboratory on algal samples from the Peterborough site was found to discolor the rock by precipitation of silver oxide (Young and Wainwright 1995:88).

Chemical contamination of biocides is not confined to the rock surface, but may leave a residue in the local environment that is toxic to plants and animals and may contaminate soil or groundwater. For example, Silver (1989) discovered a commercial biocide she was evaluating for possible conservation use contained dioxin, a substance toxic to plants. This discovery contributed to the decision not to use biocides at this site. In practice, investigators have usually sprayed or brushed on chemicals and assumed they would evaporate (particularly ammonium hydroxide) or be washed off by precipitation (Taylor *et al.* 1977: 318, Childers 1994). The chemical residue that runs down or washes off the treated surface can also contaminate sediments below the rock art panels that may contain archaeological habitation deposits. Such contamination would prevent accurate radiocarbon dating and other chemical analyses of the deposits (phosphate analysis and the wide battery of chemical tests used to identify prehistoric activity locations) and possibly might alter the preservation of some cultural materials in the soil.

A poorly recognized problem with chemical treatment is that most methods also have a mechanical component. Use of fluids causes swelling of lichens that can result in mechanical fractures of the substrate, particularly for endolithic species (Florian 1978: 97), but also for the thalli and rhizines of any species. Often researchers have not left the organisms alone to fall off naturally after applying biocides. In several cases of application to rock art, the investigator has brushed or otherwise mechanically removed the residual biomass (e.g. Childers 1994). Clarke

(1976) used ammonium hydroxide to release the grip of lichen on the surface so that the biomass could be brushed off. Wainwright (1986) applied biocides to a large monument and a year later scrubbed the lichen residue with stiff brushes. However, on delicate, finely detailed prehistoric rock art, the mechanical aspects of biocide applications are distinctly problematic.

So little testing of biocides has been conducted that actual treatments are likely to involve overkill in the number of applications or strengths of solutions. Considering that the exact mechanism for the reaction of ammonium hydroxide on lichen is not known, Clarke (1976: 16) suggests that if alkalinity is the reacting aspect of the chemical, the 10% solution he used may be stronger than necessary. Whereas Hale (1975) used a 1:5 solution of Clorox (sodium hypochlorite), treatments at the Torrey site began with that concentration, but were increased up to full strength in an effort to achieve faster or more satisfactory results, despite no available empirical evidence that stronger solutions are any more effective as biocides (Childers 1994). Although Hale (1975) reported that it takes about two years for dead lichen thalli to completely exfoliate, the biocides at Torrey were applied at more frequent intervals without the assistance of a lichenologist to determine whether or not the lichens were dead, thus making further, potentially damaging, chemical applications unnecessary. Hale (1975: 312) applied a fourth and fifth treatment to some monuments, which he thought might not have been necessary. In both of these cases, small bits of residual biota not affected by earlier treatments may have been killed by the additional treatments, but for rock art, the treatment itself can be damaging and the potential damage should be weighed against the benefits of treatment. Researchers have observed that higher concentrations of zinc or magnesium fluorosilicate were more likely to blister and exfoliate the surface (Taylor *et al.* 1977: 316). One wonders why some investigators have recommended a second application of this chemical a week later (Taylor et al. 1977: 316), since the chemical is known to have long lasting residual effects and one application may suffice for several years. These examples highlight the need for empirical testing of treatments and development of a sound methodology.

In practice, the worst untested assumptions in the application of biocides have been that they have little effect on the substrate and that they leave no residue or the residue is removed with normal precipitation. The lack of adequate testing of biocides also extends to how it affects the broader ecological context of the site. When fungicides are used to attack the fungal component of a lichen association, Florian (1978: 97) raises the issue of what happens to the algal component. Use of fungicides might select for an increase in algae, which can also damage the substrate. Several studies indicate that bacteria, in particular, increase following biocide applications (Kumar and Kumar 1999:36).

Killing lichens with biocides, with or without follow up mechanical removal, often leaves a residue of plant tissue, particularly remnants of crustose lichens (Figure 1; Wainwright 1986: 49, 51). These remnants may still obscure some of the rock art. Rock art investigators with the urge to see a "clean" surface may be tempted to apply more biocides (needlessly if the lichen is already dead) or remove the residue mechanically. Some consider the treatment not fully successful if plant tissue remains on the surface. Potentially, bits of organic matter might provide nutrients for re-colonization or some viable fragments may remain, having resisted treatment. In general, biocides used on buildings and monuments, and less often, on rock art, have proven to be unsatisfactory for rock art preservation.

3.2 Consolidant issues

The fragile eroded substrates exposed after lichen removal lead to the consideration of applying consolidants. The same problems arise: inadequate testing of consolidant chemicals and inadequate assessment of their long-term effects on the rock and the rock art. Rosenfeld (1988: 58) summarizes the situation well: "Despite the absence of long term observations on the efficacy of sealants on masonry, the consolidation of fragile rock surfaces is still often advocated for rock art, mainly by people with no direct experience in the field of rock art." Problems with early consolidants included discoloration of the surface, alteration under ultraviolet exposure, chemical instability or reaction with the substratum, and creation of an impermeable surface coating that caused exfoliation because of water or salts trapped behind the barrier. An early application of polymethyl methacrylate to prehistoric paintings in India was removed with great difficulty after it discolored and hardened, and the surface was retreated with polyvinyl acetate (Lal 1970: 145). Polyvinyl acetate, however, produces a glossy surface that alters the perceived paint colors and tends to swell when wet (Lal 1970: 146).

More recently silicon-based consolidants have been tried to avoid some of the problems of the chemicals used earlier. Rosenfeld (1988: 60-63) summarizes the properties of the silicone consolidants and some of their problems for application on rock art. Clarke tested a commercially available silicone in Australia and experienced considerable practical problems in application, including too rapid evaporation of solvents in the hot climate, evaporation carrying the polymer back to the surface, and softening of the impregnated silicone under desert heat conditions (Rosenfeld 1988: 62). Grisafe (1996) tested ethyl silicate on sandstone in Kansas, USA, and reported increased durability of the stone and a reduction in the absorption rate of about one-fifth to one-tenth of the pre-treatment rate. Chiari (this volume, Chapter 7) has experimented with ethyl silicate on similar sandstone from a Wyoming petroglyph site (samples collected from the ground, not from rock art panels) and reported a comparable reduction in porosity. The assumption that the reduced porosity will not affect preservation of the rock art is, as yet, untested. The Wyoming site is prone to intense summer rainstorms and is in a canyon prone to flash floods. A laboratory simulation of how the treated rock would behave under such conditions is a necessary prerequisite to considering consolidation treatment.

The Wyoming petroglyphs have already survived under natural conditions for more than 11,000 years (Dorn 2003, Tratebas 1993). The high porosity of the sandstone may be a key factor in the environmental equilibrium that has allowed the site to survive this long. Some petroglyph substrates at the site are case hardened by iron films, which are less permeable than silica, varnish, or oxalate (Dorn 2003). Backscattered electron microscopy shows that the iron films trap and hold water under the outer few millimeters of the substrate, resulting in decay of the underlying quartz grains. Applying consolidants would increase this effect of trapping water and eroding the outer rock surface.

Several other researchers have found ethyl silicate to be problematical. Boustead (1970: 133) reported that penetration was poor and left a surface skin that prevented water flow, and the treatment effect was temporary and needed to be repeated every few years. Experimental treatment of historic inscriptions that were in imminent danger of eroding was unsatisfactory as it was difficult to achieve good penetration of the ethyl silicate, the adjacent untreated areas continued to erode and threatened to undercut and spall the treated inscriptions, and the causes of the deterioration could not be brought under control (Oliver 2002). Oliver cautions that the lifespan of ethyl silicate treatment can be so short that its use may not be warranted in terms of cost, toxicity, and change in stone composition and appearance.

Although consolidants developed or tested recently may provide solutions to problems caused by earlier chemicals, including discoloration, brittleness, alteration by ultraviolet light, biologic growth, chemical instability, and incompatibility with the properties of the stone, (see Haskovec 1991: 99 for properties required of a consolidant for use on rock art), they may still be problematical as water and salt barriers. In addition, long-term tests of their behavior in the natural environment where they might be used are imperative prior to using them on rock art. Laboratory simulations might help assess long-term effects, but are likely to omit significant factors in the natural environment that researchers do not perceive as important. The oldest test of consolidants on a rock art site that I have encountered is the 1920s application of five consolidants to a sandstone face (lacking rock art or historic inscriptions) at El Morro National Monument, New Mexico, USA. Numbers and the words "COLORLESS COVERINGS SAVE OLD INSCRIPTIONS" were incised on the cliff for the test, and each word was treated with a different consolidant (Padgett and Barthuli 1995). It is clear today that some areas fared better over time than others. The consolidant selected from this test was "Driwall," a paraffin treatment. The more significant historic inscriptions were treated with Driwall until the 1940s (Padgett and Barthuli 1995). In the 1960s, "Pencapsula," a synthetic polyurethane resin, was tested first on prehistoric glyphs and the following year on two historic inscriptions. Treatment records suggest that the Pencapsula was not monitored beyond the first year. Finally, in the 1990s, El Morro developed a conservation program for annual condition monitoring of inscriptions and rated each in terms of the likelihood that it would erode completely. Comparison with photographs taken in 1955 showed that some inscriptions had already been lost. Several of the more threatened inscriptions were treated with ethyl silicate.

A problem with long-term tests is that advances in chemistry are likely to generate newer and better consolidants, making older consolidants obsolete. For example consolidants used in laboratory tests on limestone and sandstone samples in the 1920s (including a sandstone sample from El Morro) included varieties of petroleum distillates such as paraffin and China wood oil, aluminum soap mixtures, varnishes, inorganic salts, glue, and soluble soaps (Kessler 1923). These consolidants do not include any we would consider using today on rock art. Nevertheless, we should set up long-term tests using the most acceptable current consolidants, not directly on rock art surfaces, but on the same rock types off site. Detailed records of the test locations and results should be carefully maintained so that future researchers can effectively evaluate the efficacy of these consolidants over time.

Besides the fact that no one has adequately tested the effect of reducing the rock porosity, consolidants, like biocides, can also introduce foreign chemistry to the rock that would contaminate it for scientific studies, including dating the rock art and analyzing prehistoric paint formulas. While adding silica to a rock type that contains or is largely silica appears to be compatible, the consolidant also requires a solvent and usually a catalyst, which could leave additional chemical residues. Grisafe (1996: 375) indicates that ethyl silicate introduces no soluble foreign ions into the stone, raising the issue of what problems might be created for future scientific studies from insoluble ions that would be added or traces of chemicals left from an incomplete chemical reaction. Ethyl silicate—formula $Si(C_2H_5O)_4$ or $(C_2H_5)_4SiO_4$—could introduce carbon into the rock that would prevent or make questionable all future attempts to date rock art using the accelerator carbon 14 method (see Research Issues below). In a fully successful reaction, the silicon ester should hydrolyze to produce silica gel and alcohol that would evaporate (Rosenfeld 1988: 60); however, problems in field tests suggest that reactions are unlikely to be fully successful and may leave foreign chemicals in the rock. Treating surfaces newly exposed by lichen removal with consolidants would also hinder or prevent natural processes that accrete protective rock coatings.

Grisafe (1996) made the decision to consolidate petroglyphs in Kansas despite the lack of long-term testing of ethyl silicate because he felt that loss of the glyphs was imminent (see also Grisafe 2000). Such decisions are currently based on subjective evaluations of site condition. In some cases, single site visits or short-term site monitors may suggest imminent loss when longer-term monitors show the rock art still intact. We need objective, scientific criteria for establishing the likelihood of imminent loss to back up decisions concerning treatment intervention. A detailed database on site condition, including a time series of photographs, like that begun at El Morro, suggests a promising methodology for evaluating and making treatment decisions.

As with biocides, most uses of consolidants have been on historic buildings and monuments and almost no research has been conducted for rock art. To intervene now, without more knowledge and testing, would likely lead to further problems in rock art site preservation.

4. RESEARCH ISSUES AND CONCERNS

Removing lichen and other microorganisms from rock art surfaces destroys valuable scientific data and removes the mechanism that produces some of these data.

4.1 Dating rock art

New methods of dating rock coatings and prehistoric paints have revolutionized rock art studies (Dorn 2001, Rowe 2001a). Although these methods are still experimental or undergoing refinement, they allow researchers for the first time to place rock art in useful chronological contexts. Any chemical applied to rock art surfaces contaminates them for all future dating analyses, including methods not yet conceived or developed. Even if a researcher could demonstrate that no foreign chemicals were detected in a sample, the research community would view any dates derived from the sample as suspect.

Methods for dating petroglyphs include cation-ratio calculations using the formula, $K+Ca/Ti$, to establish a leaching curve (Dorn 2001: 175) and accelerator carbon 14 dating of carbon in oxalate coatings, or less accurate, encapsulated in the weathering rind beneath rock coatings (Dorn 2001: 173, 178-182). Any chemical treatment that includes carbon or would alter the cation leaching rate or cations used to calculate it damages petroglyphs for dating. For prehistoric paintings, carbon 14 can be used to date any organic constituents in the paint, such as binders, ground charcoal or charcoal crayons often used for black pigments (Rowe 2001a, Clottes 2001: 469-470, Watchman 1993a). Again, chemicals containing carbon would contaminate the paintings for dating.

Recent blind tests have confirmed the accuracy of varnish microlaminations for dating rock art and geologic surfaces (Liu 2003, Marston 2003). The varnish strata are useful for relative dates, as well as dates older or younger than the Younger Dryas and Heinrich events (Liu 2003, Liu and Dorn 1996). The moister climates of ice advances favor manganese-rich varnish that shows in ultra-thin sections as black strata in contrast to orange iron-rich varnish deposited during dryer climatic episodes. Both biodeterioration and mechanical damage to the rock surface from lichen removal can damage the varnish strata needed for dating. Because bacteria are essential to varnish formation, any biocide that acts on bacteria would interrupt the formation of varnish.

12. Biodeterioration of Prehistoric Rock Art

Researchers have also calculated limiting dates for rock art from radiocarbon analysis of carbon in calcium oxalate coatings overlying paintings (Russ *et al.* 1995, Watchman 1993b). Thick laminated crusts, overlying or sandwiching prehistoric paint layers in Australia, contain both oxalate layers and charcoal-rich laminations which have provided limiting dates for temporal sequences of paintings (Watchman 1993b). Research on oxalates in painted rockshelters of the Pecos River area in Texas, USA, have revealed features that under magnification show the morphology of lichen thalli, in particular, *Aspicilia calcarea,* and microcolonial fungi (Russ *et al.* 1995, 1996, 1999). This result supports previous research concluding that lichens are largely responsible for the formation of oxalate coatings on rock surfaces (Dorn 1998: 275-278, Del Monte *et al.* 1987, Seaward Chapter 2, this volume). Removal of lichens from rock art surfaces consequently destroys the mechanism that produces the means for dating the rock art. Oxalate accretions sandwiching the Pecos paintings may be the most important factor in their preservation (Kaluarachchi *et al.* 1995). The accretions protect the paintings from chemical and physical weathering. Rock art researchers may view the accretion as a nuisance because it makes the paint appear faded, however, removing it would likely accelerate biological attack and lead to increased dissolution of the underlying limestone, particularly by frost action.

As with paintings, oxalate coatings can also help preserve petroglyphs. Calcium oxalate deposition has case hardened a sandstone surface containing petroglyphs in eastern Wyoming (Dorn 1998: 88-89). Oxalate has penetrated into the sandstone pore spaces. The overlying rock coating includes a basal layer of manganiferous rock varnish, which is overlain by a thick layer of calcium oxalate, and capped with a mixture of oxalate, silica glaze, and varnish. Killing the organisms producing the calcium oxalate would eliminate this mechanism for case hardening and coating rock surfaces. Similarly, killing the organisms that contribute to varnish formation destroys the mechanism for producing the data contained in varnish. Iron films can also case harden rock surfaces and help preserve rock art, and they have some potential for dating (Dorn 1998: 146-180). Lichens are known to precipitate iron oxides (Jones and Wilson 1985), and bacteria have been implicated in the formation of iron films (Dorn 1998: 180-185).

The wealth of surface dating techniques developed recently (Dorn 2001, Rowe 2001a, Beck 1994) suggests that we can expect many innovative methods in the future. Dating with cosmogenic nuclides (Kurz and Brook 1994), for example, is so new that researchers have begun only initial attempts at using these techniques to date rock art (Phillips *et al.* 1997). Any treatments applied to rock art must take into consideration the requirements of as yet unconceived scientific analysis methods and must not chemically contaminate the rock art for those analyses or damage rock coatings and substrata either chemically or mechanically.

4.2 Paleoenvironmental data

Calcium oxalate crusts also contain paleoenvironmental data. In the Pecos River area, research suggests that the desert lichen, *Aspicilia calcarea*, produced the calcium oxalate crusts found in numerous rock shelters (Russ et al. 1996, 1999). This species should be abundant during xeric climatic episodes and lacking under mesic conditions. Radiocarbon dates on calcium oxalate crusts should therefore identify xeric climatic episodes. The samples run to date correlate well with other data on the paleoclimate of the region (Russ et al. 1996, 1999). Another method of extracting paleoenvironmental data from calcium oxalate coatings and rock varnishes is stable isotope analysis of carbon contained in the coatings (Dorn and DeNiro 1985). The $\delta^{13}C$ values should reflect the relative abundance of C_3 plants versus C_4 and CAM plants in the nearby vegetation. Considering that C_4 and CAM plants are more abundant in warm arid environments, while C_3 plants favor cool moist environments, the $\delta^{13}C$ values of the organics in varnish and oxalates should provide data on the local environment at the time of varnish or oxalate formation. Dorn and DeNiro (1985) also demonstrate the utility of the method for identifying ecological change caused by European colonization where the $\delta^{13}C$ values in varnish do not match the current vegetation. In the stratified rock coatings sampled from Northern Australian rock art sites, layers of calcium oxalate appear to reflect moister climates, while bands of mixed charcoal, clay, and quartz were deposited during droughts (Watchman 1993b). Dates for these layers correlate well with drought cycles derived from other paleoenvironmental data.

In arid regions of western North America alternating bands of black, manganese-rich and orange, iron-rich varnish have been correlated with the moister climates of glacial advances versus drier Holocene and interglacial periods (Liu 2003, Liu et al. 2000). Varnish layers with a high manganese content also correspond with high water levels of Great Basin lakes (Fleisher et al. 1999, Liu et al. 2000). In the central Sahara, varnish microlayers reflect a climatic shift from a wet environment towards increased desertification (Cremaschi 1996). Varnish strata consequently provide an excellent paleoclimate record.

Theoretically, prehistoric paints may contain paleoenvironmental data where identifiable plant parts, such as fibers, are incorporated (Watchman and Cole 1993), or DNA or blood antigen studies can identify the source of organics in the paint (Rowe 2001b). In most cases, however, paint analyses provide data about cultural practices, but not new data on past environments.

Chemical contamination of rock coatings with biocides or consolidants would damage the paleoenvironmental data they contain. Even worse, killing lichen and other surface microorganisms would destroy the mechanism that produces the data. Although researchers have suggested various origins for oxalate coatings, convincing evidence points to deposition by lichen. After years of research and discussion on varnish formation, the best explanation is a combination of biotic action by bacteria, followed by abiotic processes (Dorn 1998: 231-247).

4.3 Lichenometry and other research data derived from lichens

When rock art recorders or investigators have removed lichens from rock art, they have rarely given any consideration to the data lichens themselves may contain. A handful of researchers around the world have attempted to date rock art using lichenometry (Florian 1978: 95-96, Taylor *et al.* 1977: 301-305, Bednarik 2000). These experiments have not been extensive enough to evaluate whether the technique may be useful in certain settings or circumstances. Fully developing a method requires a greater effort. Bednarik (2000) suggests that the indirect method of calculating growth rates empirically, using surfaces of known age, may be more useful for rock art dating. An example of the indirect method is a lichen growth curve developed for the Rocky Mountains in Colorado from lichens on historic quarry walls, cairns constructed in 1910, a radiocarbon dated prehistoric game drive rock wall, and radiocarbon dated geologic surfaces (Benedict 1967). In a successful application to rock art, lichenometry demonstrated that petroglyphs in Newfoundland were old and not a recent forgery (Taylor *et al.* 1977: 303). The direct method establishes a growth curve by measuring increases in thalli size over a period of years.

Matthews (1994) stresses that lichenometry does not consist of just direct and indirect dating curves – but rather a suite of approaches to be used in different circumstances. For example, size-frequency strategies may work better in dating disturbances on polygenetic surfaces, while methods that require searches for the largest thalli over a large area of a glacial moraine may be unworkable for petroglyph dating, where surface area may be limited. Other strategies include characterizing the entire lichen population in order to limit the potential for anomalous thalli (Locke et al. 1979). Ongoing method development (for example, McCarthy and Zamiewski 2001) and investigator confidence have generated continued use of lichenometry (Bull and Brandon 1998, Sancho et al. 2001, Solomina and Calkin 2003), including some success in temperate climates (Gob et al. 2003). Lichenometry should be explored if the researcher is working with large clasts or surfaces in a latest Holocene time frame because (1) it yields usable results in a time frame that is problematic for cosmogenic nuclides, (2) it can be applied where prior exposure histories would cause problems for cosmogenic nuclides (Winchester and Chaujar 2002), (3) it can have an accuracy of ±10%, and (4) as a field-based method it is low cost.

In addition to its use for dating, lichenometry can also identify fire kill events (Bull and Brandon 1998: 71). For sites with trees, investigating fire history using tree rings would be easier. Studies of past fire frequency and severity can help site managers devise strategies for preventing future fire damage.

Ecological successions in lichen communities (Florian 1978) may also provide data on climate changes, the presence of site disturbances or alterations, or the dating of recent rock art. Surface erosion patterns adjacent to areas of present lichen growth on petroglyphs in eastern Wyoming suggest that lichens formerly covered more of the surface. Certain climatic conditions may initiate growth or expansion of lichens (Danin et al. 2002). It can be instructive to analyze lichen growth that was more expansive in the past and then correlate growth with known paleoenvironmental conditions in the region (Danin 1985). Such data might prove helpful with future efforts to control lichen growth at critical sites.

Although lichens are generally abundant in locations where they present serious deterioration problems for rock art, rock art specialists must also consider the possibility that some lichens, being evaluated for removal, may be new to science or perhaps extremely rare. Early conservation specialists were seriously mistaken in assuming that lichens and other microbiota inhabiting rock art surfaces were ecologically irrelevant and without scientific value (Rosenfeld 1988: 43).

5. ETHICAL ISSUES

Because of the complexity of problems in biodeterioration of rock art and conflicting issues, individual action without consulting all interested parties is unacceptable. Individual action increasingly leads to controversy (e.g. Jaffe 1996, Swartz 1997b). Decision making in considering any treatment of biodeterioration must include consultation with (1) the cultural group who made the rock art or is affiliated with the makers, (2) officials responsible for historic preservation, such as government historic preservation officers or land managers and entities that represent the public, such as the National Trust in the United States, and (3) research communities, including archaeologists, rock art researchers, rock art conservators, dating and paleoenvironmental scientists, botanists specializing in lichenology and related subjects, geologists, geographers, and any one else with a research interest in the site. Because of potentially conflicting interests, a consensus process may be essential for making the most appropriate decisions.

The views of the affiliated cultural group are crucial. In Australia and other parts of the world, rock art researchers view site makers or their descendants as the site owners or custodians (Mowaljarlai and Watchman 1984, Haskovec 1991, Meehan 1995, Lim 1999). In the United States, ownership of a site resides with the private landowner or government agency, even though the official terminology for government land is administering or managing the land. Conflicts and controversies concerning site use have arisen in the United States where traditional cultural sites lie outside Indian reservations. In a recent controversy over an application to drill

12. Biodeterioration of Prehistoric Rock Art

for oil in the cultural landscape of a Montana petroglyph site, the Native Americans, not the Federal government, resolved the conflict by exchanging the oil lease for another on the Indian reservation (Levendosky 2001). The controversy led to a proposed law introduced in the US Congress that would have protected Native American sacred sites (Thorne 2002). Congress opted to take no action on the bill.

Native Americans view rock art sites as sacred (Bricker *et al.* 1999). Sacred sites are integrally tied to the natural environment. Young (1988: xvii) learned that the Zuni "included the entire environmental setting of the rock art in discussions of meaning rather than focusing on the image alone." Australian Aborigines view the area surrounding a rock art site as an integral part of the site (Haskovec 1991: 98). The link to the forces of nature is key to the power of the site. Major alterations of a site, such as biocide treatment that leaves unsightly eroded rock and dead plant residues, or other changes such as rechanneling water flow in and around the site, change the site from a natural to a man-made environment. Such alteration may result in the site losing its power. Conservation intervention can also intrude foreign cultural practices into another culture's traditional customs. Major interventions can mar ceremonial practices that give a culture definition and cohesion. Protection of the Peterborough site by constructing a building around it cut it off from its natural environment (Bahn *et al.* 1995, 1996), although windows around all sides still link the site visually to its setting. Sealing channels of the streams that flowed under and through the rock (Bullen 1996) severed the site's link to the natural force of water. The sound of flowing water, which had been an integral feature of the site, was silenced. In effect, this invasive site protection killed the site (Bullen 1996: Bahn *et al.* 1995, 1996). Consultations with North American tribes have shown that most tribes oppose the use of chemicals on rock art and prefer that the site be left in a state as original, natural, and pristine as possible (Bricker *et al.* 1999). To many Native Americans touching rock art in any way is unacceptable behavior. Australian Aborigines have seemed indifferent to inevitable decay of paintings because they felt that the essence of the painting lay within the rock itself and the existing image was merely a transitory outward manifestation (Meehan 1995: 311).

Tribes in Eastern North America had a tradition of teaching rocks. Peterborough, called Kinomagewapkong, "the rocks that teach," in Ojibwa (Wainwright 1996), is a prime example. Teaching rocks were supposed to be covered by earth and mosses (Weeks 2003). When the people needed to use the teaching rocks, they were uncovered (Weeks 2003, Bullen 1996). After use, they were covered again to protect them from weathering and from people who were not ready for the knowledge the rocks held. The building over the Peterborough site intrudes a different cultural practice of covering or protecting the site. Perhaps it should be up to the Native American custodians to decide whether this shift in methods is an acceptable change from their traditional practices.

Aesthetics is another issue for biodeterioration treatments. Biocides used in the past have left remnants of dead lichens and exposed eroded rock surfaces that some people feel damage the aesthetic qualities of the rock art site (Crotty 1989: 82). Altering a site may also mar an educational message. If a site used for public education is divorced from its natural setting, visitors will miss or misunderstand part of what should be taught.

The rationale for treating a site should be clearly justifiable. Applying biocides merely to better see and record rock art is unacceptable. Perhaps the site was not meant to be seen except when in use. For most sites, it is possible to document the site adequately in order to understand and interpret it. If coatings obscure paintings, photography using cross-polarized lenses (Henderson 1995) or computerized photographic enhancement can bring out the details (Mark and Billo 2002). Photography of petroglyphs using side lighting or photogrammetry can improve visibility. Frequently, enough details are visible to see all the glyph elements despite the lichen cover (for example, compare Figures 3 and 4 before and after treatment in Childers 1994; however, another treated panel at the site was largely invisible). If treatment will halt damage and preserve a site longer, intervention may be justified, but only after all issues are considered and all interested parties consulted.

If rock art is treated with chemicals, the treatment damage should be mitigated as far as possible. In particular, all possible research data the site may contain should be collected prior to treatment. The site may still be damaged for future research that would require samples or data that we cannot presently anticipate. Any treatment should be controlled by thorough scientific documentation and scientific assessment of its effects and should be monitored over a long time.

6. TOWARDS A SOLUTION

6.1 Improved method of microflora control

Given the complexities in biodeterioration of rock art and inadequacy of biocides and consolidants used in the past, I initiated a project to investigate the issues, framed by the following parameters. Control methods should include:

1. efforts to not chemically contaminate the stone or interfere with other kinds of scientific studies, such as dating,
2. efforts to meet conservation goals of not staining, hardening, or altering the substrate, and
3. removing the biomass gently to avoid disturbing the substratum surface.

Constance Silver began working on the project, and in collaboration with Richard Wolbers, devised a method of treating lichen, using enzymes, that is significantly

12. Biodeterioration of Prehistoric Rock Art

superior to biocides used in the past (Silver, Chapter 8 this volume). An enzyme mixture containing protease, cellulase, and chitinase can be used to break down the protein, cellulose, and chitin of the biomass. The enzyme mixture selected has a short activity life of a few months. The enzymes remove exterior lichen biomasses thoroughly, leaving a clean appearance. Applied carefully to the lichen thalli, not the rock surface, enzymes appear to kill the lichen quickly and may involve significantly less mechanical damage from swelling compared with solutions sprayed or brushed onto the rock surface. Because an organic substance is acting on organic material, the method would appear to involve less chemical contamination of the rock, although a catalyst and a gel may be needed to control the enzyme action and achieve the viscosity needed to place the enzyme on the lichen, not the rock surface or the local environment. The method has yet to be tested to determine if it contaminates the rock for dating and other scientific analyses. It is likely that enzymes would have a negative effect on the organic component of paints, and we can assume that the method could be used only on petroglyphs or inorganic paints. Because some prehistoric petroglyphs were also painted or colored with charcoal "crayons," a prerequisite for using enzyme treatment would be verifying that no paints or applied charcoal would be damaged. For older petroglyphs, paints and applied charcoal may not be visible because they are heavily weathered or covered by varnish or other accretions.

Even with a significantly better treatment method, we still have not answered the basic questions concerning whether we should remove lichen from rock art:

1. Will the rock erode more quickly if lichens and/or other microorganisms are removed?
2. Can we justify killing the organisms that produce scientific data needed for rock art studies and in some situations are responsible for rock art preservation by accreting protective coatings or case hardening the rock?
3. Would treatment of the site damage or be contrary to traditional cultural use of the site?

6.2 Altering the microenvironment

A more natural method of reducing biodeterioration could be altering the microenvironment to discourage the growth and development of microflora. Despite many researchers mentioning environmental factors that appear to enhance the growth of microorganisms, controlling biodeterioration by altering the local environment has rarely been attempted and certainly "cleaning" and application of biocides has been the most common approach. One of the most obvious methods is cutting down trees and other vegetation that shade rock art and provide the moisture retention and protection from insolation needed for microflora to flourish, especially in northern latitudes. On petroglyph panels at site 48WE33 in eastern Wyoming, lichens often occupy an area that has the same shape as the shade created by trees and shrubs. The shade from two large pine trees at the Peterborough site closely matched the area covered by an algal mat, which increased in size as the trees grew (Young and Wainwright 1995: 83). Instead of felling the trees and monitoring the effect prior to using more invasive techniques, a building was erected over the site. The effect of the building on site preservation has been debated in the rock art literature (Bahn et al. 1995, 1996; Wainwright 1996). Removing shade is especially suitable in locations where tree or shrub density has increased in historic times. Petroglyph site 48WE33 is now heavily shaded, although the young age of the trees and photographs of the region taken during early historic explorations indicate that forest density was formerly much sparser. In contrast, at sites in the lower latitudes of Australia, loss of adjacent vegetation with an increase in sunlight reaching rock art panels has resulted in increased growth of microorganisms (Lambert 1989: 24-25). Manipulating shade may be the least invasive control method available and is particularly suitable when it involves environmental restoration. Where Native people still use a site, prior consultation would clearly document that the vegetation is not a significant component of the site. In a few cases, researchers have recommended altering shade in the microenvironment as part of the control effort. Following application of biocides to Mayan monuments, Hale (1975: 312) recommended preventing regrowth by cutting adjacent trees to dry out the monuments. Walderhaug and Walderhaug (1998: 131) suggest removing trees and shrubs surrounding petroglyphs to increase evaporation rates and reduce biotic growth. Andersson (1986) also recommends cutting down trees and shrubs to diminish the growth of algae and lichens on rune-stones and carvings. Even this natural method of altering the microenvironment should be scientifically assessed prior to application to ensure that it will be effective for the species to be controlled and will not produce favorable conditions for other species.

The rock art conservation literature rarely mentions controlling shade by vegetation removal or planting, but does discuss more drastic control, such as placing building or roofs over sites (Wainwright 1985: 27). The rationale for these drastic measures usually includes also altering the site hydrology, where natural precipitation or water flow provides a favorable environment for biologic growth. In general, roofs and buildings have usually compounded preservation problems, as well as drastically altered the site setting. A structure in Sweden merely resulted in a change in lichen species, plus an increase in algae that preferred water run-off channels in the greenhouse atmosphere. The structure also caused soil and leaves to accumulate since rainwater no longer washed the surface, and resulted in salts accumulating on the surface (Bahn and Hygen 1996). A shelter built over a rock art site in Japan to protect it from seasonal temperature and humidity extremes resulted in conditions that promoted moss encroachment at the site (Ogawa 1992). An opaque, green plastic roof constructed over petroglyphs at Nanaimo Petroglyph Park in British Columbia, Canada, leaked and caused algae to proliferate. The structure

12. Biodeterioration of Prehistoric Rock Art

was later removed (Lee 1991: 50-51; Kennedy 1979: 286). A building constructed at Besovy Sledki in Russia created additional preservation problems and did not solve the preservation problems, which motivated its construction (Bahn *et al.* 1995). Reports for Peterborough indicate no expansion of the area covered by algae (Young and Wainwright 1995: 89); however, trees shading the site were also removed when the building was constructed (Roach 2004). Consequently, we cannot completely separate the factor associated with removal of the tree from the effect of the building in evaluating the treatment success, although the building itself also creates some shade on the petroglyphs. Inspection shows the algae have changed from black to gray (Young and Wainwright 1995: 89), and glyphs formerly covered by algae have become more visible (Roach 2004). If the building has succeeded in slowly killing the algae, except for small areas of new growth at the north edge of the site, the price of treatment has been high since the rock art has been partially cut off from its natural environment.

Another technique for reducing microflora, as well as mechanical and chemical erosion, is redirecting water flow (see Rosenfeld 1988: 52-54 for techniques). In Australia, Aboriginal people cared for Wandjina paintings in rockshelters by piling spinifex weighted with flat rocks on top of the shelter during the rainy season to prevent rain from trickling down the painted surface (Mowaljarlai and Watchman 1989: 151). Although this technique was for preventing erosion, rather than hindering biotic growth, it demonstrates a non-invasive technique of altering water flow. More intrusive methods, such as gutters, drainage tile, and other constructions to divert water flow, can create radical alterations of the natural setting. Major alterations of the site that result in altering the micro-ecosystem, can instigate other problems, often unanticipated. Drying of the rock art surface can result in increased deposition of soluble salts (Rosenfeld 1988: 34-35). Reduction of water also can prevent the formation of mineral accretions which act to preserve rock art (Dorn 1998; Rosenfeld 1988: 21-27). As discussed above, reducing the microflora can prevent formation of oxalates and varnish that not only protect rock art, but also contain research data. At some sites water is an integral part of the site, particularly for rock art created above springs, on cliffs along rivers and lakes, or in locations where subterranean water is linked to the site via crevices. Any alteration of these water sources would damage an integral part of the site. Caves are a special case where water is integral to the formation and character of the site.

Some species of microflora favor microsites such as cracks and holes, or deposits of dust and soil. Accumulating leaves and other dried vegetation can also provide nutrients or favored settings. Removing dust, dirt, and dried vegetation is an easy solution, but involves some (preferably minimal) contact with the rock art, as from soft brushes or compressed air. For Norwegian petroglyphs on granite, vegetation and soil may be removed by washing (Hygen 1999), which would be highly damaging for sandstone substrates or fragile rock art. Although conservators have filled cracks or spalls in sites where rock art loss is imminent (Rosenfeld 1988: 55-

58), such techniques would be excessively invasive if the only need was for biologic control and would chemically contaminate the site. More appropriate would be first monitoring microflora that prefer cracks and holes to determine if they threaten to expand onto the adjacent surface. For petroglyphs, the depressions themselves provide microsites for microflora, which often selectively invade these surfaces. Consequently, it is not possible to remove all favorable microsites. In situations where lichens occur at or near the soil surface, it may be possible to control lichen development by lowering the soil surface through excavation. If the soil is covering part of the rock art, the investigator must first assess the effect of exposing the buried rock art. In some cases, soil conditions or plant roots may have deteriorated the glyphs and exposure will result in more rapid erosion (Lambert 1989: 18-20).

Another control method that fits into the category of altering the microenvironment is placing a black plastic sheet over the rock art to eliminate sunlight needed for biotic growth (Bednarik 2003). Although a number of investigators have thought of this common sense technique, I could not find any published reference to its trial and effectiveness. The method could create other problems, such as growth of algae or mold because of increased heat and humidity. Application procedures, such as how long to keep the plastic in place, would need to be determined by experimenting on surfaces without rock art. Supporting the plastic so that it shades the rock art, but does not touch it would avoid mechanical abrasion of the glyphs and rock face, but may not work as well because some light would still reach the organisms. A variant of this method has been used in Scandinavia where horizontal outcrops have been covered with a thin layer of sand to kill lichens (Bednarik 2003). Covering with sand could have potential problems, such as abrasion of the glyphs during application or mechanical damage in later removing the sand.

In cases where human visitors have caused microflora to flourish and damage rock art, the solution has been straightforward. Lascaux, Altamira and other European caves are now closed to general visitation. Only restricted numbers are allowed into the caves and the impacts of visitation are carefully monitored. Increased humidity, carbon dioxide, dust, and mirco-organisms introduced by visitors, plus artificial lights, created the conditions for microflora to invade Upper Paleolithic painted caves in Europe (Levèvre 1974, Brunet and Vidal 1980, Fernández et al. 1986, Somavilla *et al.* 1978). Extensive research was needed to bring the microorganisms under control. In Lascaux, an aerial biocide was tried initially, but had no lasting effect, and it was necessary to use formol as a biocide after considerable testing of chemicals in various concentrations (Lefèvre 1974). Formol was also used in several other Upper Paleolithic caves in France (Brunet and Vidal 1980). Additional microenvironmental controls, besides reduced visitation, included changes in cave ventilation and lighting.

The initiation of microfloral growth requires the right conditions of water, warmth, light, and nutrients (Beschel 1961: 1044). Many microenvironmental control methods have manipulated water by reducing it, while others have altered light. Shade removal reduces water when it accelerates evaporation, but may also work by removing protection from excessive insolation. Light, water, and warmth are intricately linked. Some cases of controlling one factor, for example, water, have increased another, such as warmth or light, that resulted in encouraging growth of a different microorganism. Where microenvironmental controls cannot be reversed, prior off-site tests would increase the likelihood of success.

Using microenvironmental controls also requires considering the issue of whether removing lichen and other microflora will result in faster deterioration of the substrate because the biota are now stabilizing the damaged surface. A significant difference, compared with using biocides, is that the process is much slower and allows time for natural processes to act in restabilizing the surface. Another difference compared with biocides is that a factor encouraging biotic growth is removed and periodic retreatment that continues to damage the site is not required to prevent regrowth.

7. CONCLUSIONS

Ethical issues in rock art treatment are paramount concerns that have often been neglected in planning and conducting rock art conservation. Conserving rock art is not like restoring a painting. Rock art exists in a complex ceremonial context, and its environmental setting may be as important as the images. Many native peoples believe that rock art should return to nature if its ceremonial purpose has ceased. Intervention may be inappropriate within the cultural context of such sites. Rock art researchers should not apply rock art conservation methods designed to produce a clean, fresh looking building or to make inscriptions readable from a distance, without carefully assessing issues specific to rock art. A rock art researcher must weigh conflicting issues and have justifiable goals for any treatment proposed.

No treatment methods presently available have been adequately tested for us to know their long-term effects on the rock art and the substrate. If we want to apply conservation treatments to rock art sites in the future, scrupulously scientific testing of treatment methods must be conducted. Evaluation of treatment effects must be based on scientific studies, not subjective assessments. Treatment tests should be monitored over time to better evaluate results not anticipated in post-treatment scientific tests. We should not necessarily expect to carry out intervention treatment immediately, or perhaps not in our lifetimes, if adequate methods have not been developed and definitively tested. Detailed records of tests and documentation of all sites treated should be carefully preserved for review by future researchers.

At present, a minimal intervention approach is best for most sites, particularly because we have not yet answered the fundamental question of whether the rock art and substrate will erode more quickly if the organisms that have damaged it are removed. We do not fully understand the changes in rock chemistry and structure under lichen and what the changes mean for preservation. We have not yet developed all the scientific analyses that may be important in the future and cannot know what kinds of chemicals or treatment methods may damage the rock art for these studies. The fields of dating and paleoenvironmental studies, for example, are rapidly developing new techniques. In a minimalist approach, if we are concerned about biodeterioration at a site, we should initially monitor the microflora for a reasonable period of time to determine if there is a real risk to the site. One or two site visits cannot tell us if the long-term trend is expansion or decline of the microflora. In many regions, microorganisms grow slowly and their removal may not require emergency action. Especially for more ancient rock art, nature may be in a state of equilibrium. Treatment that disturbs the stasis could be more damaging than leaving the site alone. A minimalist approach can also include altering the microenvironment to discourage lichen, particularly where the alteration is not invasive to the site setting and, even better, where it restores the former environment of the site.

Any intervention treatment should have clear goals and must mitigate for conflicting interests. For example, if any foreign substances are put on the rock art, the investigator must first collect all scientific data presently available. All conservation actions should first be subject to peer review (Bahn *et al.* 1997). Increasingly, rock art is of global interest and concern; consequently, we need an international consensus on proper rock art conservation methodologies (Swartz 1997a). Consultation with all concerned parties should be carried to a consensus prior to any treatments. Rock art sites exist in a state of nature, and it would be better to leave them there, than to apply untested treatments that result in unacceptable effects.

Not all sites contain research data for dating, paleoenvironmental, and other studies, although we cannot know what scientific studies will be developed in the future. If these data are not present and imminent lost of the rock art is certain, concerned investigators may chose to treat biodeterioration. Where the natural environment is an integral aspect of the site, any treatment should preserve the aspects of the environment vital to the site and allow the site to retain a natural appearance. Otherwise, the site should not be treated. Where Native peoples have cultural ties to a site, the first response should be to leave the site in a natural state, unless through consultation, a consensus is reached on a type of treatment.

8. ACKNOWLEDGEMENTS

This paper has benefited from discussions with Ronald Dorn, Debra Dandridge, Larry St. Clair and a research team assembled by Constance Silver. Debra Dandridge, Ronald Dorn, and Antoinette Padgett reviewed the draft manuscript.

Ronald Dorn, Antoinette Padgett, Lisa Roach, Anne Oliver and Debra Dandridge provided references and information about specific sites and issues.

REFERENCES

Anati, E. (1964) *Camonica Valley*. (Translated by L. Asher). Jonathan Cape, London.

Anati, E. (1977) *Methods of Recording and Analysing Rock Engravings*. Studi Camuni, Volume 7. (Translated by L. Diamond and R. Lawson). Edizioni del Centro, Capo di Ponte.

Andersson, T. (1986) Preservation and Restoration of Rock Carvings and Rune-Stones. In: *Case Studies in the Conservation of Stone and Wall Paintings* (N. S. Brommelle and P. Smith, ed.): 133-137. International Institute for Conservation of Historic and Artistic Works, London.

Bahn, P. G., Bednarik, R. G. and Steinbring, J. (1995) The Peterborough Petroglyph site: reflections on massive intervention in rock art. *Rock Art Research* 12: 29-41.

Bahn, P. G., Bednarik, R. G. and Steinbring, J. (1996) Peer review of massive intervention in rock art management practice. *Rock Art Research* 13: 54-60.

Bahn, P. G., Bednarik, R. G. and Steinbring, J. (1997) Hear nothing, see nothing, say nothing? *Rock Art Research* 14: 55-58.

Bahn, P. G. and Hygen, A.-S. (1996) More on massive intervention: the Aspeberget structure.*Rock Art Research* 13: 137-138.

Beck, C., ed. (1994) *Dating in Exposed and Surface Contexts*. University of New Mexico Press, Albuquerque.

Bednarik, R. G. (2000) Lichenometry and rock art. *Rock Art Research* 17: 133-135.

Bednarik, R. G. (2003) Biological deterioration. http://mc2.vicnet.au/home/conserv/web/allbio.html.

Benedict, J. B. (1967) Recent glacial history of an alpine area in the Colorado Front Range, U.S.A. I. Establishing a lichen-growth curve. *Journal of Glaciology* 6: 817-832.

Beschel, R. E. (1961) Dating rock surfaces by lichen growth and its application to glaciology and physiography (Lichenometry). In: *Geology of the Arctic* (G. O. Faasch, ed.): 1044-1062. University of Toronto Press, Toronto.

Boustead, W. M. (1970) Museum conservation of anthropological material. In: *Aboriginal Antiquities in Australia* (F. D. McCarthy, ed.): 127-134. Australian Aboriginal Studies no. 22, AIAS, Canberra.

Bricker, F., Holcomb, T. and Dean, J. C. (1999) A native American's thoughts on the preservation and conservation of rock art. In: *Images Past, Images Present: The Conservation and Preservation of Rock Art* (J. C. Dean, ed.): 7-10. IRAC Proceedings, Volume 2. American Rock Art Research Association, Tucson.

Brunet, J. and Vidal, P. (1980) Les oeuvres rupestres préhistoriques: étude des problèmes de conservation. *Studies in Conservation* 25: 97-107.

Bull, W. B. and Brandon, M. T. (1998) Lichen dating of earthquake-generated regional rockfall events, Southern Alps, New Zealand. *Geological Society of America Bulletin* 110: 60-84.

Bullen, M. (1996) Who's right? Whose right? *Rock Art Research* 13: 47-48.

Childers, B. B. (1994) Long-term lichen-removal experiments and petroglyph conservation: Fremont County, Wyoming, Ranch Petroglyph Site. *Rock Art Research* 11: 101-112.

Clarke, J. (1976) Lichen control experiment at an aboriginal rock engraving site, Bolgart, Western Australia. *ICCM Bulletin* 2 (3):15-17.

Clottes, J. (2001) Paleolithic Europe. In: *Handbook of Rock Art Research* (D. S. Whitley, ed.): 459-481. AltaMira Press, Walnut Creek, California.

Cremaschi, M. (1996) The rock varnish in the Messak Settafet (Fezzan, Libyan Sahara), age, archaeological context, and paleo-environmental implication. *Geoarchaeology* 11: 393-421.

Crotty, H. K., ed. (1989) *Preserving Our Rock Art Heritage*. American Rock Art Research Association, Occasional Paper 1. San Miguel, California.

Dandridge, D. E. (2001) The Degradation of Rock Art by Lichen. Paper presented at the 66th Annual Meeting of the Society for American Archaeology, New Orleans.

Dandridge, D. E. and Meen, J.K. (2003) The degradation effects of rock art by lichen processes. *American Indian Rock Art* 29: 43-52.

Danin, A. (1985) Palaeoclimates in Israel: Evidence from Weathering Patterns of Stones in and near Archaeological Sites. *Bulletin of the American Schools of Oriental Research* 259:33-43.

Danin, A., R. Gerson, K. Marton, and J. Garty. (1982) Patterns of Limestone and Dolomite Weathering by Lichens and Bluegreen Algae and Their Palaeoclimatic Significance. *Palaeogeography, Palaeoclimatology, Palaeoecology* 37:221-233.

Del Monte, M., Sabbioni, C. and Zappia, G. (1987) The origin of calcium oxalates on historical buildings, monuments and natural outcrops. *The Science of the Total Environment* 67: 17-39.

Dorn, R. I. (1994) Dating petroglyphs with a three-tier rock varnish approach. In: *New Light on Old Art* (D. S. Whitley and L. L. Loendorf, eds.): 13-36. Institute of Archaeology Monograph no. 36. University of California, Los Angeles.

Dorn, R. I. (1998) *Rock Coatings*. Elsevier, Amsterdam.

Dorn, R. I. (2001) Chronometric techniques: engravings. In: *Handbook of Rock Art Research* (D. S. Whitley, ed.): 167-189. AltaMira Press, Walnut Creek, California.

Dorn, R. I. (2003) Second Preliminary Report: Initial Results. Report for contract KAP030013, on file at Bureau of Land Management, Newcastle Field Office, Newcastle, Wyoming.

Dorn, R. I. and DeNiro, M. J. (1985) Stable carbon isotope ratios of rock varnish organic matter: a new paleoenvironmental indicator. *Science* 227: 1472-1474.

Fernández, P. L., Gutierrez, I., Quindós, L. S., Soto, J. and Villar, E. (1986) Natural ventilation of the paintings room in the Altamira Cave. *Nature* 321: 586-588.

Fleisher, M., Liu, T. and Broecker, W. S. (1999) A clue regarding the origin of rock varnish. *Geophysical Research Letters* 26: 103-106.

Florian, M. L. E. (1978) A review: the lichen role in rock art - dating, deterioration and control. In: *Conservation of Rock Art* (C. Pearson, ed.): 95-98. Proceedings of the International Workshop on the Conservation of Rock Art, Perth, September 1977. Institute for the Conservation of Cultural Materials.

Gob, F., F. Petit, J. -P. Bravard, A. Ozer, and A. Gob. (2003) Lichenometric Application to Historical and Subrecent Dynamics and Sediment Transport of a Corsican Stream (Figarella River—France). *Quaternary Science Reviews* 22:2111-2124.

Grisafe, D. A. (1996) Preserving native American petroglyphs on porous sandstone. *Plains Anthropologist* 41: 373-382.

Grisafe, D. A. (2000) Improvement of sandstone durability: evaluating a treatment's suitability for preserving historic inscriptions. *Technology and Conservation*, Spring 1999-2000: 20-26.

Hale, M. E. (1975) Control of Biological Growths on Mayan Archeological Ruins in Guatemala and Honduras. *National Geographic Society Research Reports*, 1975 Projects: 305-321.

Haskovec. I. P. (1991) On some non-technical issues of conservation. In: *Rock Art and Posterity: Conserving, Managing and Recording Rock Art* (C. Pearson and B. K. Swartz, eds.): 97-99. Occasional AURA Publication No. 4. Australian Rock Art Research Association, Melbourne.

Henderson, J. W. (1995) An improved procedure for the photographic enhancement of rockpaintings. *Rock Art Research* 12 (2): 75-85.

Hygen, A.-S. (1999) Preservation problems concerning rock art in the County of Østfold, South-East Norway. In *NEWS95—International Rock Art Congress Proceedings* (Dario Seglie, ed.). [Html-CD ROM edition.] Centro Studi e Museo D'Arte Preistorica, Pinerolo, Italy.

Jaffe, L. (1996) Systematic vandalism and improper conduct in the Côa Valley Rock Art Area. *AURA Newsletter* 13 (2):12-13.

Jones, D. (1988) Lichens and pedogenesis. In: *Handbook of Lichenology*, Volume III. (M. Galun, ed.): 109-124. CRC Press, Boca Raton.

Jones, D., and Wilson, M. J. (1985) Chemical activity of lichens on mineral surfaces: a review. *International Biodeterioration* 21: 99-104.

Jones, D., Wilson, M.J. and McHardy, W. J. (1981) Lichen weathering of rock-forming minerals: application of scanning electron microscopy and microprobe analysis. *Journal of Microscopy* 124: 95-104.

Kaluarachchi, W., Labadie, J. H. and Russ, J. (1995) A close look at the rock art of Amistad National Recreation Area, Texas. *Park Science* 15 (4): 15-17.

Kennedy, B. (1979) Initial attempts at conservation work at rock art sites in British Columbia. *British Columbia Provincial Museum Heritage Record* 8: 285-292.

Kessler, D. W. (1923) Exposure Tests on Colorless Waterproofing Materials. *Technologic Papers of the Bureau of Standards* 18: 1-33. Washington, DC.

Kumar, R. and Kumar, A.V. (1999) *Biodeterioration of Stone in Tropical Environments: an overview*. The Getty Conservation Institute, Los Angeles.

Kurz, M. D. and Brook, E. J. (1994 Surface exposure dating with cosmogenic nuclides. In: *Dating in Exposed and Surface Contexts* (C. Beck, ed.): 139-159. University of New Mexico Press, Albuquerque.

Lal, B. B. (1970) Indian rock paintings and their preservation. In: *Aboriginal Antiquities in Australia* (F. D. McCarthy, ed.): 139-146. Australian Aboriginal Studies No. 22, AIAS, Canberra.

Lambert, D. (1989) *Conserving Australian Rock Art: A Manual for Site Managers*. Aboriginal Studies Press, Canberra.

Lee, G. (1991) *Rock Art and Cultural Resource Management*. Wormwood Press, Calabasas, California.

Levendosky, C. (2001) Oil drilling threatens sacred sites. *The Charlotte Observer*, 27 August 2001.

Lefèvre, M. (1974) La 'Maladie Verte' de Lascaux. *Studies in Conservation* 19: 126-156.

Lim, I. L. (1999) Rock art as cultural heritage: strategies for administration. In: *Images Past, Images Present: The Conservation and Preservation of Rock Art* (J. C. Dean, ed.): 11-20. IRAC Proceedings, Volume 2. American Rock Art Research Association, Tucson.

Liu, T. (2003) Blind testing of rock varnish microstratigraphy as a chronometric indicator: results on Late Quaternary lava flows in the Mojave Desert, California. *Geomorphology* 53: 209-234

Liu, T., and Broeker, W. S. (2000) How fast does varnish grow? *Geology* 28: 183-186.

Liu, T., Broeker, W. S., Bell, J. W. and Mandeville, C. W. (2000) Terminal Pleistocene wet event recorded in rock varnish from Las Vegas Valley, Southern Nevada. *Palaeogeography, Palaeoclimatology, Palaeoecology* 161: 423-433.

Liu, T. and Dorn, R. I. (1996) Understanding the spatial variability of environmental change in drylands with rock varnish microlaminations. *Annals of the Association of American Geographers* 86: 187-212.

Lock, W. W., J. T. Andrews, and P. J. Webber. (1979) Manual for Lichenometry. *British Geomorphological Research Group, Technical Bulletin* 26:1-47.

Mark, R. and Billo, E. (2002) Application of digital image enhancement in rock art recording. *American Indian Rock Art* 28: 121-128.

Marston, R. A. (2003) Editorial Note. *Geomorphology* 53:197.

Martin, A. K. and Johnson, G. C. (1992) Chemical control of lichen growths established on building materials: a compilation of the published literature. *Biodeterioration Abstracts* 6: 101-117.

Matthews, J. A. (1994) Lichenometric Dating: A Review with Particular Reference to 'Little Ice Age' Moraines in Southern Norway. In *Dating in Exposed and Surface Contexts*, edited by C. Beck, pp. 185-212. University of New Mexico Press, Albuquerque.

McCarthy, D. P, and K. Zaniewski. (2001) Digital Analysis of Lichen Cover: A Technique for Use in Lichenometry and Licheonology. *Arctic, Antarctic, and Alpine Research* 33:107-113.

Meehan, B. (1995) Aboriginal views of the management of rock art sites in Australia. In: *Perceiving Rock Art: Social and Political Perspectives* (K. Helskog and B. Olsen, ed.): 295-316. Institute for Comparative Research in Human Culture, Oslo.

Mowaljarlai, D. and Watchman, A. (1989 An aboriginal view of rock art management. *Rock Art Research* 6: 151-153.

Oliver, A. (2002) The variable performance of ethyl silicate: consolidated stone at Three National Parks. *Association for Preservation Technology Bulletin* 33 (2-3) : 39-44.

Padgett, A. and Barthuli, K. (1995) Paso por Aqui: a history of site management at El Morro National Monument, New Mexico, U.S.A. In: *Management of Rock Art Imagery* (G. Ward and L. Ward, ed.): 26-33. Occasional AURA Publication No. 6, Australian Rock Art Research Association, Melbourne.

Phillips, F. M., Flynch, M. Elmore, D. and Sharma, P. (1997) Maximum ages for the Côa Valley (Portugal) engravings measured with ^{36}Cl. *Antiquity* 71: 100-104.

Rebricova, N. L. and Ageeva, E. N. (1995) An evaluation of biocide treatments on the rock art of Baical. In: *Methods of Evaluating Products for the Conservation of Porous Building Materials in Monuments*, pp. 69-74. Preprints of the International Colloquium, Rome, 19-21 June 1995. ICCROM.

Roach, L. (2004) Personal communication

Rosenfeld, A. (1988) *Rock Art Conservation in Australia.* Special Australian Heritage Publication Series No. 2. 2nd ed. Australian Government Publishing Service, Canberra.

Rowe, M. W. (2001a) Dating by AMS radiocarbon analysis. In: *Handbook of Rock Art Research* (D. S. Whitley, ed.): 139-166. AltaMira Press, Walnut Creek, California.

Rowe, M. W. (2001b) Physical and chemical analysis. In: *Handbook of Rock Art Research* (D. S. Whitley, ed.): 190-220. AltaMira Press, Walnut Creek, California.

Russ, J., Kaluarachchi, W. D., Drummond, L. and Edwards, H. G. M. (1999) The nature of a whewellite-rich rock crust associated with pictographs in Southwestern Texas. *Studies in Conservation* 44: 91-103.

12. Biodeterioration of Prehistoric Rock Art

Russ, J., Palma, R. L., Loyd, D. H., Bouttton, T. W. and Coy, M. A. (1996) Origin of the whewellite-rich rock crust in the Lower Pecos Region of Southwest Texas and its significance to paleoclimate reconstructions. *Quaternary Research* 46: 27-36.

Russ, J., Palma, R. L., Loyd, D. H., Farwell, D. W. and Edwards, H. G. M. (1995) Analysis of the rock accretions in the Lower Pecos Region of Southwest Texas. *Geoarchaeology* 10: 43-63.

Sancho, L. G., D. Palacios, J. de Marcos, and F. Valladares. (2001) Geomorphological Significance of Lichen Colonization in a Present Snow Hollow: Hoya del Cuchillar de las Navajas, Sierra de Gredos (Spain). *Catena* 43:323-340.

Seaward, M. R. D. (1997) Major impacts made by lichens in biodeterioration processes. *International Biodeterioration and Biodegradation* 40: 269-273.

Silver, C. (1989) Rock art conservation in the United States: wish or reality. In: *Preserving Our Rock Art Heritage* (H. K. Crotty, ed.): 3-15. American Rock Art Research Association, Occasional Paper 1. San Miguel, California.

Silver, C. (2002) Research on Lichen-Induced Deterioration of Rock Art and Prospects for Conservation Treatment. Draft report for contract DACA63-99-P-1250. Report on file, U. S Bureau of Land Management, Newcastle Field Office, Newcastle, Wyoming.

Solomina, O, and P. E. Calkin. (2003) Lichenometry Applied to Moraines in Alaska, U.S.A., and Kamchatka, Russia. *Arctic, Antarctic, and Alpine Research* 35:129-143.

Somavilla, J. F., Khayyat, N. and Arroyo, V. (1978) A comparative study of the microorganisms present in the Altamira and La Pasiega Caves. *International Biodeterioration Bulletin* 14 (4) :103-109.

St. Clair, L. (2000) Report Concerning Lichen Encroachment on Petroglyph Panels. Preliminaryreport for contract number DACA63-99-P-1250. Report on file, U. S. Bureau of Land Management, Newcastle Field Office, Newcastle, Wyoming.

Swartz, B. K. (1997a) An evaluation of rock art conservation practices at Foz Côa, Northern Portugal. *Rock Art Research* 14: 73-75.

Swartz, B. K. (1997b) An investigation of the Portuguese government policies on the management of the Foz Côa sites. *Rock Art Research* 14: 75-76.

Syers, J. K. and Iskandar, I. K. (1973) Pedogenic significance of lichens. In: *The Lichens* V. Ahmadjian and M. E. Hale, eds.): 225-248. Academic Press, New York.

Taylor, J. M., Bokman, W. and Wainwright, I. N. M. (1977) Rock art conservation: some realities and practical considerations. In: *CRARA '77, Papers from the Fourth Biennial Conference of the Canadian Rock Art Research Associates* (D. Lundy, ed.): 293-324.

Thorne, C. (2002) House bill seeks to protect sacred Indian sites. *San Francisco Chronicle*, 18 July 2002.

Tratebas, A. M. (1993) Stylistic chronology versus absolute dates for early hunting style rock art on the North American Plains. In: *Rock Art Studies: the Post-Stylistic Era* (M. Lorblanchet and P. Bahn, eds.): 163-177. Oxbow Monograph 35, Oxbow Press, Oxford.

Tratebas, A. M. and Chapman, F. (1996) Ethical and conservation issues in removing lichens from petroglyphs. *Rock Art Research* 13: 129-133.

Wainwright, I. N. M. (1985) Rock art conservation research in Canada. *Bollettino del Centro Camuno di Studi Preistorici* 22: 15-46.

Wainwright, I. N. M. (1986) Lichen removal from an engraved memorial to Walt Whitman. *Association for Preservation Technology Bulletin* 18 (4) :46-51.

Wainwright, I. N. M. (1996) Structure protects rock art. *Rock Art Research* 13: 52-53.

Walderhaug, O. and Walderhaug, E. M. (1998) Weathering of Norwegian rock art - a critical review. *Norwegian Archaeological Review* 31: 119-139.

Watchman, A. (1993a) Perspectives and potentials for absolute dating prehistoric rock paintings. *Antiquity* 67: 58-65.

Watchman, A. (1993b) Evidence of a 25,000-year-old pictograph in Northern Australia. *Geoarchaeology* 8: 465-473.

Watchman, A. and Cole, N. (1993) Accelerator radiocarbon dating of plant-fibre binders in rock paintings from Northeastern Australia. *Antiquity* 67: 355-358.

Weeks, R. (2003) Discovering Monongahela Rock-Art: The Ancient Teaching Rocks of the Upper Ohio Valley. Paper presented at the 68th Annual Meeting of the Society for American Archaeology, Milwaukee, Wisconsin.

Winchester, V., and R. K. Chaujar. (2002) Lichenometric Dating of Slope Movements, Nant Ffrancon, North Wales. *Geomorphology* 47:61-74.

Young, G. S. and Wainwright, I. N. M. 1995. The control of algal biodeterioration of a marble petroglyph site. *Studies in Conservation* 40: 82-92.

Young, M. J. (1988) *Signs from the Ancestors: Zuni Cultural Symbolism and Perceptions of Rock Art*. University of New Mexico Press, Albuquerque.

Chapter 13

RAMAN SPECTROSCOPY OF ROCK BIODETERIORATION BY THE LICHEN LECIDEA TESSELLATA FLÖRKE IN A DESERT ENVIRONMENT, UTAH, USA

HOWELL G.M EDWARDS[1], SUSANA E. JORGE VILLAR[1], MARK R.D. SEAWARD[2] and LARRY L. ST.CLAIR[3]
[1]*Department of Chemical and Forensic Sciences, University of Bradford, Bradford BD7 1DP, UK;* [2]*Department of Environmental Science, University of Bradford, Bradford BD7 1DP, UK;* [3] *Department of Integrative Biology, Brigham Young University, Provo Utah 84602, USA*

Abstract: Samples of the crustose rock lichen *Lecidea tessellata* from south eastern Utah, USA, were analysed using FT-Raman spectroscopy. Data from Raman spectra indicate significant accumulations of several biomolecules including strong bands of calcium oxalate monohydrate, with some calcium oxalate dihydrate found only in apothecia. In addition, the UV radiation protectant scytonemin was detected in rock substrata immediately adjacent to lichen thalli. A white crystalline encrustation beneath thalli demonstrated high concentrations of calcium oxalate monohydrate, while a brownish region of the crystalline encrustation contained both calcium oxalate monohydrate and haematite (iron (III) oxide). Haematite accumulation may function as a UV radiation protectant; intense summer insolation in this region may suggest a critical need for its control. The source of calcium ions to complex with the oxalic acid produced by the lichen appears to be the calcium-rich cement in the Dakota sandstone.

1. INTRODUCTION

Lichens are one of the primary colonizers of geological formations and their ability to withstand extremes of terrestrial environments is a vital part of pedogenesis. The interface between lichen thalli and rock substrata are regions of intense biogeochemical activity, which often results in severe erosion and bioweathering of rock surfaces. Lichen secondary chemistry involves a complex interplay of waste metabolic products and biomolecules. These function as survival strategies evolved by many lichens against drought, low or high temperatures, particle-borne wind abrasion and levels of damaging solar radiation, especially in the UVA, UVB and UVC regions from 360 to 250 nm.

The presence of heavy metals in rock substrata is not problematic to many lichen species which have the ability to accumulate (in many cases, hyperaccumulate) levels toxic to most other organisms (Nash 1990); in this respect, the chelating properties of the aromatic polyphenolic lichen acids, such as depsides, are also believed to play an important protective role (Culberson 1969; Huneck and Yoshimura 1996; Edwards *et al.* 2003a, 2003b).

In the most hostile environments of the Antarctic dry valleys, lichens cannot survive on rock surfaces; instead, they adopt a chasmolithic or endolithic strategy which insulates them from temperatures as low as $-50°$ C, the abrasive action of dry katabatic winds from the polar plateau and also low wavelength UV radiation reaching the terrestrial surface due to a locally depleted ozone layer (Wynn-Williams *et al.* 2000a, 2001; Wynn-Williams and Edwards 2002). Production of cryogenic, desiccative and radiation protectants by many lichens now assumes a major strategic importance for species colonizing such environments; some interesting comparisons have been made with the chemical strategies of Antarctic lichens, especially those found at more hospitable maritime localities, with similar species occurring in more temperate climates, such as the Mediterranean (Edwards *et al.* 1998; Holder *et al.* 2000).

In previous studies, we have examined lichen colonies on diverse substrata such as sandstone, limestone, dolomite, granite (Edwards and Seaward 1997; Prieto *et al.* 1999a, 1999b, 2000; Seaward and Edwards 1995, 1997; Seaward *et al.* 1995) and even epiphylls (de Oliveira *et al.* 2002). In all cases, the availability of calcareous substrate material for reaction with oxalic acid metabolically formed by the lichen mycobiont is seen to be a prime factor for colony development.

In a recent study of the Moroccan desert lichen *Aspicilia caesiocinerea* agg., which was responsible for severe erosion of sandstone outcrops, a remarkable strategy involving the creation of concentric escarpments was generated by undercutting (Edwards *et al.* 2002). Here, we have extended these studies by an FT-Raman investigation of *Lecidea tesselata* Flörke, a frequent lichen colonizer of rocks in desert environments of western North America, which is clearly having a dramatic biodeteriorative impact on its substratum. Through Raman spectroscopy we have been able to identify key biomolecules involved in the lichen's survival strategy and effect a comparison with the chemical biodeterioration processes observed in other hot and cold desert environments (Wynn-Williams *et al.* 2000b).

2. EXPERIMENTAL

2.1 Specimens

Lecidea tessellata Flörke, a crustose lichen with a whitish grey areolate thallus profusely covered with black apothecia (Fig. 13-1), occurs commonly on a variety of rock substrates (quartzite, sandstone, granite, and basalt) throughout the western United States. This species is widely distributed, reported from a wide range of habitats including arid and semi-arid sites from warm deserts to alpine tundra.

Figure 13-1. Lecidea tessellata Flörke on sandstone on Colorado Plateau, 10 km N of Monticello, Utah, USA

2.2 Study site

South eastern Utah is located in the geological region know as the Colorado Plateau. The canyon-plateau landscape in this region is formed by sedimentary rocks, often vertically offset by faulting and folding, and extensively downcut by the Colorado River and its tributaries. Much of the development of the Plateau country has occurred in the last 10 million years. Low precipitation and thin, sandy soils generally result in sparse vegetation dominated by desert shrubs including Pinyon pine (*Pinus edulis* Engelm.), Utah juniper (*Juniperus osteosperma* (Torr.) Little), Big sagebrush (*Artemisia tridentata* Nutt.), and Rabbitbrush (*Chrysothamnus nauseosus* var. *bigelovii* (Gray) Hall); ravine and canyon locations also support Gambel's oak (*Quercus gambelii* Nutt., Skunkbush (*Rhus aromatica* var.*trilobata* (Nutt.) Gray), as well as the tree species Ponderosa pine (*Pinus ponderosa* Lawson), and Quaking aspen (*Populus tremuloides* Michx.).

Lichen samples for this study were collected at a site 10 km north of Monticello, Utah along U. S Highway 191 (San Juan Co.; 37° 58.136' N, 109° 21.473' W; alt. 2055m). This site, slightly lower in elevation than nearby Monticello, is in a relatively shallow, short canyon with its northern extremity dropping steeply into the Lisbon Valley area. This portion of U.S. Highway 191 is bordered on both sides by the Dakota sandstone formation, the oldest Cretaceous formation in the Plateau country. All samples were collected from Dakota sandstone in an area dominated by Pinyon pine, Utah juniper, Gambel's oak, Big sagebrush, and Skunkbush.

At the study site, *Lecidea tessellata* covers extensive areas of sandstone; where fully developed, significant portions of the thalli have become detached to reveal fully eroded sandstone beneath. At the margins of the craters generated by these detachments, crystalline encrustations up to 5 mm in thickness are evident (Fig. 13-2), the top surface of which is stained deep brown in color. Biodeterioration of the substratum caused by lichen invasion is evidenced by the extent of the erosion and the friability of the specimens. Numerous collections were made by two of us (MRDS & LLStC) and representative herbarium material (MRDS 111311) was chosen for more detailed investigation using FT-Raman Spectroscopy.

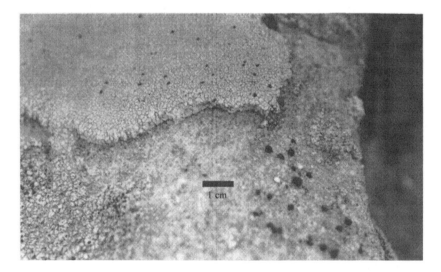

Figure 13-2. The major features of *Lecidea tessellata* subjected to the analytical spectroscopic study: black apothecium on greenish-grey areolate thallus with several mm thickness of white crystalline encrustation, the top surface of which in immediate contact with the lichen is deep chocolate brown in color

2.3 Raman spectroscopy

FT-Raman spectra were obtained using a Bruker IFS66 instrument with FRA106 Raman module attachment and dedicated Raman microscope facility; excitation was effected using 1064 nm radiation in the near-infrared from a Nd^{3+}/YAG laser and an InGaAs liquid-nitrogen cooled detector. In the macroscopic mode the spectral "footprint was 100 μm and laser powers of up to 100 mW were used; in the microscopic mode, a typical spectral "footprint" was about 40 μm using a 40x microscope lens objective, which could be decreased to about 10 μm using a 100 x lens objective. In the microscopic mode, laser powers of 20 mW or less were used to prevent specimen damage. Some 2000-4000 spectral scans were accumulated at 4 cm^{-1} spectral resolution to produce a spectrum which was typical of those illustrated here; each specimen was sampled spectroscopically several times to check for heterogeneity.

3. RESULTS AND DISCUSSION

The FT-Raman spectrum (Fig. 13-3) of the substratal rock sampled macroscopically in regions remote from the lichen biodeterioration indicates a quartz spectral signature, with major bands at 465, 207 and 129 cm^{-1} and weaker features at 1161, 795, 696 cm^{-1}. This correlates with the geological attribution of the rock as sandstone.

The spectrum of the rock adjacent to the green colored biological colonization (Fig. 13-4) is interesting in that this too can be assigned as quartz, but now a new band appears at 1590 cm^{-1} of medium-weak intensity; this is characteristic of scytonemin, a UV-radiation protectant biomolecule which is synthesized by cyanobacteria in highly-stressed environments. We have identified scytonemin previously in mats of the cyanobacterium *Nostoc commune* on exposed Antarctic lacustrine beaches, where survival of the organism is critically dependent on its capability of dealing with excessive dosage of low wavelength UV radiation. Scytonemin is a key biomolecule in this respect as it absorbs UVA, UVB and UVC radiation and is present in the outer sheaths of *Chroococcodiopsis* at the limits of life survivability in Antarctic habitats. The presence of scytonemin here indicates that there had almost certainly been cyanobacterial colonization of the eroded rock surface prior to lichen development at this point; the longevity of survival of scytonemin following the death of the host organism is currently a subject receiving much attention from geochemists with regard to its detection in biomodified geological environments.

The spectrum of the crystalline white encrustations (Fig. 13-5) shows the presence of quartz and new features at 1490, 1463, 897 and 503 cm^{-1} which are all characteristic of calcium oxalate monohydrate (whewellite); its presence in the encrustation is significant in that it is a major product of several lichen biodeteriorators of calcareous habitats, such as *Dirina massiliensis* forma *sorediata*, which can produce up to 50% of its biomass as hydrated calcium oxalate (Edwards *et al.* 1997).

The question then arises as to where the lichen has obtained the calcium ions to complex with the oxalic acid produced in the metabolic cycle of the mycobiont; in previous studies of non-calcareous substrata (Prieto *et al.* 1999a, 1999b), the presence of spectral signatures of hydrated calcium oxalate is generally attributable to wind-borne dust from neighbouring limestone outcrops, to on-shore winds from a marine or lacustrine environment or to ground water run-off. It is a matter of conjecture as to what may be responsible for the calcium oxalate production in the present case but the most likely source is the calcium-enriched cement of the sandstone. Dakota sandstone was deposited on beaches and sandbars near the shore of an ancient Cretaceous sea.

Microscopic Raman examination of the chocolate brown region of the crystalline encrustation, the white region of which analyzed above as calcium oxalate monohydrate and quartz, revealed strong evidence of calcium oxalate monohydrate and quartz (Fig. 13-6), but also weaker signals from haematite, iron (III) oxide, at 294 and 400 cm^{-1} (Fig. 13-7). A novel feature emerged in sampling this region and Figure 13-8 shows the presence of two strong bands at 440 and 607 cm^{-1}, along with weaker calcium oxalate monohydrate bands at 1464 and 1490 cm^{-1}; these new features are characteristic of rutile, TiO_2. Like quartz, SiO_2, the titanium (IV) oxide has extremely strong metal-oxygen bonds, which are not available for attack by biodeteriorative organisms. Hence, particles of TiO_2 are identified here, along with quartz in the encrustation (Fig. 13-8).

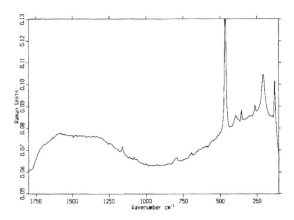

Figure 13-3. FT-Raman spectrum of the rock substratum, recorded from a region remote from lichen biodeterioration; 1064 nm excitation, 4 cm-1 spectral resolution, 2000 spectral scans accumulated

13. Raman Spectroscopy of Rock Biodeterioration 235

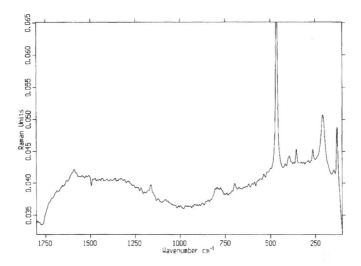

Figure 13-4. FT-Raman spectrum of biodeterioration area adjacent to green colonies; conditions as for Figure 13-3

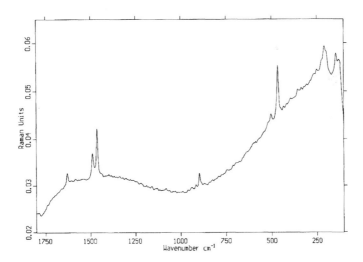

Figure 13-5. FT-Raman spectrum of white crystalline encrustation; conditions as for Figure 13-3; note the presence of calcium oxalate monohydrate bands

Biodeterioration of Rock Surfaces

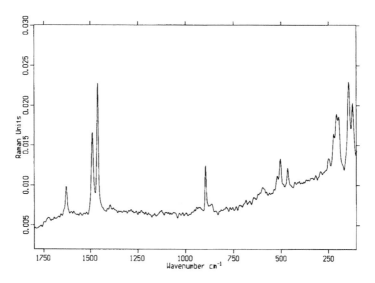

Figure 13-6. FT-Raman spectrum of chocolate brown colored region of lichen encrustation; conditions as for Figure 13-3

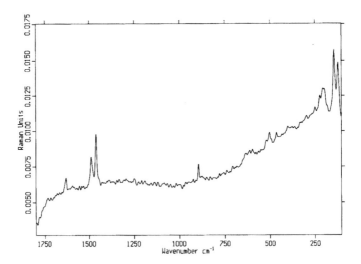

Figure 13-7. FT-Raman spectrum of chocolate brown colored region of lichen encrustation, showing the presence of iron (III) oxide bands

13. Raman Spectroscopy of Rock Biodeterioration

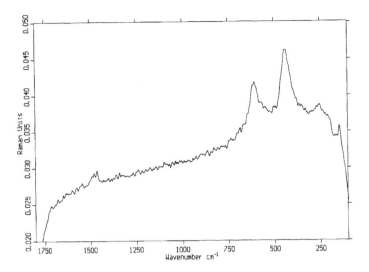

Figure 13-8. . Detail of FT-Raman spectrum of red particulate matter in lichen encrustation, showing the presence of titanium (IV) oxide bands

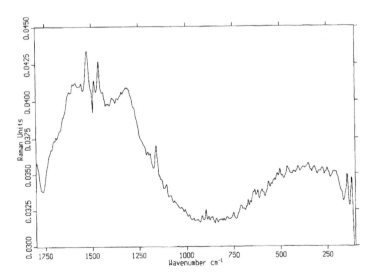

Figure 13-9. Cortex containing photobiont: FT-Raman spectrum showing presence of scytonemin, beta-carotene, whewellite and chlorophyll; conditions as for Figure 13-3

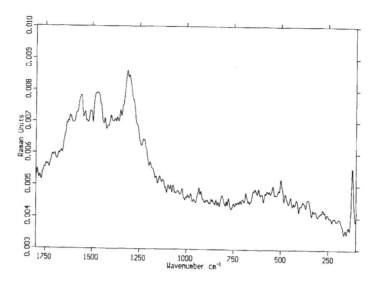

Figure 13-10. Black apothecium: FT-Raman spectrum showing presence of chlorophyll, weddelite and unidentified lichen substance(s)

The Raman spectra of the cortex occupied by the photobiont (Fig. 13-9) shows the presence of beta-carotene (1525, 1159 cm^{-1}), calcium oxalate monohydrate (1490, 1463, 897 cm^{-1}), chlorophyll (1320 cm^{-1}) and scytonemin (1590, 1560, 1391 cm^{-1}). The black apothecium also shows characteristic bands of chlorophyll and some unidentified organic material, with features at 1618, 1559, 1230, 934 cm^{-1} which are assignable to aromatic compounds such as polyphenolic lichen acids (Fig. 13-10). There is no spectral signature for scytonemin or beta-carotene in the black apothecia, but a broad feature at 1470 cm^{-1} is identifiable, with supporting bands at 900 and 500 cm^{-1}, as calcium oxalate dihydrate (weddellite). The fact that the dihydrate form of the calcium oxalate is limited only to the apothecium could be relevant for its anti-herbivoral properties, as has been suggested previously (Seaward *et al.* 1998).

In summary, the severe biodeteriorative effect of this lichen/cyanobacterial system on the rock substratum is as a consequence of the ability to assimilate calcium ions to immobilize the oxalic acid produced by the colonies into a significant encrustation, which consists mainly of calcium oxalate monohydrate with quartz and rutile particles. The brown coloration of the upper surface of this encrustation is attributable to the presence of haematite – and its localization in this region is worthy of note. We have found a similar pattern in Antarctic cryptoendolithic systems, where mobilization of the iron (III) oxide has occurred from within the rock to the surface; a possible explanation of this phenomenon for the Antarctic Dry Desert region is the UV-radiation protection afforded by haematite (Benton Clark 1998). Perhaps a similar situation pertains in the Utah desert, and the lichen/cyanobacterial system studied here needs to provide similar protection against high levels of insolation. A study of the recently eroded rock surface adjoining the green colonies reveals the presence of scytonemin; in the living colonies adjacent to this region there are clear spectral signatures of scytonemin and other materials produced by the organisms – the residual scytonemin remaining in the biodeteriorated surface following the death and detachment of the organism is expected from the longevity of the survival of this biomolecule in ancient geological environments.

ACKNOWLEDGEMENT

Susana Jorge Villar is grateful to the University of Burgos for a Visiting Fellowship, during the tenure of which this work was carried out.

REFERENCES

Benton Clark, C. (1998) Surviving the limits to life at the surface of Mars. *Journal of Geophysical Research* 103: 28545-28555.

Culberson, C.F. (1969) *Chemical and Botanical Guide to Lichen Products*. University of North CarolinaPress, Chapel Hill.

Edward, H.G.M., Farwell, D.W. and Seaward, M.R.D. (1997) FT-Raman spectroscopy of *Dirina massiliensis* f. *sorediata* encrustations growing on diverse substrata. *The Lichenologist* 29: 83-90.

Edwards, H.G.M., Holder, J.M., Seaward, M.R.D. and Robinson, D.A. (2002) A Raman spectroscopic study of lichen-assisted weathering of sandstone outcrops in the High Atlas Mountains, Morocco. *Journal of Raman Spectroscopy* 33: 449-454.

Edwards, H.G.M., Holder, J.M. and Wynn-Williams, D.D. (1998) Comparative FT-Raman spectroscopy of *Xanthoria* lichen-substratum systems from temperate and Antarctic habitats. *Soil Biology and Biochemistry* 30: 1947-1953.

Edwards, H.G.M., Newton, E.M. and Wynn-Williams, D.D. (2003a) Molecular structural studies of lichen substances II: Atranorin, gyrophoric acid, fumarprotocetraric acid, rhizocarpic acid, calycin, pulvinic dilactone and usnic acid. *Journal of Molecular Structure* 651-653: 27-37.

Edwards, H.G.M., Wynn-Williams, D.D., Newton, E.M. and Coombes, S.J. (2003b) Molecular structural studies of lichen substances I: Parietin and emodin. *Journal of Molecular Structure* 648: 49-59.

Holder, J.M., Wynn-Williams, D.D., Rull Perez, F. and Edwards, H.G.M. (2000) Raman spectroscopy of pigments and oxalates in situ within epilithic lichens: *Acarospora* from the Antarctic and Mediterranean. *New Phytologist* 145: 271-280.

Huneck, S. and Yoshimura, I. (1996) *Identification of Lichen Substances*. Springer-Verlag, Berlin.

Nash, T.H. (1990) Metal tolerance in lichens. In: *Heavy Metal Tolerance in Plants: Evolutionary Aspects* (A.J.Shaw ed.): 119-131. CRC Press, Boca Raton. de Oliveira, L.F.C., Edwards, H.G.M., Feo Manga, J.C., Seaward, M.R.D. and Lücking, R. (2002). FT-

Raman spectroscopy of three foliicolous lichens from the Costa Rican rain forests. *The Lichenologist* 34: 259-266.

Prieto, B., Edwards, H.G.M. and Seaward, M.R.D. (2000) A Fourier-Tranform Raman spectroscopic study of lichen strategies on granite monuments. *Geomicrobiology Journal* 17: 55-60.

Prieto, B., Seaward, M.R.D., Edwards, H.G.M., Rivas, T. and Silva, B. (1999a) Biodeterioration of granite monuments by *Ochrolechia parella* (L.) Mass.: an FT-Raman spectroscopic study. *Biospectroscopy* 5: 53-59 (1999)

Prieto, B., Seaward, M.R.D., Edwards, H.G.M., Rivas, T. and Silva, B. (1999b) An FT-Raman spectroscopic study of gypsum neoformation by lichens growing on granitic rocks. *Spectrochemica Acta* 55A: 211-217.

Seaward, M.R.D. and Edwards, H.G.M. (1995) Lichen-substratum interface studies with particular reference to Raman microscopic analysis. 1. The deterioration of works of art by *Dirina massiliensis* forma *sorediata*. *Cryptogamic Botany* 5: 282-287.

Seaward, M.R.D. and Edwards, H.G.M. (1997) Biological origin of major chemical disturbances on ecclesiastical architecture studied by Fourier-Transform Raman microscopy. *Journal of Raman Spectroscopy* 28: 691-696.

Seaward, M.R.D., Edwards, H.G.M. and Farwell, D.W. (1995) FT-Raman microscopic studies of *Haematomma ochroleucum* var. *porphyrium*. *Bibliotheca Lichenologica* 57: 395-407.

Seaward, M.R.D., Edwards, H.G.M. and Farwell, D.W. (1998) Fourier-Transform Raman microscopy of the apothecia of *Chroodiscus megalophthalmus* (Müll.Arg.) Vĕzda & Kantvilas. *Nova Hedwigia* 66: 463-472.

Wynn-Williams, D.D. and Edwards, H.G.M. (2002) Environmental UV radiation: biological strategies for protection and avoidance. In: *Astrobiology: The Quest for the Conditions of Life* (G.Horneck and C.Baumstark-Khan ed.): 245-260. Springer-Verlag, Berlin.

Wynn-Williams, D.D., Edwards, H.G.M. and Newton, E.M. (2000a) Raman spectroscopy of microhabitats and microbial communities: Antarctic deserts and Mars analogues. In: *Abstracts of the Proceedings of the Lunar and Planetary Sciences Institute Conference XXXI* (S.Clifford ed.), Houston.

Wynn-Williams, D.D., Holder, J.M. and Edwards, H.G.M. (2000b) Lichens at the limits of life: past perspectives and modern technology. *Bibliotheca Lichenologica* 75: 275-288.

Wynn-Williams, D.D., Newton, E.M. and Edwards, H.G.M. (2001) The role of habitat structure for biomolecule integrity and microbial survival under extreme environmental stress in Antarctica (and Mars?): ecology and technology. In: *Proceedings of the First European Workshop on Exo/Astro-Biology* (P.Ehrenfreund, O.Angerer and B.Battrick ed.): 225-237. ESA Special Publications, Frascati.

Chapter 14

LICHENS AND MONUMENTS
An Analtyical Bibliography

ROSANNA PIERVITTORI[1], ORNELLA SALVADORI[2] and MARK R.D. SEAWARD[3]
[1]*Dipartimento di Biologia Vegetale, Università di Torino viale Mattioli 25, I-10125 Torino, Italy* [2]*Soprintendenza Speciale per il Polo Museuale Veneziano-Laboratorio Scientifico, Cannaregio 3553, I-30131 Venice, Italy;* [3]*Department of Environmental Science, University of Bradford, Bradford, BD7 1DP, UK*

Since 1994, periodic bibliographic reviews have been published in *The Lichenologist* (see numbers 550, 451 and 452 in the list) with abstracts and analytical index referring to major themes. These form the basis of the following analytical bibliography.

Ackerman, R.E. (1964) Lichens and the patination of chert in Alaska. *American Antiquity* 29: 386-387. [2]

Adamo, P. (1996) Ruolo dell'attività dei licheni nell'alterazione di substrati rocciosi e nella neogenesi di entità mineralogiche. In: Atti del XIII Convegno Nazionale della Società Italiana di Chimica Agraria: 13-26. Patron Editore, Bologna. [2]

Adamo, P. (1997) Bioalterazione di rocce di natura vulcanica, dolomitica e mafica indotta da licheni. In: Abstracts del Convegno annuale della Società Lichenologica Italiana *Licheni e Ambiente* (D. Ottonello, ed.), (Palermo, 9-12 Dicembre 1995). *Notiziario della Società Lichenologica Italiana* 10: 63-64. [2]

Adamo, P. (2000) Lo studio della bioalterazione di substrati minerali indotta da licheni: applicazione di nuove metodologie analitiche. In: Abstracts del Convegno annuale della Società Lichenologica Italiana *Licheni e Ambiente* (D. Ottonello, ed.), (Palermo, 9-12 Dicembre 1995). *Notiziario della Società Lichenologica Italiana* 13: 62. [2]

Adamo, P. and Violante, P. (1989) Bioalterazione di roccia dolomitica operata da una specie lichenica del genere *Lepraria*. *Agricoltura Mediterranea* 119: 460-464. [2]

Adamo, P. and Violante, P. (1991) Bioalterazione di rocce diabasiche, gabbriche e serpentinose operata da licheni. In: Atti XXX Convegno annuale della Società Italiana di Genetica Agraria *Fertilità del Terreno e Biomassa Microbica*: 77-81. Congedo Editore, Galatina (Lecce). [2]

Adamo, P. and Violante, P. (1991) Weathering of volcanic rocks from Mt Vesuvius associated with the lichen *Stereocaulon vesuvianum*. *Pedobiologia* 35: 209-217. [2]

Adamo, P. and Violante, P. (1992) Biological weathering of volcanic and serpentine rocks. In: Book of Abstracts, *Mediterranean Clay Meeting (MCM '92), Gruppo Italiano A.I.P.E.A.* (Aeolian Islands, 27-30 September 1992): 7-8. Lipari, Italy. [2]

Adamo, P. and Violante, P. (2000) Weathering of rocks and neogenesis of minerals associated with lichen activity. *Applied Clay Science* 16: 229-256. [2, 7]

Adamo, P., Marchetiello, A. and Violante, P. (1990) I licheni *Stereocaulon vesuvianum* Pers. e *Parmelia conspersa* (Ach.) Ach. agenti della bioalterazione di roccia vulcanica. In: Atti VIII Convegno Nazionale Società Italiana di Chimica Agraria (S.I.C.A.), (Bari, 25-27 Settembre 1990). Congedo, Lecce, Italy. [2]

Adamo, P., Marchetiello, A. and Violante, P. (1993) The weathering of mafic rocks by lichens. *The Lichenologist* 25: 285-297. [2]

Adamo, P., Colombo, C. and Violante, P. (1995) Occurrence of poorly ordered Fe-rich phases at the interface between the lichen *Stereocaulon vesuvianum* and volcanic rock from Mt. Vesuvius. In: Book of Abstracts, *Euroclay '95, Clays and Clay Mineral Sciences* (A. Elsen, P. Grobet, M. Keung, H. Leeman, R. Schoonheydt and H. Toufar, eds.): 341-342. Leuven, Belgium. [2]

Adamo, P., Terribile, F. and Violante, P. (1996) Micromorphological study of lichens activity on volcanic rock. In: Book of Abstracts, 10th International Working Meeting on *Soil Micromorphology* (Moscow, 8-13 July 1996): 13. Moscow State University, Moscow. [2]

Adamo, P., Colombo, C. and Violante, P. (1997) Iron oxides and hydroxides in the weathering interface between *Stereocaulon vesuvianum* and volcanic rocks. *Clay Minerals* 32: 453-461. [2]

Adamo, P., Vingiani, S. and Violante, P. (2000) I licheni *Stereocaulon vesuvianum* Pers. e *Lecidea fuscoatra* (L.) Ach. ed il muschio *Grimmia pulvinata* agenti di bioalterazione di una tefrite fonolitica dell'Etna (Sicilia). In: Abstracts del Convegno annuale della Società Lichenologica Italiana *Licheni e Ambiente* (D. Ottonello, ed.), (Palermo, 9-12 Dicembre 1995). *Notiziario della Società Lichenologica Italiana* 13: 3-64. [2]

Affini, A., Favali, M.A., Pedrazzini, R. and Fossati, F. (1996) Metodologie di intervento e di conservazione dei monumenti del Giardino Ducale di Parma. *Giornale Botanico Italiano* 130: 430. [4]

Agrawal, O.P., ed. (1972) Conservation in the Tropics. In: Proceedings of Asia-Pacific Symposium *Conservation of Cultural Property*. New Delhi. [4]

Aires-Barros, L. and Mauricio, A. (1992) Topoclimatic characterization of south down gallery in the Jeronimos Monastery cloister, Lisboa. In: Proceedings of the International Congress *Deterioration and Conservation of Stone* (J. Delgado Rodrigues, F. Henriques and F. Telmo Jeremias, eds.): 335-344. LNEC, Lisbon. [1]

Aires-Barros, L. and Mauricio, A. (1995) Contaminaçao liquénica no claustro do mosteiro dos Jeronimos e correlaçao com factores de alteraçao meteorica das suas rochas. *Monumentos* 3: 51-61. [1]

Aires-Barros, L., Basto, J., Graça, R.C., Alves, L.M., Esteves, L., Carmo, A. and Ribeiro, I. (1989) The decay of limestone caused by saxicolous lichens: the case of the Monastery of Jeronimos (Lisbon, Portugal). In: Proceedings of 1st International Symposium *La Conservazione dei monumenti nel bacino del Mediterraneo* (F. Zezza, ed.): 225-229. Grafo Edizioni, Bari. [2]

Aires-Barros, L., Alves, L. and Esteves, L. (1991) New data on saxicolous lichenic vegetation in monuments as bioindicator of pollution. In: Proceedings 2nd International Symposium *The Conservation of Monuments in the Mediterranean Basin* (D. Decrouez, J. Chamay and F. Zezza, eds.): 115-120. Genève. [5]

Alessandrini, G. and Realini, M. (1990) Le pellicole ad ossalati: origine e significato nella conservazione. Note conclusive sul Convegno. *Arkos* 9/10: 19-36. [3]

Alessandrini, G. Bonecchi, R., Peruzzi, R. and Toniolo, L. (1989) Caratteristiche composizionali e morfologiche di pellicole ad ossalato: studio comparato su substrati lapidei di diversa natura. In: Atti del Convegno *Scienza e Beni Culturali, Le pellicole ad ossalato: origine e significato nella conservazione delle opere d'arte* (V. Fassina, ed.): 137-150. Centro del C.N.R. "Gino Bozza", Milano. [3]

Alessi, P. and Visintin, D. (1988) Protective agents as a possible substrate for biogenic cycles. *Studia Geobotanica* 8: 99-112. [2]

Allsopp, D. and Seal, K.L. (1986) *Introduction to Biodeterioration*. Arnold, London. [1]

Alstrup, V. (1992) Ceased grazing: the consequences of overgrowing on the flora of soil and stone inhabiting lichens. *International Journal of Mycology and Lichenology* 5: 1-2. [1]

Alstrup, V. (1992) Effects of pesticides on lichens. *Bryonora* 9: 2-4. [4]

Altieri, A., Giuliani, M.R., Nugari, M.P., Pietrini, A.M., Ricci, S. and Roccardi, A. (1998) Il degrado di origine biologica delle opere d'arte: diagnosi e interventi. In: *Diagnosi e progetto per la conservazione dei materiali dell'architettura* (Istituto Centrale per il Restauro, ed.):125-137. De Luca Edizioni, Roma. [1, 4]

Altieri, A., Laurenti, M.C. and Roccardi, A. (1999) The conservation of archaeological sites: materials and techniques for short-term protection of archaological remains. In: Proceedings of the 6th International Conference, *Non-destructive Testing and Microanalysis for the Diagnosticts and Conservation of the Cultural and Environmental Heritage* (Rome, 17-19 May 1999): 673-687. Roma. [4]

Altieri, A., Mazzone, A., Pietrini, A.M., Ricci, S. and Roccardi, A. (2000) Indagini diagnostiche sul biodeterioramento delle fontane. In: *Piazza di Corte, il recupero dell'immagine berniniana* (M. Natoli, ed.): 34-57. Ministero per i Beni e le Attività Culturali, Palombi Editore, Roma. [1, 4]

Altieri, A., Pietrini, A.M., Ricci, S., Roccardi, A. and Piervittori, R. (2000) The temples of the archaeological area of Paestum (Italy): a case study on biodeterioration. In: Proceedings of 9th International Congress *Deterioration and Conservation of Stone* (V. Fassina, ed.), (Venice, 19-24 June 2000): 433-451. Elsevier, Amsterdam. [1]

Alunno Rossetti, V. and Laurenzi Tabasso, M. (1973) Distribuzione degli ossalati di calcio $CaC_2O_4.H_2O$ e $CaC_2O_4.2H_2O$ nelle alterazioni delle pietre di monumenti esposti all'aperto. In: *Problemi di Conservazione* (G. Urbani, ed.): 375-386. Editrice Compositori, Bologna. [3]

Alvarez, A., Argemì M. de Laorden, V., Domènech, X., Gerbal, J., Navarro, A., Prada, J.L., Pugés, M., Rocabayera, R. and Vilaseca, L. (1994) Physical, chemical and biological weathering detected in the romanic portal of the Sant Quirze de Pedret church (XIIc.). In: Proceedings of the 3rd International Symposium *The Conservation of Monuments in the Mediterranean Basin* (V. Fassina, H. Ott and F. Zezza, eds.): 365-369. Elsevier, Amsterdam. [2]

Amadori, L., Mecchi, A.M., Monte, M., Musco, S. and Salvatori, A. (1989) La conoscenza dei materiali e delle strutture per un progetto di restauro nel Parco Archeologico di Gabii. In: Atti del Convegno *Scienza e Beni Culturali, Il Cantiere della conoscenza, il Cantiere del Restauro* (G. Biscontin, M. Dal Col and S. Volpin, eds.): 295-308. Libreria Progetto Editrice, Padova. [1]

Amoros i Gurrera, M., Duprè i Raventos, X. Escon arré i Bou, V., Prada Perez, J.L., Rocabayera Vinas, R. and Valenciano Horta, A. (1994) Degradation forms and weathering mechanisms in the Berà Arch (Terragona, Spain). In: Proceedings of the 3rd International

Symposium *The Conservation of Monuments in the Mediterranean Basin* (V. Fassina, H. Ott & F. Zezza, eds): 673-679. Soprintendenza ai beni artistici e storici, Venezia. [2]

Anagnostidis., K., Gehrmann, C.K., Gross, M., Krumbein, W.E., Lisi, S., Pantazidou, A., Urzì, C. and Zagari, M. (1991) Biodeterioration of marbles of the Parthenon and Propylacea, Acropolis, Athens. Associated organisms, decay and treatment suggestions. In: Proceedings of the 2nd International Symposium *The Conservation of Monuments in the Mediterranean Basin* (J. Decrouez, J. Chamay and F. Zezza, eds.): 305-325. Genève. [1, 4]

Andersson, T. (1985) The investigation and conservation of middle age stone sculpture of the island of Gotland, Sweden. In: Proceedings of the 5th International Congress *Deterioration and Conservation of Stone*: 1035-1043. Presses Polytechniques Romandes, Lausanne. [1, 4]

Andreoli, C., Caniglia, G., Salvadori, O. and Scapin, C. (1982) Studi preliminari su due forme licheniche sviluppantesi sulle colonne della Basilica di Santa Maria Assunta a Torcello (VE). *Giornale Botanico Italiano* 116: 166-167. [1, 4]

Anonymous (1996) Lichen growth rates used to determine building age. *ASPP Newsletter* 23: 12. [1]

Anom, I.G.N. and Samidi, S. (1993) Conservation of stone monuments in Indonesia. In: Proceedings of the International RILEM/UNESCO Congress *Conservation of Stone and Other Materials* (M.J. Thiel, ed.): 853-859. Spon, London. [1, 4]

Appolonia, L., Grillini, G.C. and Pinna, D. (1996) Origin of oxalate films on stone monuments: I. Nature of films on unworked stone. In: Proceedings of 2nd International Symposium *The oxalate films in the conservation of works of art* (M. Realini, L. Toniolo, eds.), (Milan, 25-27 March 1996): 257-268. Centro del C.N.R. "Gino Bozza", Milano, Italy. [3]

Aptroot, A. and Bran, A.M. (1996) Lichenen van de voorjaarsexcursie 1995 naar Bramsche, Niedersachen. *Buxbaumiella* 39: 41-46. [1]

Aptroot, A. and James, P.W. (2002) Monitoring lichens on monuments. In: *Monitoring with Lichens – Monitoring Lichens* (P.L. Nimis, C. Scheidegger and P.A. Wolseley, eds.): 239-253. Kluwer Academic Publishers, Printed in the Netherlands. [5]

Aptroot, A. and Spier, L. (1995) Lichenen op de portogees Israelietische Begraafplats te Ouderkerk a/d Amstel. *Buxbaumiella* 38: 53-54. [1]

Aptroot, A., Bakke, P., Van den Boo, C., Her, V. and Spie, L. (1995) Lichenen op hunebedden. *Buxbaumiella* 38: 16-24. [1]

Ariño, X. and Saiz-Jimenez, C. (1994) Granite weathering by the lichen *Rhizocarpon geographicum*. In: *Granite Weathering and Conservation* (E. Bell and T.P. Cooper, eds.): 33-35. Director of Buildings' Office, Trinity College, Dublin. [2, 1]

Ariño, X. and Saiz-Jimenez, C. (1996) Colonization and deterioration processes in Roman mortars by cyanobacteria, algae and lichens. *Aerobiologia* 12: 9-18. [2]

Ariño, X. and Saiz-Jimenez, C. (1996) Biological diversity and cultural heritage. *Aerobiologia* 12: 279-282. [1]

Ariño, X. and Saiz-Jimenez, C. (1996) Factors affecting the colonization and distribution of cyanobacteria, algae and lichens in ancient mortars. In: Proceedings of the International Congress *Deterioration and Conservation of Stone* (J. Riederer, ed.): 725-730. Möller, Berlin. [1]

Ariño, X. and Saiz-Jimenez, C. (1996) Lichen deterioration of consolidants used in the conservation of stone monuments. *The Lichenologist* 28: 391-394. [4]

14. Lichens and Monuments

Ariño, X. and Saiz-Jimenez, C. (1997) Deterioration of the Elephant Tomb (Necropolis of Carmona, Seville, Spain). *International Biodeterioration and Biodegradation* 40: 233-239. [1, 2]

Ariño, X., Gomez-Bolea, A., Hladun, N. and Saiz-Jimenez, C. (1995) La colonizacion del Foro Romano de Baelo Claudia (Cadiz): aplicacion del tratamiento de imagenes al estudio de las comunidades liquenicas. In: Proceedings of the XI Simposio Nacional de Botanica Criptogamica: 221-223. Santiago de Compostela. [1]

Ariño, X., Ortega-Calvo, J.J., Gomez-Bolea, A. and Saiz-Jimenez, C. (1995) Lichen colonization of the Roman pavement at Baelo Claudia (Cadiz, Spain): biodeterioration vs. bioprotection. *The Science of the Total Environment* 167: 353-363. [1]

Ariño, X., Gomez-Bolea, A., Hladun, N. and Saiz-Jimenez, C. (1996) Lichen colonization of mortars in archaeological areas. Application of imaging analysis. In: Abstracts of the International Symposium IAL 2 *Progress and Problems in Lichenology in the Nineties* (R. Türk, ed.): 143. Universität Salzburg, Salzburg. [1]

Ariño, X., Gomez-Bolea, A., Hladun, N. and Saiz-Jimenez, C. (1996) *Roccella phycopsis* Ach.: colonization and weathering processes on granite. In: Proceedings of the EC Environmental Workshop *Degradation and Conservation of Granitic Rocks in Monuments* (M.A. Vicente, J. Delgado-Rodrigues and J. Acevedo, eds.): 399-404. Protection and Conservation of the European Cultural Heritage, Research Report No. 5. [1, 2]

Ariño, X., Gomez-Bolea, A. and Saiz-Jimenez, C. (1997) Lichens on ancient mortars. *International Biodeterioration and Biodegradation* 40: 217-224. [1]

Armstrong, R.A. (1976) The influence of the frequency of wetting and drying on the radial growth of three saxicolous lichens in the field. *New Phytologist* 77: 719-724. [1]

Armstrong, R.A. (1977) The response of lichen growth to transplantation to rock surfaces of different aspect. *New Phytologist* 78: 473-478. [1]

Armstrong, R.A. (1978) The colonization of a slate rock surface by a lichen. *New Phytologist* 81: 85-88. [1]

Armstrong, R.A. (1981) Field experiments on the dispersal, establishment and colonization of lichens on a late rock surface. *Environment and Experimental Botany* 21: 115-120. [1]

Armstrong, R.A. (1990) The influence of calcium and magnesium on the growth of the lichens *Parmelia saxatilis* and *Xanthoria parietina* on slate substratum. *Environmental and Experimental Botany* 30: 51-57. [1]

Ascaso, C. (1985) Structural aspects of lichens invading their substrata. In: *Surface Physiology of Lichens* (C. Vicente, D.H. Brown and M.E. Legaz, eds.): 87-113. Universidad Complutense, Madrid. [2]

Ascaso, C. (2000) Lichens on rock substrates: observation of the biomineralization processes. In: *New Aspects in Cryptogamic Research. Contributions in Honour of Ludger Kappen* (B. Schroeter, M. Schlensog and T.G.A. Green, eds.): 127-135. J. Cramer, Berlin-Stuttgart. [2, 1, 7]

Ascaso, C. and Galvan, J. (1976) Studies on the pedogenetic action of lichen acids. *Pedobiologia* 16: 321-331. [2]

Ascaso, C. and Ollacarizqueta, M.A. (1991) Structural relationship between lichen and carved stonework of Silos Monastery, Burgos, Spain. *International Biodeterioration* 27: 337 - 349. [2]

Ascaso, C. and Ollacarizqueta, M.A. (1992) Biodeterioro por liquenes. In: Book of Preprints of Simposio Internacional sobre *Biodeterioro*: 75. Madrid. [1]

Ascaso, C. and Wierzchos, J. (1994) Nuevas aplicaciones de las técnicas submicroscopicas en el estudio del biodeterioro producido por los talos liquénicos. *Microbiologia SEM* 10: 103-110. [2]

Ascaso, C. and Wierzchos, J. (1994) Structural aspects of the lichen-rock interface using back-scattered electron imaging. *Botanica Acta* 107: 251-256. [2]

Ascaso, C. and Wierzchos, J. (1995) Estudio de la interfase talo liquénico-substrato litico con microscopia electronica de barrido en modo de electrones retrodispersados. In: *Flechten Follman Contributions to Lichenology in Honour of Gerhard Follmann* (F.J.A. Daniels, M. Schul and J. Peine, eds.): 43-54. Botanical Institute, University of Cologne, Cologne. [2]

Ascaso, C. and Wierzchos, J. (1995) Study of the biodeterioration zone between the lichen thallus and the substrate. *Cryptogamic Botany* 5: 270-281. [2]

Ascaso, C. and Wierzchos, J. (1996) Study of the weathering processes of granitic micaceous minerals by lichen activity. In: Proceedings of the EC Workshop *Degradation and Conservation of Granitic Rocks in Monument* (M.A. Vicente, J. Delgado-Rodrigues and J. Acevedo, eds.): 411-416. Protection and Conservation of the European Cultural Heritage, Research Report No. 5. [2]

Ascaso, C., Galvan, J. and Ortega, C. (1976) The pedogenetic action of *Parmelia conspersa*, *Rhizocarpon geographicum* and *Umbilicaria pustulata*. *The Lichenologist* 8: 151-171. [2, 1]

Ascaso, C., Galvan, J. and Rodriguez-Pascal, C. (1982) The weathering of calcareous rocks by lichens. *Pedobiologia* 24: 219-229. [2]

Ascaso, C., Sancho, L.G. and Rodriguez-Pascal, C. (1990) The weathering action of saxicolous lichens in maritime Antarctica. *Polar Biology* 11: 33-39. [2]

Ascaso, C., Vizcayno, C. and Garcìa-Gonzalez, T. (1992) Biodeterioration produced by *Lecanora albescens* (Hoffm.) Branth & Rostr. on the abaci of the Silos Monastery. In: *Alteracion de granitos y rocas afines* (A. Vicente, E. Molina and V. Rives, eds.): 181-186. CSIC, Madrid, Spain. [2]

Ascaso, C., Wierzchos, J. and De Los Rios, A. (1995) Cytological investigations of lithobiontic microorganisms in granitic rocks. *Botanica Acta* 108: 474-481. [2]

Ascaso, C., De Los Rios, A. and Wierzchos, J. (1996) *In situ* investigations of lichen invading rocks at cellular and enzymatic level. In: Proceedings of the International Symposium IAL 2 on *Progress and Problems in Lichenology in the Nineties* (R. Türk, ed.): 12. Universität Salzburg, Salzburg. [2]

Ascaso, C., Wierzchos, J. and Castello, R. (1998) Study of the biogenic weathering of calcareous litharenite stones caused by lichen and endolithic microorganisms. *International Biodeterioration and Biodegradation* 42: 29-38. [2]

Ascaso, C., Wierzchos J., De Los Rios, A. (1998) *In situ* cellular and enzymatic investigations of saxicolous lichens using correlative microscopial and microanalytical techniques. *Symbiosis* 24: 221-234. [2]

Ascaso, C., Wierzchos, J., Delgado Rodrigues, J., Aires-Barros, L., Henriques, F.M.A. and Charola, A.E. (1998) Endolithic microorganisms in the biodeterioration of the Tower of Belem. *Internationale Zeitschrift für Bauinstandsetzen* 4: 627-640. [2, 1]

Ascaso, C., Wierzchos, J., Souza-Egipsy, V. and Delgado Rodrigues, J. (1999) Application of electron microscopy in the study of monumental stones. In: Book of Abstracts of the International Conference on Microbiology and Conservation (ICM '99), *Of microbes and art: The role of microbial communities in the degradation and protection of Cultural Heritage*: 113-117. Tribuna di Galileo, Museo della Specola, Florence. [2]

Ascaso, C., Wierzchos, J., Souza-Egipsy, V., de los Rios, A. and Delgado Rodrigues, J. (2002) *In situ* evaluation on the biodeteriorating action of microorganisms and the effects of biocides on carbonate rock of the Jerominos Monastery (Lisbon). *International Biodeterioration and Biodegradation* 49: 1-12. [2, 4]

Ashurst, J. (1977) *Control of Organic Growth 1 - Algal Slimes, Lichens, Mosses.* DAMHB Technical Note. [1]

Aslan, A., Demircioglu, N. and Karagöz, Y. (2002) Likenlerin tas yüzeyler ve eski anitlar üzerine etkileri. In: 1. Ulusal Cevre Sorunlari Sempozyumu, (Ekim 16-18 2002): 31-38. Atakürk Üniversitesi, Erzurum, Turkey. [1, 2]

Asta, J. and Lachet, B. (1977) Analyses de rélation entre la teneur en carbonate de calcium des substrates et divers groupements phyto sociologiques de lichens saxicoles. *Oecologia Plantarum* 13: 193-206. [1]

Atzeni, C., Cabiddu, M.G., Massidda, L., Sanna, U. and Sistu, G. (1994) Degradation and conservation of sandstone and pyroclastic rocks used in the prehistoric complex Genna Maria (Villanovaforru, Sardegna, Italy). In: Proceedings of the 3rd International Symposium *The Conservation of Monuments in the Mediterranean Basin* (V. Fassina V., H. Ott and F. Zezza, eds.), (Venice, 22-25 June 1994): 533-539. Soprintendenza ai beni artistici e storici, Venezia. [1, 4]

Awasthi, D.D. (1989) Lichens and monuments. In: Proceedings of International Conference on *Biodeterioration of Cultural Property* (O.P. Agrawal and S. Dhawan, eds.), (Lucknow, 20-25 February 1989): 175-180. Macmillan, Delhi. [1, 4]

Bachmann, E. (1890) Die Beziehungen der Kalkflechten zu ihrem substrat. *Berichte der deutschen botanischen Gesellschaft* 8: 141 - 145. [2]

Bachmann, E. (1892) Der Thallus der Kalkflechten likens. *Berichte der deutschen botanischen Gesellschaft* 10: 30-37. [2]

Bachmann, E. (1904) Die Beziehungen der Kalkflechten zu ihrem substrat. *Berichte der deutschen botanischen Gesellschaft* 22: 101-104. [2]

Bachmann, E. (1911) Die Beziehungen der Kalkflechten zu ihrem substrat. *Berichte der deutschen botanischen. Gesellschaft* 29: 261-273. [2]

Bachmann, E. (1917) Die Beziehungen der Kieselflechten zu ihrer Unterlage. III Bergkristall und Flint. *Berichte der deutschen botanischen Gesellschafft* 35: 464 - 476. [2]

Bajpai, P. K., Bajpai, R.P. and Shinha, G.P. (1992) A fast technique for investigating the effect of biocides on lichens. *International Biodeterioration and Biodegradation* 30: 1-8. [4]

Bajpai, P.K., Bajpai, R.P. and Sinha, G.P. (1993) Study of the effect of biocides on lichens using biophoton emission. In: Proceedings of the 2nd International Conference *Biodeterioration of Cultural Property*: 619-623 (Yokohama, 5-8 October 1992). Tokyo. [4]

Banfield, J.F., Barker, W.W., Welch, S.A. and Taunton, A. (1999) Biological impact on mineral dissolution: application of the lichen model to understanding mineral weathering in the rhizosphere. *Proceedings of the National Academy of Science* 96: 3404-3411. [2]

Barker, W.W. and Banfield, J.F. (1996) Biologically versus inorganically mediated weathering reactions: relationships between minerals and extracellular microbial polymers in lithobiontic communities. *Chemical Geology* 132: 55-69. [2]

Barker, W.W. and Banfield, J.F. (1998) Zones of chemical and physical interaction at interfaces between microbial communities and minerals: a model. *Geomicrobiology* 15: 223-244. [2]

Barker, W.W., Welch, S.A. and Banfield, J.F. (1998) Biogeochemical weathering of silicate minerals. In: *Geomicrobiology: Interactions Between Microbes and Minerals* (J.F. Banfield and K.H. Nealson eds.). Reviews in Mineralogy and Geochemistry 35: 391-428. Mineralogical Society of America, Washington DC. [2]

Baroni, E. (1893) Notizie e osservazioni sui rapporti dei licheni calcicoli col loro sostrato. *Bullettino della Società Botanica Italiana* 1893: 136-140. [1]

Barquín Sainz de la Maza, P. and Terrón, A.A. (1997) Colonización liquénica de la Catedral de León. *PH Boletín* 20: 52-59. [1]

Barquín Sainz de la Maza, P. and Terrón, A.A. (1997) Lichen communities in the cathedral of Leon. *Aerobiologia* 13: 191-197. [1]

Barquín Sainz de la Maza, P. and Terrón, A.A. (1999) Análisis ecológico de las comunidades liquénicas de la Catedral de León (España). *Cryptogamie, Mycologie* 20: 41-47. [1]

Barquín Sainz de la Maza, P., Terrón, A.A., Fernández-Salegni, A.B. and Marcos, R. (1999) Procesos de biodeterioro liquénico en rocas monumentales: granitos, dolomías y gneis. In: Proceedings of the *XIII Simposio de Botánica Criptogámica*: 131. Universidad Complutense, Madrid. [2]

Barr, A.R.M. (1977) Comparative studies on the inhibition of lichens and algal growth on asbestos paving slabs. In: Working Group of International Biodegradation Research Grouping Constructional Materials. [1]

Bartoli, A. (1990) I licheni della Peschiera dei Tritoni nell'Orto Botanico di Roma, Villa Corsini. *Giornale Botanico Italiano* 125: 87. [1]

Bartoli, A. (1992) Flora e vegetazione lichenica del Porto di Traiano. In: *Il parco archeologico-naturalistico del porto di Traiano. Ministero Beni Culturali Ambientali, Soprintendenza Archeologica di Ostia*: 167-171. Edizioni Gangemi, Roma. [1]

Bartoli, A. (1997) I licheni del Colosseo. In: Abstracts del Convegno annuale della Società Lichenologica Italiana *Licheni e Ambiente* (D. Ottonello, ed.), (Palermo, 9-12 Dicembre 1995). *Notiziario della Società Lichenologica Italiana* 10: 67. [1]

Bartoli, A. (1997) I licheni del Colosseo. *Allionia* 35: 59-67. [1]

Bartoli, A., Massari, G. and Ravera, S. (1997) The lichens of Munazio Planco's Mausoleum (Gaeta). In: Proceedings of the 3rd IAL Symposium *Progress and Problems in Lichenology in the Nineties* (R. Türk and R. Zorer, eds.). *Bibliotheca Lichenologica* 68: 145. [1]

Bartoli, A., Massari, G. and Ravera, S. (1998) The lichens of the Mausoleum of *Manatius Plancu*s (Gaeta). *Sauteria* 9: 53-60. [1]

Basile, G., Chilosi, M.G. and Martellotti, G. (1987) La facciata della cattedrale di Termoli: un esempio di manutenzione programmata. *Bollettino d'Arte* 41: 283-304. [4]

Bates, J.W. (1978) The influence of metal availability on the bryophyte and macrolichen vegetation of four rock types on Skye and Rhum. *Journal of Ecology* 66: 457-481. [1]

Bech-Andersen, J. (1984) Biodeteriotion of natural and artificial stone caused by algae, lichens, mosses and higher plants. In: Proceedings of the 6th International Biodeterioration Symposium *Biodeterioration 6* (S. Barry, D.R. Houghton, G.C. Llewellyn and C.E. O'Rear, eds.): 126-131. C.A.B. International, Slough. [1, 2]

Bech-Andersen, J. (1987) Oxalic acid production by lichens causing deterioration of natural and artificial stones. In: Proceedings of the Biodeterioration Society Meeting on *Biodeterioration of Constructional Materials* (L.H.G. Morthon, ed.): 9-13. Delft. [3]

Bech-Andersen, J. and Christensen, P. (1983) Studies on lichen growth and deterioration of rocks and buildings materials using optical methods. In: *Biodeterioration 5* (T.A. Oxeley and S. Barry, eds.): 568-572. Wiley, Chichester. [2]

Bell, E., Dowding, P. and Cooper, T. P. (1991) The effect of a biocide treatment and a silicone treatment on the weathering of limestone. *Environmental Technology* 13: 687-693. [4]

Bernardini, C. (1993) Biocidi e prevenzione microbiologica: alcune osservazioni in cantiere. *Kermes* 16: 12-19. [4]

Bettini, C. (1984) Gli interventi di restauro sulla decorazione della Villa di Papa Giulio II. *Bollettino d'Arte* 27: 127-131. [4]

Bettini, C. and Villa, A. (1981) Description of a method for cleaning tombstones. In: Proceedings of 2nd International Symposium on *The Conservation of Stone* (R. Rossi Manaresi, ed.): 523-534. Centro per la Conservazione delle Sculture all'Aperto, Bologna. [4]

Bewley, J.J.D. (1979) Physiological aspects of desiccation tolerance. *Annual Review of Plant Physiology* 30: 195-238. [1]

Bewley, J.J.D. and Krochko, J.E. (1982) Desiccation tolerance. In: *Encyclopedia of Plant Physiology* 12 B: 325-378. Springer, Berlin. [1]

Biscontin, G. (1983) La conservazione dei materiali lapidei: trattamenti conservativi. In: Atti del Convegno Internazionale *La Pietra: interventi, conservazione, restauro* (A. Cassiano, O. Curti and G. Delli Ponti, eds.), (Lecce, 6-8 Novembre 1981): 111-120. Congedo Editore, Lecce. [4]

Bjelland, T. and Ekman, S. (2000) On the occurrence of endolithic hyphae beneath *Ophioparma ventosa*. In: Book of Abstracts, The Fourth IAL Symposium *Progress and Problems in Lichenology at the Turn of the Millenium* (Barcelona, 3-8 September 2000): 38. Universitat of Barcelona, Barcelona. [2]

Bjelland, T., Sæbø, L. and Thorseth, I.H. (2002) The occurrence of biomineralization products in four lichen species growing on sandstone in western Norway. *The Lichenologist*, 34: 429-440. [3]

Bjerke, J.W. (2000) Bevaring og tilrettelegging av helleristninger sett fra et biologisk stasted. *Riksantikvarens Rapporter* 29: 75-86. [1, 4]

Blazquez, F., Calvet, F. and Vendrell, M. (1995) Lichenic alteration and mineralization in calcareous monuments of north-eastern Spain. *Geomicrobiology Journal* 13: 223-247. [2]

Bolivar, F., Garcia-Rowe, J., Manzano, E. and Saiz-Jimenes, C. (1991) Estudio de la comunidad de liquenes de la fuente del claustro principal del hospital de San Juan de Dios, Granada. In: Preprints of the Simposio International sobre *Biodeterioro*: 73. Madrid. [1]

Bolle, E. (1995) El rol de los liquenes en la conservacion de sitios arqueologicos. Aministracion y conservacion de sitios de arte rupestre. *Contribuciones al estudio del arte rupestre sudamericano* 4: 22-28. Sociedad de Investigacion del Arte Rupestre de Bolivia, La Paz. [1]

Braconnot, H. (1825) De la présence de l'oxalate de chaux dans le régne mineral; existance du meme sel en quantité enorme dans les plantes de la famille des lichens, et moyen advantageux d'en extraire l'acide oxalique. *Annales de Chimie et de Physique* 28: 318-322. [3]

Brady, P.V., Dorn, R. I., Brazel, A.J., Clark, J., Moore, R.B. and Glidewell, T. (1999) Direct measurement of the combined effects of lichen, rainfall, and temperature on silicate weathering. *Geochimica and Cosmochimica Acta* 63: 3293-3300. [2, 1]

Bratt C. (1997) Lichens of Cabrillo National Monument, Point Loma, San Diego. *Bulletin of the California Lichen Society* 4: 1-2. [1]

Brightman, F. H. and Laundon, J. R. (1993) Lichens in churchyards. *Newsletter of the Association for Gravestone Studies* 17: 8-9. [1]

Brightman, F.H. and Seaward, M.R.D. (1977) Lichens of man-made substrates. In: *Lichen Ecology* (M.R.D. Seaward, ed.): 253-293. Academic Press, London. [1]

Brown, D.H. (1991) Lichen mineral studies - currently clarified or confused? *Symbiosis* 11: 207-223. [2]

Brown, D.H. and Beckett, R.P. (1984) Minerals and lichens: acquisition, localization and effect. In: Proceedings of Congress *Surface Physiology of Lichens* (C. Vicente, D.H. Brown and M.E. Legaz, eds.): 127-149. Universitad Complutense, Madrid. [2]

Brown, S.K. and Martin, A.K. (1993) Chemical control and encapsulation of organic growths on weathered asbestos-cement roofing. In: Proceedings of the International RILEM/UNESCO Congress *Conservation of Stone and Other Materials* (M.J. Thiel, ed.): 758-767. Spon, London. [4]

Buchet, G. (1890) Les lichens attaquent le verre et, dans les vitraux, semblement préférer certaines couleurs. *Céreal Séance Société Biologique* 2: 13. [1]

Building Research Establishment Digest (1972) Control of lichens, moulds and similar growths. *BRE Digest* 139: 1-4. [4]

Cabrera Garrido, J.M. (1979) Causas de alteración y métodos de conservatión aplicables a los monumentos hechos con piedra. In: *Materiales de Construccion*: 174. C.S.I. e Istituto Torroja, Madrid. [1, 4]

Cadot-Leroux, L. (1996) Rural nitrogenous pollution and deterioration of granitic monuments: results of a statistical study. In: Proceedings of the International Congress *Deterioration and Conservation of Stone* (J. Riederer, ed.): 301-309. Möller, Berlin. [5, 2]

Camuffo, D. (1993) Reconstructing the climate and air pollution of Rome during the life of the Trajan column. *The Science of the Total Environment* 128: 205-226. [5]

Caneva, G. (1993) Ecological approach to the genesis of calcium oxalate patinas on stone monuments. *Aerobiologia* 9: 149-156. [3]

Caneva, G. and Roccardi, A. (1989) Harmful flora in the conservation of Roman Monuments. International Conference *Biodeterioration of Cultural Property* (O.P. Agrawal and S. Dhawan, eds.), (Lucknow, 20-25 February 1989): 212-218. Macmillan, Delhi. [1]

Caneva, G. and Salvadori, O. (1987) I pesticidi nel controllo del biodeterioramento dei monumenti: problemi tecnici e sanitari. In: Atti del Convegno *Inquinamento in ambienti di vita e di lavoro: esperienze e linee di intervento* (F. Candura and A. Messineo, eds.): 81-91. Edizioni Acta Medica, Roma. [4]

Caneva, G. and Salvadori, O. (1988) Biodeterioration of stone. In: *The Deterioration and Conservation of Stone* (L. Lazzarini and R. Pieper eds.). *Studies and Documents on the Cultural Heritage* 16: 182-234. UNESCO, Venezia. [1, 2]

Caneva, G. and Salvadori, O. (1988) I pesticidi nel diserbo dei monumenti. *Verde Ambiente* 1: 79-84. [4]

Caneva, G. and Salvadori, O. (1989) Sistematica e sinstematica delle comunità vegetali nella pianificazione degli interventi di restauro. In: Atti del Convegno *Scienza e Beni Culturali, Il Cantiere della Conoscenza, il Cantiere del Restauro* (G. Biscontin, M. Dal Col and S. Volpin, eds.): 235-335. Libreria Progetto Editore, Padova. [1]

Caneva, G., Roccardi, A., Marenzi, A. and Napoleone, I. (1985) Proposal for a data base on biodeterioration of stone artworks. In: Proceedings of the 5th International Congress on *Deterioration and Conservation of Stone*, vol. 2 (G. Félix, ed.): 587-596. Presses Polytechniques Romandes, Lausanne. [1]

Caneva, G., Roccardi, A., Marenzi, A. and Napoleone, I. (1989) Correlation analysis in the biodeterioration of stone artworks. *International Biodeterioration* 25: 161-167. [1, 2]

Caneva G., Nugari M.P. and Salvadori O. (1991) *Biology in the Conservation of Works of Art.* ICCROM, Roma. [1, 4]
Caneva, G., Gori, E. and Danin, A. (1992) Incident rainfall in Rome and its relation to biodeterioration of buildings. *Atmosperic Environment* 26B, 255-259. [5]
Caneva, G., Nugari, M. P., Ricci, S. and Salvadori, O. (1992) Pitting of marble in Roman monuments and the related microflora. In: Proceedings of the International Congress on *Deterioration and Conservation of Stone* (J. Delgado Rodrigues, F. Henriques and F. Telmo Jeremias, eds.): 521-530. LNEC, Lisbon. [2]
Caneva, G., Nugari, M.P. and Salvadori, O. (1994) *La Biologia nel Restauro.* Nardini Editore, Firenze. [1, 2, 3, 4]
Caneva, G., Danin, A., Ricci, S. and Conti, C. (1994) The pitting of Trajan's column, Rome: an ecological model of its origin. *Contributi del Centro Linceo Interdisciplinare "Beniamino Segre"* 88: 77-102. Accademia Nazionale dei Lincei, Roma. [2]
Caneva, G., Gori, E. and Montefinale, T. (1995) Biodeterioration of monuments in relation to climatic changes in Rome between 19th-20th centuries. *The Science of the Total Environment* 167: 205-214. [2, 1, 5]
Caneva, G., Nugari, M.P., Pinna, D. and Salvadori, O. (1996) *Il Controllo del Degrado Biologico. I biocidi nel restauro dei materiali lapidei.* Nardini Editore, Firenze. [4]
Caneva, G., Piervittori, R. and Roccardi, A. (1998) Ambienti esterni: problematiche specifiche. In: *Aerobiologia e Beni culturali. Metodologie e tecniche di misura* (P. Mandrioli and G. Caneva, eds.): 247-251. Nardini Editore, Firenze. [6]
Caniglia, G., Cornale, G. and Salvadori, O. (1993) The lichens on the statues in the orangery of "Villa Pisani" at Stra (Venezia). *Giornale Botanico Italiano* 127: 620. [1]
Cao, J. and Wang, F. (1998) Reform of carbonate rock subsurface by crustose lichens and its environmental significance. *Acta Geologica Sinica* 72: 94-99. [2, 3]
Cao, J. and Wang, F. (1998) Relationships of biokarst microforms of algae and lichens and the terrestrial environment. *Geological Review* 44: 656-661. [1]
Capponi, G. and Meucci, C. (1987) Il restauro del paramento lapideo della facciata della Chiesa di S. Croce a Lecce. *Bollettino d'Arte* 41: 263-282. [4]
Caramiello, R., Piervittori, R., Papa, G. and Fossa, V. (1991) Estrazione di pollini e spore da talli lichenici. Giornale Botanico Italiano, 125 (3): 331. [6]
Caramiello, R., Siniscalco, C. and Piervittori, R. (1991) The relationship between vegetation and pollen deposition in soil and in biological traps. *Grana* 30: 291-300. [6]
Carballal, R., Paz-Bermudez, G., Sánchez-Biezma, M.J. and Prieto, B. (2001) Lichen colonization of coastal churches in Galicia: biodeterioration implaction. *International Biodeterioration and Biodegradation* 47: 157-163. [1]
Cardilli Aloisi, L. (1985) La restauration de la Porta del Popolo a Rome. In: Proceedings of the 5th International Congress on *Deterioration and Conservation of Stone* (G. Félix, ed.): 1083-1091. Presses Polytechniques Romandes, Lausanne. [4]
Carrol, D. (1970) *Rock Weathering.* Plenum Press, New York. [2]
Carter, N. (2000) Small ecological project progress report: control on lichen species distribution, community structure and species richness on limestone heritage buildings in Oxford and the Cotswolds. *British Lichen Society Bulletin* 86: 33-35. [1]
Casares-Porcel, M. and Gutierrez-Carretero, L. (1993) Synthesis of the thermo- and mesomediterranean gypsicolous lichen vegetation in the Iberian Peninsula. *Cryptogamie, Bryologie et Lichénologie* 14: 361-388. [1]
Cavaletti, R., Strazzabosco, G., Manoli, N. and Toson, P. (1990) Relazione tecnica di restauro della statua n. 85 'Andrea Briosco'. In: Atti della Giornata di Studio, *Il Prato della Valle e*

le opere di pietra calcarea collocate all'aperto: 313-331. Libreria Progetto Editore, Padova. [4]
Cengia-Sambo, M. (1939) Licheni che intaccano i mosaici fiorentini. *Nuovo Giornale Botanico Italiano* 46: 141-145. [1]
Chatterjiee, S., Sinha, G.P., Upreti, D.K. and Singh, A. (1996) Preliminary observations on lichens growing over some indian monuments. *Flora and Fauna (India)* 2: 1- 4. [1]
Chen, J. and Blume, H.P. (1999) Biotic weathering of rocks by lichens in Antarctica. *Chinese Journal of Polar Science* 10: 25-32. [2]
Chen, J., Blume, H.P. and Beyer, L. (2000) Weathering of rocks induced by lichen colonization – a review. *Catena* 39: 121-146. [2, 3, 7]
Cherido, M.M. (1990) Relazione tecnica di progetto per il restauro della statua n. 41 "Cesare Piovene". In: Atti Giornata di Studio, *Il Prato della Valle e le opere di pietra calcarea collocate all'aperto*: 355-359. Libreria Progetto Editore, Padova. [4]
Childers, B.B. (1994) Long-term lichen-removal experiments and petroglyph conservation: Fremont County, Wyoming, Ranch Petroglyph Site. *Rock Art Research* 11: 101-112. [4]
Chisholm, J.E., Jones, G.C. and Purvis, O.W. (1987) Hydrated copper oxalate, moolooite, in lichens. *Mineralogical Magazine* 51: 715-718. [3]
Chu, F.J., Seaward, M.R.D. and Edwards, H.G.M. (1998) Application of FT-Raman spectroscopy to an ecological study of the lichen *Xanthoparmelia scabrosa* in the supralittoral zone, Hong Kong. *Spectrochimica Acta* 54A: 967-982. [2]
Ciarallo, A., Festa, L., Piccioli, C. and Raniello, M. (1985) Microflora action in the decay of stone monuments. In: Proceedings of the 5th International Congress on *Deterioration and Conservation of Stone*, vol. 2 (G. Félix, ed.): 607-616. Press Polytechniques Romandes, Lausanne. [1, 2]
Ciccarone, C. and Pinna, D. (1993) Calcium oxalate films on stone monuments: microbiological observations. *Aerobiologia* 9: 33-37. [3]
Cipriani, C. and Franchi, L. (1958) Sulla presenza di whewellite fra le croste di alterazione dei monumenti romani. *Bollettino del Servizio Geologico* 79: 555-564. [3]
Clarke, J. (1976) A lichen control experiment at an aboriginal rock engraving site, Bolgart, Western Australia. *ICCM Bulletin* 2: 15-17. [4]
Clarke, J. (1978) Conservation and restoration of painting and engraving sites in Western Australia. In: *Conservation of Rock Art* (C. Pearson, ed.): 89-94. ICCM, Sydney. [4]
Cooks, J. and Otto, E. (1990) The weathering effects of the lichen *Lecidea* aff. *sarcogynoides* (Koerb) on Magaliesberg quartzite. *Earth Surface Processes Landforms* 15: 491-500. [2]
Copper, R. and Rudolph, E.D. (1953) Role of lichens in soil formation and plant succession. *Ecology* 34: 805 - 807. [2]
Crespo, A. and Bueno, A.G. (1982) Flora y vegetacion liquénicas de la Casa del Campo de Madrid (Espana). *Lazaroa* 4: 327-356. [1]
Curri, S. (1986) Processi litoclastici di origine biologica: diagnosi e trattamenti. In: *Tecniche della Conservazione* (A. Bellini, ed.): 144-206. Franco Angeli, Milano. [2]
Daniels, F.J.A. and Harkema, M. (1992) Epilithic lichen vegetation on man-made, calcareous substrates in The Netherlands. *Phytocoenologia* 21: 209-235. [1]
Danin, A. (1989) Remnants of biogenic weathering as a tool for studying paleoclimates. *Braun-Blanquetia* 3: 257-262. [2]
Danin, A. (1993) Biogenic weathering of marble monuments in Didim, Turkey, and in Trajan's column, Rome. *Water Science and Technology* 27: 557-563. [2]
Danin, A. (1993) Pitting of calcareous rocks by organisms under terrestrial conditions. *Israel Journal of Earth Science* 41: 201-207. [2]

14. Lichens and Monuments

Danin, A. and Caneva, G. (1990) Deterioration of limestone walls in Jerusalem and marble monuments in Rome caused by cyanobacteria and cyanophilous lichens. *International Biodeterioration* 26: 397-417. [2]

Danin, A. and Garty, J. (1983) Distribution of cyanobacteria and lichens on hillsides of the Negev Highlands and their impact on biogenic weathering. *Zeitschrift für Geomorphologie NF* 27: 423-444. [1]

Danin, A., Gerson, R., Marton, K. and Garty, J. (1982) Patterns of limestone and dolomite weathering by lichens and blue-green algae and their palaeoclimatic significance. *Palaeogeography, Palaeoclimatology, Palaeoecology* 37: 221-233. [2, 1]

Danin, A., Gerson, R. and Garty, J. (1983) Weathering patterns on hard limestone and dolomite by endolithic lichens and cyanobacteria: supporting evidence for eolian contribution to Terra Rossa soil. *Soil Science* 136: 213-217. [2, 1]

De Henau, P. (1982) Altération des marbres en Belgique et leur traitement. *Annales des Mines de Belgique* 9: 857-860. [2, 4]

De Los Rios, A., Ascaso, C. and Wierzchos, J. (1999) Study of lichens with different state of hydration by the combination of low temperature scanning electron and confocal laser scanning microscopies. *International Microbiology* 2: 251-257. [1]

Degelius, G. (1962) Uber Verwitterung von Kalk und Dolomitgestein durch Algen und Flechten. Eine Ubersicht. In: *Chemie im Dienst der Archaeologie Bautechmik Denkmalpflege* (J.A.. Hedvall, ed.). Hakam Ohlssons, Lund. [1]

Del Monte, M. (1989) I monumenti in pietra e i licheni. *Rassegna dei Beni Culturali* 3: 12-17. [1]

Del Monte, M. (1990) Microbioerosions and biodeposits on stone monuments: pitting and calcium oxalate patinas. In: Proceedings of the Advanced Workshop, *Analytical Methodologies for the Investigation of Damaged Stones* (F. Veniale, U. Zezza, eds.), 27 pp. Pavia, Italy. [2, 3]

Del Monte, M. (1991) Trajan's column: lichens don't live here any more. *Endeavour* 15: 86-93. [1]

Del Monte, M. and Ferrari, A. (1989) Patine di biointerazione alla luce delle superfici marmoree. In: Atti del Convegno *Scienza e Beni Cukturali, Le pellicole ad ossalato: origine e significato nella conservazione delle opere d'arte* (V. Fassina, ed.): 171-182. Centro del C.N.R. "Gino Bozza", Milano. [3]

Del Monte, M. and Sabbioni, C. (1983) Weddellite on limestone in the Venice environment. *Environmental Science and Technology* 17: 518-522. [3]

Del Monte, M. and Sabbioni, C. (1986) Chemical and biological weathering of an historical building: Reggio Emilia cathedral. *The Science of the Total Environment* 50: 165-182. [2]

Del Monte, M. and Sabbioni, C. (1987) A study of the patina called "Scialbatura" on imperial Roman marbles. *Studies in Conservation* 32: 114-121. [3]

Del Monte, M., Sabbioni, C. and Zappia, G. (1987) The origin of calcium oxalates on historical buildings, monuments and natural outcrops. *The Science of the Total Environment* 67: 17-39. [3]

Del Monte, M., Rattazzi, A., Romão, P.M.S. and Rossi, P. (1996) The role of lichens in the weathering of granitic building. In: Proceedings of the EC Environmental Workshop on *Degradation and Conservation of Granitic Rocks in Monuments* (M.A. Vicente, J. Delgado-Rodrigues and J. Acevedo, eds.): 301-306. European Commission, Brussels. [2, 3]

Deruelle, S. (1988) Effets de la pollution atmosphérique sur la végétation lichénique des monuments historiques. *Studia Geobotanica* 8: 23 - 31. [5]

Deruelle, S., Lallemant, R. and Roux, C. (1979) La végétation lichénique de la basilique de Notre-Dame de l'Epine (Marne). *Documents Phytosociologiques* 4: 217–234. [1]

Di Benedetto, L. and Grillo, M. (1995) Contributo alla conoscenza dei biodeteriogeni rilevati nel complessso archeologico del Teatro greco-romano ed Anfiteatro romano di Catania. *Quaderni Botanica Ambientale Applicata* 6: 61-66. [1]

Di Francesco, C., Grillini, G.C., Pinna, D. and Tucci, A. (1989) Il restauro come occasione di studio e documentazione nel caso del campanile del Duomo di Ferrara. In: Atti del Convegno *Scienza e Beni Culturali, Il cantiere della conoscenza, il Cantiere del Restauro* (G. Biscontin, M. Dal Col and S. Volpin, eds.): 251-264. Libreria Progetto Editore, Padova. [4]

Dillon, P., Skeggs, S. and Goodey, C. (1992) Some investigations on habitats of lichens on sarsen stones at Fyfield Down, Wiltshire. *The Wiltshire Archaeological and Natural History Magazine* 85: 128-139. [1]

Dobson, F.S. (1997) *Lichens on Man-made Surfaces. Encouragement and Removal.* British Lichen Society, London. [1]

Dubernat, P.J. and Pezerat, H. (1974) Fautes d'empilement dans les oxalates dihydratés des métaux divalents de la séries magnésienne (Mg, Fe, Co, Ni, Zn, Mn). *Journal of Applied Crystallography* 7: 387-394. [3]

Earland-Bennett, P. (1999) Colchester walls. *British Lichen Sociey Bulletin* 85: 1-4. [1]

Easton, R.M. (1994) Lichens and rocks: a review. *Geoscience Canada* 21: 59-76. [1, 2, 7]

Edwards, H.G.M. (1998) Raman spectroscopy of fresco fragment substrates. *Asian Journal of Physics* 7: 383-389. [2]

Edwards, H.G.M. and Rull Perez, F. (1999) Lichen biodeterioration of the Convento de la Peregrina, Sahagún, Spain. *Biospectroscopy* 5: 47-52. [3, 2]

Edwards, H.G.M. and Russell, N.C. (1998) Vibrational spectroscopic study of iron(II) and iron(III) oxalates. *Journal of Molecular Structure* 443: 223-231. [3, 2]

Edwards, H.G.M. and Seaward, M.R.D. (1993) Raman microscopy of lichen-substratum interfaces. *Journal of the Hattori Botanical Laboratory* 74: 303-316. [3, 2]

Edwards, H.G.M. and Seaward, M.R.D. (1993) Raman spectroscopy and lichen biodeterioration. *Spectroscopy Europe* 5: 16-20. [3, 2]

Edwards, H.G.M. and Seaward, M.R.D. (1994) FT-Raman spectroscopic studies of lichen-substratum interfaces in biodeterioration studies. In: *Biodeterioration and Biodegradation 9* (A. Bousher, M. Chandra and R. Edyvean, eds.): 199-203. Institution of Chemical Engineers, Rugby. [3, 2]

Edwards, H.G.M., Farwell, D.W. and Seaward, M.R.D. (1991) Raman spectra of oxalates in lichen encrustations on Renaissance frescoes. *Spectrochimica Acta* 47A: 1531-1539. [3, 2]

Edwards, H.G.M., Farwell, D.W., Seaward, M.R.D. and Giacobini, C. (1991) Preliminary Raman microscopic analysis of a lichen encrustation involved in the biodeterioration of Renaissance frescoes in central Italy. *International Biodeterioration* 27: 1-9. [1, 3]

Edwards, H.G.M., Farwell, D.W., Jenkins, R. and Seaward, M.R.D. (1992) Vibrational Raman spectroscopic studies of calcium oxalate monohydrate and dihydrate in lichen encrustations on Renaissance frescoes. *Journal of Raman Spectroscopy* 23: 185-189. [3]

Edwards, H.G.M., Farwell, D.W., Lewis, I.R. and Seaward, M.R.D. (1993) FT-Raman spectroscopic studies of encrustations formed at the lichen-substratum interface. In: Proceedings of the 9th International Conference on *Fourier Transform Spectroscopy*, Calgary, Canada (J.E. Bertie and H. Wieser, eds.): 256-257. SPIE Publishing, Washington DC. [3]

Edwards, H.G.M., Farwell, D.W., Lewis, I.R., Seaward, M.R.D., Turner, P. and Whithley, A. (1993) FT-Raman microscopy and lichen biodeterioration. *Bruker Report* 139: 8-11. [3]

Edwards, H.G.M., Farwell, D.W. and Seaward, M.R.D. (1994) FT-Raman-spectroscopic studies of lichen encrustations on Renaissance frescoes. In: 4th International Conference on Non-destructive Testing of Works of Art (Berlin, 3-8 October 1994). *Deutsche Gesellschaft für Zerstörrungsfreie Prüfüng* 45: 743-752. [3, 2]

Edwards, H.G.M., Farwell, D.W., Russell, N.C. and Seaward, M.R.D. (1994) Fourier-transform Raman microscopic studies of colored lichen encrustations on biodeteriorated rock substrata. In: Proceedings of XIVth International Conference *Raman Spectroscopy* (N-T. Yu and X-Y. Li, eds.), (Hong Kong, August 1994): 856-857. Wiley, Chicester. [3]

Edwards, H.G.M., Farwell, D.W., Giacobini, C., Lewis, I.R. and Seaward, M.R.D. (1994) Encrustations of the lichen *Dirina massiliensis* forma *sorediata* on Renaissance frescoes: an FT-Raman spectroscopic study. In: Proceedings of XIVth International Conference *Raman Spectroscopy* (N-T. Yu and X-Y. Li, eds.), (Hong Kong, August 1994): 896-897. Wiley, Chichester. [3]

Edwards, H.G.M., Edwards, K.A.E., Farwell, D.W., Lewis, I.R. and Seaward, M.R.D. (1994) An approach to stone and fresco lichen biodeterioration through Fourier transform Raman microscopic investigation of thallus-substratum encrustations. *Journal of Raman Spectroscopy* 25: 99-103. [3]

Edwards, H.G.M., Russell, N.C., Seaward, M.R.D. and Slarke, D. (1995) Lichen biodeterioration under different microclimates: an FT Raman spectroscopic study. *Spectrochimica Acta* 51A: 2091-2100. [2]

Edwards, H.G.M., Russell, N.C. and Seaward, M.R.D. (1997) Calcium oxalate in lichen biodeterioration studied using FT-Raman spectroscopy. *Spectrochimica Acta* 53A: 99-105. [3]

Edwards, H.G.M., Farwell, D.W. and Seaward, M.R.D. (1997) Raman spectroscopy of *Dirina massiliensis* f. *sorediata* encrustations growing on diverse substrata. *The Lichenologist* 29: 83-90. [3]

Edwards, H.G.M., Gwyer, E.R. and Tait, J.K.F. (1997) Fourier transform Raman analysis of paint fragments from biodeteriorated Renaissance frescoes. *Journal of Raman Spectroscopy* 28: 677-684. [3]

Edwards, H.G.M., Drummond, L. and Russ, J. (1998) Fourier-transform Raman spectroscopic study of pigments in native American Indian rock art: Seminole Canyon. *Spectrochimica Acta* 54A: 1849-1856. [3]

Edwards, H.G.M., Farwell, D.W., Rull Perez, F. and Villar, S.J. (1999) Spanish mediaeval frescoes at Basconcillos del Tozo: a Fourier transform Raman spectroscopic study. *Journal of Raman Spectroscopy* 30: 307-311. [3, 2]

Edwards, H.G.M., Holder, J.M. and Wynn-Williams, D.D. (1999) Comparative FT-Raman spectroscopy of *Xanthoria* lichen-substratum systems from temperate and antarctic habitats. *Soil Biology and Biochemistry* 30: 1947-1953. [2]

Egea, J.M. and Llimona, X. (1987) La comunidades de liquenes de las rocas siliceas del SE de Espana. *Acta Botanica Barcinonensia* 36: 1-123. [1]

Egea, J.M. and Llimona, X. (1991) Phytogeography of silicicolous lichens in Mediterranean Europe and NW Africa. *Botanika Cronika* 10: 179-198. [1]

Enteco, S.R.L. (1990) Relazione tecnica di restauro della statua n. 86 "Albertino Papafava di Carrarresi". In: Atti Giornata di Studio *Il Prato della Valle e le opere di pietra calcarea collocate all'aperto*: 299-311. Libreria Progetto Editore, Padova. [4]

Farmer, V.C. and Mitchell, B.D. (1963) Occurrence of oxalates in soil clays following hydrogen peroxide treatment. *Soil Science* 96: 221-229. [3]

Fassina, V., ed. (2000) Proceedings of the 9th International Congress on *Deterioration and Conservation of Stone* (Venice, 19-24 June 200), 2 vols. Elsevier, Amsterdam. [1, 2, 3,4]

Favali, M.A., Fossati, F. and Realini, M. (1989) Studio della natura delle pellicole osservate sul Duomo e sul Battistero di Parma. In: Atti del Convegno *Scienza e Beni Culturali, Le pellicole ad ossalato: origine e significato nella conservazione delle opere d'arte* (V. Fassina, ed.): 261-270. Centro del C.N.R. "Gino Bozza", Milano. [3]

Favali, M.A., Fossati, F., Mioni, A. and Realini, M. (1995) Biodeterioramento da licheni crostosi dei calcari selciferi lombardi. In: Atti del Convegno *Scienza e Beni Culturali, La pulitura delle superfici dell'architettura* (G. Biscontin and G. Driussi, eds.): 201-209. Libreria Progetto Editore, Padova. [2, 1]

Feilden, B.M. (1982) Botanical, biological and microbiological causes of decay. In: *Conservation of Historic Buildings*: 131 -134. Butterworth, London. [1]

Florian, M.L.E. (1978) A review: the lichen role in rock art - dating deterioration and control. In: *Conservation of Rock Art* (C. Pearson, ed.): 95 - 98. ICCM, Sydney. [1, 2, 7]

Forero, L.E. (1986) Investigación biologica en el parque arqueólogico de San Agustin (Huila). Erradiación y control de liquenes, hepaticas y musgos que deterioran la estatus agustiniana. *Restauracion Hoy* 1: 5-9. [1, 4]

Franceschi, V.R. and Horner, H.T. (1980) Calcium oxalate crystal in plants. *Botanical Review* 46: 361-427. [3]

Franzini, M., Gratzui, C. and Wicks, E. (1984) Patine ad ossalato di calcio sui monumenti marmorei. *Rendiconti Società Italiana di Mineralogia e Petrografia* 39: 59-70. [3]

Frey, T., von Reis, J. and Barov, Z. (1993) An evaluation of biocides for control of the biodeterioration of artifacts at Hearst Castle. In: Preprints of 10th Triennal Meeting (Washington-DC, 22-27 August 1993): 875-881. ICOM Committee for Conservation, Paris. [4]

Frey-Wyssling, A. (1981) Crystallography of the two hydrates of crystalline calcium oxalate in plants. *American Journal of Botany* 68: 130-141. [3]

Friedman, E.I. (1982) Endolithic organisms in the Antarctic cold desert. *Science* 215: 1045-1053. [1, 2]

Friedmann, E. I. and Weed, R. (1987) Microbial trace-fossil formation, biogenous, and abiotic weathering in the Antarctic cold desert. *Science* 236: 703-705. [1, 2]

Fry, E.J. (1922) Some types of endolithic limestone lichens. *Annals of Botany* 36: 541-562. [1]

Fry, E.J. (1924) A suggested explanation of the mechanical action of lithophytic lichens on rocks (shale). *Annals of Botany* 38: 175-196. [2]

Fry, E.J. (1927) The mechanical action of crustaceous lichens on substrate of shale, schist, gneiss, limestone and obsidian. *Annals of Botany* 41: 437-460. [2]

Fry , M.F. (1985) The problems of ornamental stonework: lichen. *Stone Industries* 20: 22-25. [1]

Gabrielli, N. (1986) Restauro di opere lapidee. *Rassegna dei Beni Culturali* 8/9: 44-50. [3]

Gairola, T.R. (1968) Ejemplos de preservation de monumentos en la India. *Museum and Monuments* 1: 150-164. UNESCO, Paris. [4]

Gallo, L.M. and Piervittori, R. (1991) La flora lichenica rupicola dei Monti Pelati di Baldissero (Canavese, Piemonte). In: Atti del Convegno *I Monti Pelati di Baldissero. Importanza Paesistica e Scientifica* (P.M. Giachino, ed.): 25-31. Feletto (Torino). [1]

Gallo, L.M. and Piervittori, R. (1993) Lichenometry as a method for holocene dating: limits in its applications and realibility. *Il Quaternario* 6: 77-86. [7]

Gallo L.M., Piervittori R. and Montacchini, F. (1989) Rapporti con il substrato litologico degli esemplari del gen. *Rhizocarpon* presente nelle Collezioni crittogamiche dell'*Herbarium Horti Taurinensis* (TO). *Bollettino Museo Regionale di Scienze Naturali di Torino* 7: 129-156. [1]

Galsomies, L., Robert, M. and Orial, G. (1996) Biological factors in the weathering of historical monuments in granite (Brittany – France). In: Proceedings of the EC Workshop *Degradation and Conservation of Granitic Rocks in Monuments* (M.A. Vicente, J. Delgado-Rodrigues and J. Acevedo, eds.): 95-101. Protection and Conservation of the European Cultural Heritage, Research Report No. 5. [2]

Galsomies, L., Robert, M. and Orial, G. (1999) Interaction lichens-roche sur monument historique en granite. *Bulletin d'Informations de l'Association Française de Lichénologie, Mémoires* 3: 35-42. [1, 2]

Galvan, J., Rodriguez, C. and Ascaso, C. (1981) The pedogenic action of lichens in metamorphic rock. *Pedobiologia* 21: 60-73. [2]

Garcia-Rowe, J. and Saiz-Jimenez, C. (1988) Colonization of mosaics by lichens: the case study of Italica (Spain). *Studia Geobotanica* 8: 65-71. [1]

Garcia-Rowe, J. and Saiz-Jimenez, C. (1989) Colonizacion y alteracion de mosaicos por liquenes y briofitos. In: Coloquio Nacional de Conservacion de Mosaicos: 61-84. Diputacion Provincial, Palencia. [1]

Garcia-Rowe, J. and Saiz-Jimenez, C. (1991) Colonizacion y alteracion de la piedra por liquenes, briofitos y plantas superiores en las catedrales de Salamanca, Sevilla y Toledo. In: Proceedings of the Jornadas Restauracion y Conservacion de Monumentos: 71-79. Instituto de Conservacion y Restauracion de Bienes Culturales, Madrid. [1, 2]

Garcia-Rowe, J. and Saiz-Jimenez, C. (1991) Lichens and bryophytes as agents of deterioration of buildings materials in Spanish cathedrals. *International Biodeterioration, Special Issue Biodeterioration of Cultural Property* 28: 151-163. [1, 2]

Garcia-Rowe, J. and Saiz-Jimenez, C. (1992) A case study on the corrosion of stone by lichens: the mosaics of the Roman remains of Italica. In: Proceedings of the 2nd EFC Workshop *Microbial Corrosion* (C.A.C. Sequeira and A.K. Tiller, eds.): 275-281. The Institute of Materials, London. [2]

Garcia-Rowe, J.G., Saiz-Jimenez, C. and Sameno, M. (1995) Colonizacion y alteracion de mosaicos romanos por liquenes y briofitos. *Boletin informativo, Instituto Andaluz del Patrimonio Historico* 3: 37-39. [1, 2]

Garcia-Velles, M.T., Krumbein, W.E., Urzì, C. and Vendrell-Saz, M. (1996) Biological pathways leading to the formation and transformation of oxalate-rich layers on monument surfaces exposed to Mediterranean climate. In: Proceedings of the II International Symposium *The Oxalate Films in the Conservation of Works of Art*: 319-334. Centro del C.N.R. "Gino Bozza", Milano. [3]

Garg, K.L., Dhawan, S. and Agrawal, O.P. (1988) *Deterioration of Stone and Buildings Materials by Algae and Lichens: a Review*. Natural Research Laboratory Conservation of Cultural Property, Lucknow. [1, 2, 7]

Gargani, G. (1971-1972) Fattori biologici nel degradamento delle opere d'arte. In Vita e decadenza delle opere d'arte. *Atti dell'Accademia di Scienze di Ferrara* 49: 133-142. [1]

Garty, J. (1988) Some observation on the establishment of the lichen *Caloplaca aurantia* on concrete tiles in Israel. *Studia Geobotanica* 8: 13-21. [1]

Garty, J. (1992) The postfire recovery of rock inhabiting algae, microfungi and lichens. *Canadian Journal of Botany* 70: 301-312. [1]

Garty, J. and Delarea, J. (1987) Some initial stages in the formation of epilithic crustose lichens in nature: a SEM study. *Symbiosis* 3: 49-56. [2, 1]

Garty, J., Galun, M. and Gal, M. (1974) The relationship between physico-chemical soil properties and substrate choice of "multisubstrate" lichen species. *The Lichenologist* 6: 146-150. [2]

Gayathri, P. (1980) Effects of lichens on granite statues. *Birla Archaeological Cultural Research Institute, Research Bulletin* 2: 41-52. [2]

Gehrmann, C.K. and Krumbein, W.E. (1994) Interactions between epilithic and endolithic lichens and carbonate rocks. In: Proceedings of the 3rd International Symposium *The Conservation of Monuments in the Mediterranean Basin* (V. Fassina, H. Ott and F. Zezza, eds.): 311-316. Soprintendenza ai beni artistici e storici, Venezia. [2]

Gehrmann, C.K., Krumbein, W.E. and Petersen, K. (1988) Lichen weathering activities on mineral and rock surfaces. *Studia Geobotanica* 8: 33-45. [2]

Gehrmann, C.K., Petersen, K. and Krumbein, W. E. (1988) Silicicole lichens on Jewish tombstone. Interactions with the environment and biocorrosion. In: Proceedings of the 6th International Congress on *Deterioration and Conservation of Stone*: 33-38. Nicholas Copernicus University, Torun. [1, 2]

Gehrmann, C.K., Krumbein, W.E. and Petersen, K. (1992) Endolithic lichens and the corrosion of carbonate rocks. A study of biopitting. *International Journal of Mycology and Lichenology* 5: 37-48. [1, 2]

Genin, G. (1973) Control of lichens, fungi, and othere organisms. *Paint Pigments Vernis* 49: 3-6. [4]

Giacobini, C. and Bettini, C. (1978) Traitements des vestiges archeologiques détériorés par les lichens et les algues. In: Proceedings International Symposium *Deterioration and Protection of Stone Monuments*: 4-3. UNESCO/RILEM, Paris. [4]

Giacobini, C. and Roccardi, A. (1984) Indagini sui problemi biologici connessi alla conservazione dei manufatti storico-artistici. *Restauri nel Polesine*: 34-35. Electa Editrice, Rovigo. [1]

Giacobini, C. and Seaward, M.R.D. (1991) Licheni e monumenti: studi in Veneto e in Puglia. In: Atti del Convegno *Scienza e Beni Culturali, Le pietre nell'architettura: Struttura e Superfici* (G. Biscontin, ed.): 215-224. Libreria Progetto Editrice, Padova. [1]

Giacobini, C., Bettini, C. and Villa, A. (1979) Il controllo dei licheni, alghe e muschi. In: Proceedings of the 3rd International Symposium *Deterioration and Preservation of Stones*, (Venice, 24-27 October 1979): 305-312. Venezia.[4]

Giacobini, C., Nugari, M.P., Micheli, M.P., Mazzone, B. and Seaward, M.R.D. (1986) Lichenology and the conservation of ancient monuments: an interdisciplinary study. In: *Biodeterioration 6* (S. Barry, D.R. Houghton, G.C. Llewellyn and C.E. O'Rear, eds.): 386-392. C.A.B. International Mycological Institute, Slough. [1, 2, 3, 4]

Giacobini, C., Roccardi, A. and Tigliè, I. (1986) Ricerche sul biodeterioramento. In: Atti del Convegno *Scienza e Beni Culturali, Manutenzione e conservazione del costruito tra tradizione e innovazione* (G. Biscontin, ed.): 687-705. Libreria Progetto Editore, Padova. [1]

Giacobini, C., Roccardi, A., Bassi, M. and Favali, M.A. (1986) The use of electronic microscopes in research on the biodeterioration of works of art. In: Proceedings of the Symposium on *Scientific Methodologies Applied to Works of Art* (P.L. Parrini, ed.): 71-75. Florence. [2]

14. Lichens and Monuments

Giacobini, C., Pietrini, A.M., Ricci, S. and Roccardi, A. (1987) Problemi di biodeterioramento. *Bollettino d'Arte* 41: 53-64. [1, 2]

Giacobini, C., Pedica, M. and Spinucci, M. (1989) Problems and future projects on the study of biodeterioration: mural and canvas paintings. In: Proceeding of the International Conference on *Biodeterioration of Cultural Property*. (O.P. Agrawal and S. Dhawan, eds.), (Lucknow-India, 20-25 February 1989): 275-286. Macmillan, Delhi. [4, 1]

Gilbert, O.L. (1977) Lichen conservation in Britain. In: *Lichen Ecology* (M.R.D. Seaward, ed.): 415-436. Academic Press, London. [1]

Gomez-Alarcon, G. and Angeles de la Torre, M. (1994) Mechanism of microbial corrosion of stone. *Microbiologia* (*Madrid*) 10: 111-120. [2]

Gomez-Alarcon, G., Munoz, M., Ariño, X. and Ortega-Calvo, J.J. (1995) Microbial communities in weathered sandstones-the case of Carrascosa del Campo church, Spain. *The Science of the Total Environment* 167: 249-254. [1]

Gomez-Bolea, A., Ariño X., Balzarotti, R. and Saiz-Jimenez, C. (1999) Surface treatment of stones: consequences on lichenic colonization. In: Book of Abstracts, International Conference on *Microbiology and Conservation (ICM '99), Of microbes and art: The role of microbial communities in the degradation and protection of Cultural Heritage*: 233-237. Tribuna di Galileo, Museo della Specola, Florence. [4]

Gorbushina, A.A., Boettcher, M., Brumsack, H.J., Krumbein, W.E. and Vendrell-Saz, M. (2001) Biogenic forsterite and opal as a product of biodeterioration and lichen stromaltolite formation in table mountain systems (tepuis) of Venezuela. *Geomicrobiology Journal* 18: 117-132. [2]

Gorgoni, C., Lazzarini, L. and Salvadori, O. (1992) Minero-geochemical transformation induced by lichens in the biocalcarenite of the Selinuntine monuments. In: Proceedings of the International Congress on *Deterioration and Conservation of Stone* (J. Delgado Rodrigues, F. Henriques and F. Telmo Jeremias, eds.): 531-539. LNEC, Lisbon. [2]

Gratziu, C. (1986) Primi dati sulle caratteristiche petrografiche e stratigrafiche delle patine ad ossalato di calcio sui monumenti marmorei romani. In: Atti del Convegno *Scienza e Beni Culturali, Manutenzione e Conservazione del Costruito fra Tradizione e Innovazione*: 751-763. Bressanone. [3]

Green, T. and Snelgar, W. (1977) *Parmelia scabrosa* on glass in New Zealand. *The Lichenologist* 9: 170-172. [1]

Griffin, P.S., Indicator, N. and Koestler, R. J. (1991) The biodeterioration of stone: a riview of deterioration mechanisms, conservation case histories, and treatment. *International Biodeterioration, Special Issue Biodeterioration of Cultural Property* 28: 187-207. [2,4]

Guidobaldi, F., Laurenzi Tabasso, M. and Meucci, C. (1984) Monumenti in marmo di epoca imperiale a Roma: indagine sui residui di trattamenti superficiali. *Bollettino d'Arte* 24 121-134. [4]

Hale, M.E. (1975) Informe sobre el crecimiento de liquenes en los monumentos de Copan, Honduras. *Yaxkin* 1: 6-9. [1]

Hale, M.E. (1975) Control of biological growths on Mayan archaeological ruins in Guatemala and Honduras. *National Geographic Society Research Reports, 1975 Projects*: 305-321. [4]

Hale, M.E. (1979) Conservatión de monumentos arqueológicos Mayas en Copan. Honduras: el programa biológico. *Yaxkin* 3: 135-149. [4]

Hallbauer, D.K. and Jahns, H.M. (1977) Attack of lichens on quartzitic rock surfaces. *The Lichenologist* 9: 119-122. [2]

Hawksworth, D.L. (2001) Do lichens protect or damage stonework? *Mycological Research* 105: 386. [1]
Heaton, N. (1925) The deterioration of stained glass. *Journal of British Society Master Glass Painters* 2: 7. [2]
Hocke, B. and Daniels, F. (1993) Über die epilitische Flechtenflora und -vegetation im Stadtgebiet von Münster. *Natur und Heimat* 53: 41-54. [1]
Holder, J.M., Wynn-Williams, D.D., Rull Perez, F. and Edwards, H.G.M. (2000) Raman spectroscopy of pigments and oxalates in situ within epilithic lichens: *Acarospora* from the Antarctic and Mediterranean. *New Phytologist* 145: 271-280. [3]
Hueck van der Plas, E.H. (1968). The microbiology deterioration of porous building materials. *International Biodeterioration Bulletin* 4: 11-28. [2]
Hyvert, G. (1973) Borobodur, les bas-relief. Materiaux - facteurs responsables des degradations - programme de conservation. *Studies in Conservation* 18: 131-155. [2]
Iskandar, I.K. and Syers, J.K. (1971) Solubility of lichen compounds in water: pedogenetic inplications. *The Lichenologist* 5: 45-50. [2]
Iskandar, I.K. and Syers, J.K. (1972) Metal-complex formation by lichen compounds. *Journal of Soil Science* 23: 255-265. [2]
Isquith, A.J., Abbott, E.A. and Walter, P.A. (1972).Surface-bonded antimicrobial actuvity of an organo silicon quaternary ammonium chloride. *Applied Microbiology* 24: 859-863. [2]
Jackson, D.W. (1981) An SEM study of lichen pruina crystal morphology. *Scanning Electron Microscopy* 3: 279-284. [2]
Jackson, T.A. and Keller, W.D. (1970) A comparative study of the role of lichens and the "inorganic" processes in the chemical weathering of recent Hawaiian lava flows. *American Journal of Science* 269: 446-466. [2]
Jain, K.K., Mishra, A.K. and Singh, T. (1993) Biodeterioration of stone: a review of mechanisms involved. In: *Recent Advances in Biodeterioration and Biodegradation* (K.L. Garg, N. Garg and K.G. Mukerji, eds.), vol. I: 323-354. Naya Prokash, Calcutta. [2, 7]
Jaton, C. and Orial, G. (1979) Etat des recherches sur les traitements des pierres (1975-1979) - Nettoyage. In: Ministere de la Culture et de la Communication, Laboratoire de Recherche des Monuments Historiques: 20-27. [4]
Jaton, C., Orial, G. and Brunet, A. (1985) Actions des vegetaux sur le materiaux pierreux. In: Proceedings of the 5th International Congress *Deterioration and Conservation of Stone*: 577-586. Presses Polytechniques Romandes, Lausanne. [1, 2]
Jenkins, I.D. and Middleton, A.P. (1989) Paint on the Parthenon sculptures. *The Annual of the British School of Archaeology at Athens* 83: 183-207. [1]
Johansson, P. (1992) Lavfloran på Lunds domkyrka. *Svensk Botanisk Tidskrift* 87: 25-30. [1]
John, E. and Dale, M.R.T. (1991) Determinants of spatial patterns in saxicolous lichen communities. *The Lichenologist* 23: 227-236. [1]
Johnston, C.G. and Vestal, J.R. (1993) Biogeochemistry of oxalate in the Antarctic cryptoendolithic lichen-dominated community. *Microbial Ecology* 25: 305-319. [3]
Jones, R.J. (1959) Lichen hyphae in limestone. *The Lichenologist* 1: 119. [2]
Jones, D. (1988) Lichens and pedogenesis. In: *Handbook of Lichenology, Volume 3* (M. Galun, ed.): 109-124. CRC Press, Boca Raton. [2, 3]
Jones, D. and Wilson, M.J. (1985) Chemical activity of lichens on mineral surfaces: a review. *International Biodeterioration* 21: 99-104. [2, 7]
Jones, D. and Wilson, M.J. (1986) Biomineralization in crustose lichens. In: *Biomineralization in Lower Plants and Animals* (B.S.C. Leadbeater and R. Riging, eds.): 91-105. Clarendon Press, Oxford. [2]

Jones, D., Wilson, M. J. and Tait, J.M. (1980) Weathering of a basalt by *Pertusaria corallina*. *The Lichenologist* 12: 277-289. [2]

Jones, D., Wilson, M. J. and McHardy, W.J. (1981) Lichen weathering of rock-forming minerals: application of scanning electron microscopy and microprobe analysis. *Journal of Microscopy* 124: 95-104. [2]

Jones, D., Wilson, M. J. and McHardy, W.J. (1987) Effects of lichens on mineral surfaces. In: Proceedings of the 7th International Biodeterioration Symposium *Biodeterioration 7* (D.R. Houghton, E. Smith, eds.): 129-134. Elsevier, Amsterdam. [2]

Jones, D., Muehlchen, A. and Young, M.E. (1994) Calcium-rich crystals associated with apothecia of crustose lichens on sandstone. *Biomineralization Newsletter*. Masonry Conservation Research Group Publications, Aberdeen. [3]

Keen, R. (1976) Controlling algae and other growths on concrete. *Advisory note of the Cement and Concrete Association*: 45-120. London. [4]

Koestler, R.J. and Vedral, J. (1991) Biodeterioration of cultural property: a bibliography. In: *International Biodeterioration, Special Issue Biodeterioration of Cultural Property* 28: 229-340. [7]

Krumbein, W. E. (1969) Uber den Einfluss der Mikroflora auf die exogene Dynamik (verwitterung und krustenbildung). *Geologische Rundschau* 58: 333-365. [2]

Krumbein, W. E. and Urzì, C. (1991) Biologically induced decay phenomena of antique marbles, some general considerations. In: Proceedings of the 2nd International Symposium *The Conservation ef Monuments in the Mediterranean Basin* (D. Decrouez, J. Chamay and F. Zezza, eds): 219-235. Genève. [1]

Krumbein, W.E. (1993) Color changes of building stones and their direct and indirect biological causes. In: Proceedings of the 7th International Congress *Deterioration and Conservation of Stone* (J. Delgado Rodrigues, F. Henriques and F. Telmo Jeremias, eds.): 443-452. Laboratório Nacional de Engenharia Civil, Lisbon. [1]

Krumbein, W.E. and Schonborn-Krumbein, C.E. (1987) Biogene Bauschaden I und II. Anamnese, Diagnose und Therapie in Bautenschutz und Denmalpflege. *Bautenschutz Bausanierung* 10: 14-23 and 110-117. [1, 4]

Krumbein, W.E., Urzì, C.E. and Gehrmann, C.K. (1991) On the biocorrosion and biodeterioration of antique and mediaeval glass. *Geomicrobiology Journal* 9: 139-160. [2]

Kumar, R. and Kumar, A.V. (1999) Biodeterioration of stone in tropical environments. *Research in Conservation*: 1-85. Getty Conservation Institute, Los Angeles. [2, 4]

Kushmir, E., Tietz, A. and Galun, M. (1978) "Oil hyphae" of endolithic lichens and their fatty acid composition. *Protoplasma* 97: 47-60. [2]

Lal, B.B. (1970) Indian rock paintings and their preservation. *Australian Aboriginal Studies* 22: 139-146. [1, 4]

Lal, B.B. (1978) Weathering and preservation of stone monuments under tropical conditions: some case histories. In: Proceedings of the International Symposium *Deterioration and Protection of Stone Monuments*: 7-8. UNESCO/RILEM, Paris. [1, 4]

Lallemant, R. (1992) Les altérations biologiqies de la pierre: 2 - Le rôle des lichens. In: *La conservation de la Pierre Monumentale en France* (J. Philippon, D. Jeannet and R.A. Lefevre, eds.): 95-100. Presses du CNRS, Paris. [1]

Lallemant, R. and Deruelle, S. (1978) Présence des lichens sur les monuments en pierre: nuisance ou protection? In: Proceedings of the International Symposium on *Deterioration and Protection of Stone Manuments*: 4-6. UNESCO/RILEM, Paris. [1]

Lamas, B.P., Brea, M.T.R. and Hermo, B.M.S. (1995) Colonization by lichens of granite churches in Galicia (Northwest Spain). *The Science of the Total Environment* 167: 343-351. [1]

Lamprecht, I., Reller, A., Riesen, R. and Wiedemann, H. G. (1997) Ca-oxalate films and microbiological investigations of the influence of ancient pigments on the growth of lichens. *Journal of Thermal Analysis* 49: 1601-1607. [4]

Lazzarini, L. (1979) I rilievi degli arconi dei portali della Basilica di San Marco a Venezia: ricerche tecnico scientifiche. In: *Die Sculpturen Von San Marco in Venedig*: 58-65. Deutscher Kunstverlag, Munich. [1, 2]

Lazzarini, L. and Laurenzi Tabasso, M. (1986) *Il restauro della pietra*. Cedam, Padova. [4]

Lazzarini, L. and Salvadori, O. (1989) A reassessment of the formation of the patina called scialbatura. *Studies in Conservation* 34: 20-26. [3]

Lee, M.R. (1999) Organic-mineral interactions studied by controlled pressure SEM. *Microscopy and Analysis European Edition, March 1999*: 23-25. [2]

Lee, M.R. and Parsons, I. (1999) Biomechanical and biochemical weathering of lichen-encrusted granite: textural controls on organic-mineral interactions and deposition of silica-rich layers. *Chemical Geology* 161: 385-397. [2]

Lehmann, J. (1976) Quelques nouvelles recherches sur le nettoyages et la preservation des sculptures en pierre expoièes à l'exterieur en Polonie. In: Proceedings of the International Congress *The Conservation of Stone* (R. Rossi Manaresi, ed.): 477-483. Bologna. [4]

Lehmann, J. (1987) The methodology for the cleaning and desalting of stone objects in Goluchow Castle Museum. In: Proceedings of the 8th Triennal Meeting ICOM Committee for Conservation: 487- 491. Sydney. [4]

Lelikova, D.S. and Tomasevich, G.N. (1975) Protection of quarry stone and brick of architectural monuments against phyco-chemical effects end biological deterioration. In: Proceedings of the 4th Triennal Meeting ICOM, Venezia. [4]

Levin, F.M. (1949) The role of lichens in the weathering of limestone and diorites. *Vestnik Gosudarstvenuogo Mosskovskogo Universiteta* 9: 149-150. [2]

Liebig J. von (1853) Überden Thierschite. *Liebigs Annalen der Chemie und Pharmazie* 86: 113-115. [1]

Lloyd, A.O. (1972) An approach to the testing of lichen inhibitors. In: *Biodeterioration of Materials* (A. H. Walters and E. H. Heuck van der Plats, eds.), 2: 185-191. Applied Science Publishers, London. [2]

Lloyd, A.O. (1974) Lichen attack on marble at Torcello - Venice. In: Atti del Congresso *Petrolio e Ambiente*: 221-224. Editore Artioli, Modena. [1, 2]

Lloyd, A.O. (1975) Progress in studies of deteriogenic lichens. In: Proceedings of the 3rd International Biodegradation Symposium: 395-402. [2]

Lorenz, J. (1973) Developments in anti-fouling paints. *Journal of the Oil Color Chemists' Association* 56: 369-372. [1]

Lounamaa, K.J. (1965) Studies of the content of iron, manganese and zinc in macrolichens. *Annales Botanici Fennici* 2: 127-137. [2]

Lovering, T.S. (1959) Significance of accumulator plants in rock weathering. *Bulletin of the Geological Society of America* 70: 127-137. [2]

Maggi, O., Pietrini, A.M., Piervittori, R., Ricci, S. and Roccardi, A. (1998) L'aerobiologia applicata ai beni culturali. In: *Areobiologia e Beni culturali. Metodologie e tecniche di misura* (P. Mandrioli and G. Caneva, eds.): 32-36. Nardini Editore, Firenze. [6]

Marchese, E.P., Razzara, S., Grillo, M. and Galesi, R. (1990) Indagine floristica e restauro conservativo dell'Abbazia di San Nicolò l'Arena di Nicolosi (Etna). *Bollettino Accademia Gioenia Scienze Naturali* 23: 707-720. [1, 4]

Marchese, E.P., Di Benedetto, L., Luciani, F., Razzara, S., Grillo, M., Stagno, F. and Auricchia A. (1997) Biodeteriogeni vegetali di monumenti del centro storico della città di Noto. *Archivi di Geobotanica* 3: 71-80. [1, 4]

Marcos Laso, B. (1999) Biodiversidad e influencia medioambiental sobre la colonizacion liquenica de algunos monumentos en la Ciudad de Salamanca. In: XIII Simposio de Botánica Criptogámica: 129. Universidad Complutense, Madrid. [1]

Marcos Laso, B. (2001) Biodiversidad y colonización liquénica de algunos monumentos en la Ciudad de Salamanca (España). *Botanica Complutensis* 25: 93-102. [1]

Martin, A.K. and Johnson, G.C. (1992) Chemical control of lichen growths established on building materials: a compilation of the published literature. *Biodeterioration Abstracts* 6: 101-107. [4]

Martin, A.K., Johnson, G.C., McCarthy, D.F. and Filson, R.B. (1992) Attempts to control lichens on asbestos cement roofing. *International Biodeterioration and Biodegradation* 30: 261-271. [4]

Matteini, M. and Moles, A. (1986) Le patine di ossalato sui manufatti in marmo. In: *Restauro del Marmo: Opere e Problemi.* Quaderni dell'Opificio delle Pietre Dure e Laboratori di Restauro di Firenze: 65-73. Opus Libri, Firenze. [3]

Mattirolo, O. (1917) Sulla natura della colorazione rosea della calce dei muri vetusti e sui vegetali inferiori che danneggiano i monumenti e le opere d'arte. *Rivista Archeologica della Provincia e Antica Diocesi di Como* 73/75: 1-19. [1]

Mattirolo, O. (1928) I licheni e la malattia delle vetrate antiche. *Rivista Archeologica della Provincia e Antica Diocesi di Como* 94-95: 1-23. [1]

May, E., Lewis, F.J., Pereira, S., Tayler, S., Seaward, M.R.D. and Allsopp, D. (1993) Microbial deterioration of building stone. *Biodeterioration Abstracts* 7: 109-123. [1]

McCarroll, D. and Viles, H. (1995) Rock-weathering by the lichen *Lecidea auriculata* in an arctic alpine environment. *Earth Surface Processes Landforms* 20: 199-206. [2]

Mellor, E. (1921) Les lichens vitricoles et leur action mécanique sur les vitraux d'eglise. *Compte Rendu des Séances de l'Académie des Sciences* 173: 1106. [2]

Mellor, E. (1922) Les lichens vitricoles et la détérioration des vitraux d'eglise. *Révue Générale de Botanique* 34: 1-16. [2]

Mellor, E. (1923) Lichens and their action on the glass and leading of church windows. *Nature* 25: 299.300. [2]

Mellor, E. (1924) The decay of window glass from the point of view of lichenous growths. *Journal Society of Glass Technology* 8: 182-186. [2]

Mellor, E. and Davy de Virville, A. (1921) La détérioration des vitraux d'Eglise de la Mayenne par les lichens. *Bulletin de Mayenne-Sciences-Laval*: 53-67. [2]

Micheli, M.P. and Napoli, B. (1987) Interventi conservativi sulla facciata della Chiesa di Santa Maria Assunta a Ponte di Cerreto. *Bollettino d'Arte* 41: 247-262. [4]

Mitchell, B.D., Birnie, A.C. and Syers, J.K. (1966) The thermal analysis of lichens growing on limestone. *Analyst* 91: 783-789. [2]

Mitteilung, K. (1974) Selecticity in lichen-substrate relationship. *Flora* 163: 530-534. [1, 2]

Modenesi, P. and Lajolo, L. (1988) Microscopical investigation on a marble encrusting lichen. *Studia Geobotanica* 8: 47-64. [2]

Modenesi, P., Giordani, P. and Brunialti, G. (2000) Factors determining the formation of weddellite and whewellite in lichens. In: Book of Abstracts, Fourth IAL Symposium on

Progress and Problems in Lichenology at the Turn of the Millenium (Barcelona, 3-8 September 2000): 37-38. Universitat de Barcelona, Barcelona. [3]

Monte, M. (1989) I licheni su monumenti: bioindicatori ambientali. In: Proceedings of the International Congress on *Science, Technology and European Cultural Heritage*: 355-359. Bologna. [5]

Monte, M. (1991) La lichenologia applicata alla conservazione dei monumenti in pietra esposti all'aperto: problemi e prospettive. In: Atti del Convegno *Scienza e Beni Culturali, Le pietre nell'Architettura: Struttura e Superfici* (G. Biscontin, ed.): 287-298. Libreria Progetto Editore, Padova. [1, 3, 4, 5]

Monte, M. (1991) Lichens on monuments: environmental bioindicators. In: *Science, Technology and European Cultural Heritage* (N.S. Baer, C. Sabbioni and A.I. Sors, eds.): 355-359. Butterworth Heinemann, London. [5]

Monte, M. (1991) Lichens on the ruins of a Roman aqueduct. *Biodeterioration and Biodegradation 8* (H.W. Rossmoore, ed.): 394-396. Elsevier Applied Science, London. [1]

Monte, M. (1991) Multivariate analysis applied to the conservation of monuments: lichens on the Roman aqueduct Anio Vetus in San Gregorio. *International Biodeterioration* 28: 133-150. [1]

Monte, M. (1994) Licheni come bioindicatori: analisi dell'ambiente e del degrado dei monumenti. In: *Studi e Ricerche sulla Conservazione delle Opere d'Arte alla Memoria di Marcello Paribeni*: 211-221. C.N.R., Roma. [5]

Monte, M. and Ferrari, R. (1996) Biodeterioration of the temple of Segesta (Sicily). In: Proceedings of the International Congress *Deterioration and Conservation of Stone* (J. Riederer, ed.): 585-591. Möller, Berlin. [1]

Monte, M. and Nichi, D. (1997) Azione di due biocidi nell'eliminazione di licheni dai monumenti in pietra. In: Abstracts del Convegno annuale della Società Lichenologica Italiana *Licheni e Ambiente* (D. Ottonello, ed.), (Palermo, 9-12 Dicembre 1995). *Notiziario della Società Lichenologica Italiana* 10: 83. [4]

Monte, M. and Nichi, D. (1997) Effects of two biocides in the elimination of lichens from stone monuments: preliminary findings. *Science and Technology for Cultural Heritage* 6: 209-216. [4]

Monte, M. and Tretiach, M. (1992) Licheni sui Nuraghi: un cantiere di ricerca. In: Atti del Convegno *Scienze dei Materiali e Beni Culturali, Esperienze e prospettive nel restauro delle costruzioni nuragiche* (C. Atzeni and U. Sunna, eds.): 73-81. [1]

Monte Sila, M. (1986) Biodeterioramento dei materiali musivi e proposte di intervento. In: Atti del II Seminario di Studi *Metodologia e prassi della conservazione musiva* (Ravenna, 22-23 Gennaio 1986): 345-51. Longo Editore, Ravenna. [1, 4]

Moriconi, G., Castellano, M.G. and Collepardi, M. (1994) Dégradation de mortiers de murs en mançonnierie dans les édifices historiques. Un exemple: Le Mole de Vamvitelli à Ancona. *Materials and Structure* 27: 408-414. [1, 2]

Moses, C.A. and Smith, B.J. (1993) A note on the role of the lichen *Collema auriforme* in solution basin development on a carboniferous limestone substrate. *Earth Surface Processes Landforms* 18, 363-368. [1]

Mottershead, D. and Lucas, G. (2000) The role of lichens in inhibiting erosion of a soluble rock. *The Lichenologist* 32: 601-609. [2, 1]

Narduzzi, P.A. and Soccal, M. (1990) Relazione tecnica di cantiere: il restauro della statua n.36 "Galileo Galilei". In: Atti della Giornata di Studio *Il Prato della Valle e le opere di pietra calcarea collocate all'aperto*: 337-346. Libreria Progetto Editore, Padova. [4]

Nascimbene, J. (1997) Licheni e conservazione dei monumenti. Un'esperienza in campo didattico. In: Abstracts del Convegno annuale della Società Lichenologica Italiana *Licheni e Ambiente* (D. Ottonello, ed.), (Palermo, 9-12 Dicembre 1995). *Notiziario della Società Lichenologica Italiana* 10: 55-56. [1]

Nienow, J.A. and Friedmann, E.I. (1993) Terrestrial lithophytic (rock) communities. In: *Antarctic Microbiology* (E.I. Friedmann, ed.): 343-412. Wiley-Liss, New York. [1, 2]

Nimis, P.L. (1999) Opere d'arte e di storia: ecosistemi minacciati. In: *Frontiere della Vita* 4: 531-541. Istituto della Enciclopedia Italiana, Roma. [1]

Nimis, P.L. (2001) Artistic and historical monuments: threatened ecosystems. In: *Frontiers of Life*, Part 2 - *Discovery and Spoliation of the Biosphere*, sect. - *Man and the Environment*: 557-569. Academic Press, San Diego. [1]

Nimis, P.L. and Martellos, S. (2001) *Checklist of the lichens of Italy 3.0*. University of Trieste, Department of Biology, IN2.0/2 (http://dbiodbs.univ.trieste.it). [1]

Nimis, P.L. and Monte, M., eds. (1988) Lichens and monuments. *Studia Geobotanica* 8: 1-133. [1, 2, 3, 4]

Nimis, P.L. and Monte, M. (1988) The lichen vegetation on the cathedral of Orvieto (Central Italy). *Studia Geobotanica* 8: 77-88. [1]

Nimis, P.L. and Salvadori, O. (1997) La crescita dei licheni sui monumenti di un parco. Uno studio pilota a villa Manin. In: *Restauro delle sculture lapidee nel parco di Villa Manin a Passariano. Il viale delle Erme* (E. Accornero, ed.): 109-142. Centro di Catalogazione e Restauro dei Beni Culturali, Villa Manin di Passariano, Venezia. [1, 4]

Nimis, P.L. and Tretiach, M. (1996) Studies on the biodeterioration potential of lichens, with particular reference to endolithic forms. In: *Interactive Physical Weathering and Bioreceptivity Study on Building Stones, Monitored by Computerized X-ray Tomography (CT) as a Potential Non-destructive Research Tool* (M. de Cleene ed.). European Commission Environment/Protection and Conservation of European Cultural Heritage, *Research Report* 2: 63-122. University of Ghent. [2]

Nimis, P.L. and Zappa, L. (1988) I licheni endolitici calcicoli su monumenti. *Studia Geobotanica* 8: 125-133. [1, 2]

Nimis, P.L., Monte, M. and Tretiach, M. (1987) Flora e vegetazione lichenica di aree archeologiche del Lazio. *Studia Geobotanica* 7: 3-161. [1]

Nimis, P.L., Pinna, D. and Salvadori, O. (1992) *Licheni e Conservazione dei Monumenti*. CLUEB, Bologna. [1, 2, 3, 4]

Nimis, P.L., Seaward, M.R.D., Ariño, X. and Barreno, E. (1998) Lichen-induced chromatic changes on monuments: a case-study on the Roman amphitheater of Italica (S. Spain). *Plant Biosystems* 132: 53-61. [1]

Nishiura, T. and Ebisawa, T. (1993) Conservation of carved natural stone under extremely severe conditions on the top of an high mountain: elimination of lichens and protective treatment. In: Proceedings of the 2nd International Conference *Biodeterioration of Cultural Property*: 506-511 (Yokohama, 5-8 October 1992). Tokyo. [4]

Nishiura, T., Okabe, M. and Kuchitsu, N. (1994) Study on the conservation treatment of Irimizu Sanjusan Kannon - cleaning and protective treatment of a marble Buddha image. *Hozon kagaku (Science for Conservation)* 33: 67-72. [4]

Normal-3/80 (1980) *Materiali Lapidei: Campionamento*. C.N.R.-I.C.R, Roma. [1]

Normal-19/85 (1985) *Microflora Autotrofa ed Eterotrofa: Tecniche di Indagine Visiva*. C.N.R.-I.C.R, Roma. [1]

Normal-1/88 (1990) *Alterazioni Macroscopiche dei Materiali: Lessico*. C.N.R.-I.C.R, Roma. [1]

Normal-30/89 (1991) *Metodi di Controllo del Biodeterioramento*. C.N.R.-I.C.R, Roma. [4]
Not, R. and Ottonello, D. (1989) Osservazioni sulla colonizzazione lichenica e briofitica del Tempio della Vittoria (Sicilia settentrionale). *Giornale Botanico Italiano* 123: 161. [1]
Nugari, M.P., D'Urbano, M.S. and Salvadori, O. (1993) Test methods for comparative evaluation of biocide treatments. In: Proceedings of the International RILEM/UNESCO Congress *Conservation of Stone and Other Materials*: 565-572. Spon, London. [4]
Nugari, M.P., Pallecchi, P. and Pinna, D. (1993) Methodological evaluation of biocidal interference with stone materials: preliminary laboratory tests. In: Proceedings of the International RILEM/UNESCO Congress *Conservation of Stone and OtherMaterials*: 295-302. Spon, London. [4]
Orial, G. and Brunet, A. (1988) Les lichens: impact sur les matérials pierreux et recherche d'une thérapie. In: Actes des Journeés d'étude de la SFIIC *Patrimoine culturel et altérations biologiques*. Poitiers. [1, 4]
Ortega, J., Martin, A., Aparicio, A. and Garcia, J. (1988) Bioalteration of the Cathedral of Seville. In: Proccedings of the 6th International Congress *Deterioration and Conservation of Stone*: 1-8. Nicholas Copernicus University, Torun. [1]
Ottonello, D. (1990) Osservazioni preliminari sulla flora lichenica gipsicola. *Giornale Botanico Italiano* 124: 91. [1]
Ottonello, D., Alaimo, R., Calderone, S. and Montana, G. (1991) Contributo alla conoscenza del rapporto tra i licheni e i substrati litici. *Giornale Botanico Italiano* 125: 263. [1, 2]
Paleni, A. and Curri, S. (1972) Biological aggression of works of art in Venice. In: *Biodeterioration of Materials 2* (A.H. Walters and E.H. Hueck van der Plas, eds.): 392-400. Applied Science Publishers, London. [1]
Paleni, A. and Curri, S. (1972) L'aggression des algues et des lichens aux pierres et moyens pur la combattre. In: 1er Colloque International *Deterioration des Pierres en Oeuvre*: 157-166. La Rochelle, France. [1, 4]
Paleni, A. and Curri, S. (1972) La pulitura delle sculture all'aperto. *Petrolieri d'Italia* 5. [4]
Paleni, A., Curri, S. and Benassi, R. (1973) Aggressione biologica ad una statua del 700 in marmo di Carrara in ambiente collinare campestre. In: Atti del Convegno nazionale *Petrolio e Ambiente, sezione Arte*: 29-50. [1, 4]
Pallecchi, P. and Pinna, D. (1988) Azione della crescita dei licheni sulla pietra nell'area archeologica di Fiesole. *Studia Geobotanica* 8: 113-124. [1]
Pallecchi, P. and Pinna, D. (1988) Alteration of stone caused by lichen growth in the Roman Theatre of Fiesole (Firenze). In: Proceedings 6th International Congress *Deterioration and Conservation of Stone*: 30-47. Nicholas Copernicus University, Torun. [1]
Pentecost, A. and Fletcher, A. (1974) Tufa: an interesting lichen substrate. *The Lichenologist* 6: 100-101. [1]
Piervittori, R. (1992) I popolamenti lichenici rupicoli calcifughi nel settore occidentale delle Alpi (Piemonte e Valle d'Aosta). *Biogeographia* 16: 91-104. [1]
Piervittori, R. and Caramiello, R. (2002) Importance of biological elements in conservation of stonework: a case study on a Romanesque church (Cortazzone, N. Italy). In: Proceedings 3rd International Congress *Science and Technology for the Safeguard of Cultural Heritage in the Mediterranean Basin* (A. Guarino, ed.), (Alcalá de Henares-Spain, 9-14 July 2001): 891-894. Universidad de Alcalá, Spain. [1, 4]
Piervittori, R. and Laccisaglia, A. (1993) Lichens as biodeterioration agents and biomonitors. *Aerobiologia* 9: 181-186. [1, 2, 5]

Piervittori, R. and Roccardi, A. (1998) Licheni. In: *Aerobiologia e Beni culturali. Metodologie e tecniche di misura* (P. Mandrioli and G. Caneva, eds): 179-183. Nardini Editore, Firenze. [1]

Piervittori, R. and Roccardi, A. (2002) Indagini aerobiologiche in ambienti esterni: valutazione della componente lichenica. In: Book of Abstracts, X Congresso Nazionale di Aerobiologia *Aria e Salute, Sezione Beni Culturali* (Bologna, 13-15 Novembre 2002): 70. Associazione Italiana di Aerobiologia, Bologna. [6]

Piervittori, R. and Salvadori, O. (2002) Il contributo della Lichenologia alla conoscenza e conservazione dei Beni Culturali: esperienze di studio e proposta di protocollo metodologico. In: Book of Abstracts, 97th Congresso della Società Botanica Italiana (Lecce, 24-27 Settembre 2002): 9. Edizioni del Grifo, Lecce. [1, 4]

Piervittori, R. and Sampò, S. (1987-1988) Colonizzazione lichenica su manufatti litici: la facciata dell'Abbazia di Vezzolano. Asti (Piemonte). *Allionia* 28: 93-101. [1, 4]

Piervittori, R. and Sampò, S. (1988) Lichen colonization on stoneworks: examples from Piedmont and Aosta Valley. *Studia Geobotanica* 8: 73-75. [1]

Piervittori, R., Gallo, L.M. and Laccisaglia, A. (1991) Analisi qualitative dell'interfaccia lichene-substrato litico: metodologie con il microscopio polarizzatore. *Giornale Botanico Italiano* 125: 256. [2]

Piervittori, R., Laccisaglia, A., Appolonia, L. and Gallo, L.M. (1991) Aspetti floristico-vegetazionali e metodologici relativi ai licheni su materiali lapidei in Valle d'Aosta. *Revue Valdôtaine d'Histoire Naturelle* 45: 53-86. [1, 2]

Piervittori, R., Salvadori, O. and Laccisaglia, A. (1994) Literature on lichens and biodeterioration of stonework I. *The Lichenologist* 26: 171-192. [7]

Piervittori, R., Salvadori, O. and Laccisaglia, A. (1996) Literature on lichens and biodeterioration of stonework II. *The Lichenologist* 28: 471-483. [7]

Piervittori, R., Salvadori, O. and Isocrono, D., (1998) Literature on lichens and biodeterioration of stonework III. *The Lichenologist* 30: 263-277. [7]

Piervittori, R., Roccardi, A. and Isocrono, D. (2002) Aspetti della colonizzazione lichenica sui monumenti. In: Abstracts del Convegno annuale della Società Lichenologica Italiana (G. Massari and S. Ravera, eds.), (Roma, 26-28 Ottobre 2001). *Notiziario della Società Lichenologica Italiana* 15: 71-72. [6]

Piervittori, R., Sampò, S., Appolonia, L., Gallo, L.M. and Polini, V. (1990) Caractéres écologiques d'éspéces licheniques dans les châteaux de la Vallée d'Aoste (Italie). In: Book of Abstracts, VI OPTIMA (Organization for the Phyto-Taxonomic Investigation of the Mediterranean Area) Meeting (Delphi, 10-16 September 1989): 124. [1]

Piervittori, R., Valcuvia-Passadore, M. and Nola, P. (1990/91). Italian lichenological bibliography: 1568-1989. *Allionia* 30: 99-169. [7]

Piervittori, R., Valcuvia-Passadore, M. and Laccisaglia, A. (1995) Italian lichenological bibliography. First update (1989-1994) and addenda. *Allionia* 33: 153-179. [7]

Piervittori, R., Valcuvia-Passadore, M. and Isocrono, D. (1998) Italian lichenological bibliography. Second update (1995-1998) and addenda. *Allionia* 36: 67-88. [7]

Piervittori, R., Valcuvia-Passadore, M. and Isocrono, D. (2001) Italian lichenological bibliography. Third update (1999-2001) and addenda. *Allionia* 38: 81-94. [7]

Piervittori, R., Salvadori, O., Castelli, S. and Favero-Longo, S. (2002) Interazioni licheni-ofioliti in ambiente alpino. In: Abstracts del Convegnoannuale della Società Lichenologica Italiana (G. Massari and S. Ravera, eds.), (Roma, 26-28 Ottobre 2001). *Notiziario della Società Lichenologica Italiana* 15: 38. [2]

Pinna, D. (1993) Fungal physiology and the formation of calcium oxalate films on stone monuments. *Aerobiologia* 9: 157-167. [3]

Pinna D. (2002) Crescita biologica su monumenti lapidei trattati con protettivi e consolidanti. In: Abstracts del Convegno annuale della Società Lichenologica Italiana (G. Massari and S. Ravera, eds.), (Roma, 26-28 Ottobre 2001). *Notiziario della Società Lichenologica Italiana* 15: 34-35. [4]

Pinna, D. and Salvadori, O. (1992) Effects of *Dirina massiliensis* and *stenhammari* growth on various substrata. In: Book of Abstracts, 2nd International Lichenological Symposium IAL: 103. Bastad [2]

Pinna, D. and Salvadori, O. (1999) Biological growth on Italian monuments restored with organic or carbonatic compounds. In: Book of Abstracts, International Conference *Microbiology and Conservation (ICMC '99), Of Microbes and Art: The role of microbial communities in the degradation and protection of Cultural Heritage* (O. Ciferri, P. Tiano and G. Mastromei, eds.): 149-154. C.N.R., Firenze. [4]

Pinna, D. and Salvadori, O. (2000) Endolithic lichens and conservation: an underestimated question. In: Proceedings of the 9th International Congress on *Deterioration and Conservation of Stone* (V. Fassina, ed.), (Venice, 19-24 June 2000): 513-519. Elsevier, Amsterdam. [2]

Pinna, D., Biscontin, G. and Driussi, G. (1995) La pulitura e il controllo della crescita biologica sui materiali lapidei. In: Atti del Convegno di Studi *Scienza e Beni Culturali, La pulitura delle superfici dell'architettura* (G. Biscontin and G. Driussi, eds.): 619-624. Libreria Progetto Editore, Padova. [4]

Pinna, D., Salvadori, O. and Tretiach, M. (1998) An anatomical investigation of calcicolous endolithic lichens from the Trieste Karst (NE Italy). *Plant Biosystems* 132: 183-195. [2]

Piterans, A., Fisere, D. and Vulfa, L. (1992) Biologiskais apaugums uz sunakmens pieminekliem un ta likvidesanas iespejas. *Latvijas Zinatnu Akademijas Vestis* B: 71-74. [1]

Poli-Marchese, E., Razzara, S., Grillo, M. and Galesi, R. (1990) Indagine floristica e restauro conservativo dell'Abbazia di San Nicolò l'Arena di Nicolosi (Etna). *Bollettino dell'Accademia Gioenia Scienze Naturale Catania* 23: 707-720. [1, 4]

Poli-Marchese, E., Di Benedetto, L., Luciani, F., Grillo, M., Auricchia, A. and Stagno, F. (1996) Indagine sui vegetali causa di degrado dei monumenti della città di Noto (Sicilia orientale). *Giornale Botanico Italiano* 130: 457. [1]

Poli-Marchese, E., Di Benedetto, L., Luciani, F., Razzara, S., Grillo, M., Stagno, F. and Auricchia A. (1997) Biodeteriogeni vegetali di monumenti del centro storico della città di Noto. *Archivio Geobotanico* 3: 71-80. [1]

Poli-Marchese, E., Luciani, F., Razzara, S., Grillo, M., Auricchia, A. and Stagno, F. (1995) Biodeteriogeni di origine vegetale causa del degrado del complesso monumentale dei Benedettini di Catania. *Giornale Botanico Italiano* 129: 58. [1]

Poli-Marchese, E., Luciani, F., Razzara, S., Grillo, M., Auricchia, A., Stagno, F., Giacone G. and Di Martino V. (1995) Biodeteriorating plants entities on monuments and stonework in historical city centre of Catania. "Il Monastero dei Benedettini". In: Proceedings of 1st International Congress *Science and Technology for the Safeguard of Cultural Heritage in the Mediterranean Basin*: 1195-1203. Catania, Italy [1]

Politi, M.A. (1989) Datare i monumenti studiando i licheni. *Scienza and Vita Nuova* 6: 136-139. [7, 1]

Pomar, L., Esteban, M., Llimona, X. and Fontarnau, R. (1975) Action de liquenes, algas y hongos en la telodiagenesis de la rocas carbonatads de la zona litoral y prelitoral catalan. *Instituto de Investigaciones Geològicas, Universitat de Barcelona* 30: 84-117. [2, 1]

Popescu, M. (1979) Probleme de biodeteriorare la piatra. *Biharea* 15: 411-426. [1]
Prieto, B., Rivas, T. and Silva, B. (1994) Colonization by lichens of granite dolmens in Galicia (NW Spain). *International Biodeterioration and Biodegradation* 34: 47-60. [2]
Prieto, B., Rivas, M.T., Silva, B.M., Carballal, R. and Lopez de Silanes, M.E. (1995) Colonization by lichens of granite dolmens in Galicia (NW Spain). *International Biodeterioration and Biodegradation* 35: 47-60. [1]
Prieto, B., Rivas, M.T., Silva, B.M., Carballal, R. and Sanchez-Bilma, M.J. (1995) Etude écologique de la colonization lichénique des églises des environs de Saint-Jaques-de-Compostelle (NW Espagne). *Cryptogamie, Bryologie et Lichénologie* 16: 219-228. [1]
Prieto, B., Rivas, M. T., Silva, B. M. and Lopez de Silanes, M. E. (1996) Ecological characteristics of lichens colonizing granite monuments in Galicia (Northwest Spain). In: Proceedings of the EC Workshop *Degradation and Conservation of Granitic Rocks in Monuments* (M.A. Vicente, J. Delgado-Rodrigues and J. Acevedo, eds.): 295-300. Protection and Conservation of the European Cultural Heritage, Research Report No 5. [1]
Prieto, B., Rivas, M.T. and Silva, B.M. (1996) Effectiveness of biocide treatments on granite. In: Proceedings of the EC Workshop *Degradation and Conservation of Granitic Rocks in Monuments* (M.A. Vicente, J. Delgado-Rodrigues and J. Acevedo, eds.): 361-366. Protection and Conservation of the European Cultural Heritage, Research Report No 5. [4]
Prieto, B., Silva, B., Rivas, T., Wierzchos, J. and Ascaso, C. (1997) Mineralogical transformation and neoformation in granite caused by the lichens *Tephromela atra* and *Ochrolechia parella*. *International Biodeterioration and Biodegradation* 40: 193-199. [2]
Prieto, B., Rivas, T. and Silva, B. (1999) Environmental factors affecting the distribution of lichens on granitic monuments in the Iberian Peninsula. *The Lichenologist* 31: 291-305. [1]
Prieto, B., Seaward, M.R.D., Edwards, H.G.M., Rivas, T. and Silva, B. (1999) Biodeterioration of granite monuments by *Ochrolechia parella* (L.) Mass.: an FT Raman spectroscopic study. *Biospectroscopy* 5 53-59. [3, 2]
Prieto, B., Seaward, M.R.D., Edwards, H.G.M., Rivas, T. and Silva, B. (1999) An Fourier transform-Raman spectroscopic study of gypsum neoformation by lichens growing on granitic rocks. *Spectrochimica Acta* 55A: 211-217. [2, 3]
Prieto, B., Edwards, H.G.M. and Seaward, M.R.D. (2000) A Fourier transform-Raman spectroscopic study of lichen strategies on granite monumentss. *Geomicrobiology Journal* 17: 55-60. [3, 2]
Prieto Lamas, B., Rivas Brae, M.T. and Silva Hermo, B.M. (1995) Colonization by lichens of granite churches in Galicia (Northwest Spain). *The Science of the Total Environment* 167: 343-351. [1, 2, 3]
Prudon, T., Labine, C. and Flaherty, C. (1980) Removing stains from masonry. *The Hold-House Journal Compendium*: 97-98. The Overlook Press, Woodstock. [4]
Puckett, K.J., Nieboer, E., Gordzynsky, M.J. and Richardson, D.H.S. (1973) The uptake of metal ions by lichens: a modified ion-exchange process. *New Phytologist* 72: 329-342. [2]
Puertas, F., Blanco-Varela, M.T., Palomo, A., Ariño, X., Ortega-Calvo, J.J. and Saiz-Jimenez, C. (1994) Characterization of mortars from the mosaics of Italica: causes of deterioration. In: Proceedings of the 3rd International Symposium *The Conservation of Monuments in the Mediterranean Basin* (V. Fassina, H. Ott and F. Zezza, eds.): 577- 583. Soprintendenza ai beni artistici e storici, Venezia. [1]
Puertas, F., Blanco-Varela, M.T., Palomo, A., Ariño, X., Hoyos, M. and Saiz-Jimenez, C. (1995) Causes and forms of decay of stuccos and concretes from the Roman city of Baelo Claudia (Southern Spain). In: *Architectural Studies, Materials and Analysis* (C.A. Brebbia

and B. Leftheris, eds.): 171-178. Structural Studies of Historical Buildings IV. Computational Mechanics Publications. [2]

Puertas, F., Blanco-Varela, M.T., Palomo, A., Ortega-Calvo, J.J., Ariño, X. and Saiz-Jimenez, C. (1995) Decay of Roman and repair mortars in mosaics from Italica, Spain. *The Science of the Total Environment* 153: 123-131. [2]

Puertas, F., Blanco, M.T., Palomo, A., Ortega-Calvo, J.J., Ariño, X. and Saiz-Jimenez, C. (1995) Characterization of mortar from Italica mosaics: causes of deterioration. In: Proceedings of the 5th Conference of the International Committee for the *Conservation of Mosaics*: 197-202. Conimbriga, Spain. [2]

Purvis, O.W. (1984) The occurrence of copper oxalate in lichens growing on copper sulphide-bearing rocks in Scandinavia. *The Lichenologist* 16: 197-204. [2, 3]

Purvis, O.W. and Halls, C. (1996) A review of lichens in metal-enriched environments. *The Lichenologist* 28: 571-601. [2, 3, 7]

Purvis, O.W., Elix, J.A., Broomhead, J.A. and Jones, G.C. (1987) The occurrence of copper-norstitic acid in lichens from cupriferous substrate. *The Lichenologist* 19: 193-203. [2, 3]

Purvis, O.W., Elix, J.A. and Gaul, K.L. (1990) The occurence of copper-psoromic acid in lichens from cupriferous substrata. *The Lichenologist* 22: 345-354. [2]

Realini, M., Mioni, A., Favali, M.A. and Fossati, F. (1994) Lichen-stone surface interaction under different environmental conditions. *Giornale Botanico Italiano* 128: 363. [2]

Rebricova, N.L. and Ageeva E.N. (1995) An evaluation of biocide treatments on the rock art of Baical. In: Proceedings of the International Colloquium on *Methods of Evaluating Products for the Conservation of Porous Building Materials in Monuments*: 69-74. ICCROM, Roma. [4]

Rechenberg, W. (1972) The avoidance and control of algae and other growths on concrete. *Betontechnische Berichte* 22: 249-251. [1]

Redazione (a cura di) (1990) Il restauro della Fontana dei Tritoni. *Rassegna dei Beni Culturali* 6: 49-52. [4]

Richardson, B.A. (1973) Control of biological growths. *Stone Industries* 8: 2-6. [4]

Richardson, B.A. (1976) Control of moss, lichen and algae on stone. In: Proceedings International Congress *The Conservation of Stone* (R. Rossi Manaresi, ed.): 225-231. Bologna. [4]

Richardson, B.A. (1977) *Colonization of structural surfaces by algae, lichens and other organisms*. British Association for the Advancement of Science Annual Meeting, University of Aston. [1]

Richardson, B.A. (1987) Control of Microbial Growths on Stone and Concrete. In: Proceedings of the 7th International Biodeterioration Symposium *Biodeterioration 7* (D.R. Houghton, R.N.Smith and H.O.W. Eggings, eds.): 101-106. Elsevier Applied Science, London. [4]

Richardson, D.H.S. (1991) Lichens and man. In: *Frontiers in Mycology* (D.L. Hawksworth, ed.). CAB International, Wallingford. [1, 5]

Riederer, J. (1986) Protection from weathering of building stone in tropical countries. In: Proceedings of the II C Congress *Case Studies in the Cconservation of Stone end Wall Painting*: 51-54. Bologna. [4, 2]

Rigoni, C. (1990) Le statue di Villa Cordellina. Problemi e indagini sui licheni. In: *Atti della Giornata di Studio su Il Prato della Valle e le opere di pietra calcarea collocate all'aperto*: 227-232. Libreria Progetto Editore, Padova. [1]

Rinne, D. (1976) *The Conservation of Ancient Marble*. J. Paul Getty Museum, Los Angeles. [4]

Ritchie, T. (1978) Cleaning of brickwork. *Canadian Building Digest* 194: 1-4. [4]
Robert, M. and Berthelin, J. (1986) Role of biological and biochemical factors in soil mineral weathering. In: *Interactions of Soil Minerals with Natural Organics and Microbes* (P.M. Huang and M. Schnitzer, eds.): 453-495. Soil Science Society of America, Madison. [2]
Robert, M., Vicente Hernandez, M. A., Molina Ballesteros, E. and Rives Arnau, V. (1993) The role of biological factors in the degradation of stone and monuments. In: Actas Workshop, *Alteracion de granitos y rocas afines, empleados como materiales de construccion: deterioro de monumentos historicos* (M.A. Vicente, E. Molina, V. Rives, eds.), (Avila-Spain, 1993): 103-115. Consejo Superior de Investigaciones Cientificas, Madrid. [2]
Robinson, D.A. and Williams, R.B.G. (2000) Accelerated weathering of a sandstone in the High Atlas Mountains of Morocco by an epilithic lichen. *Zeitschrift für Geomorphologie* 44: 513-528. [2]
Roccardi, A. and Bianchetti, P. (1988) The distribution of lichens on some stoneworks in the surroundings of Rome. *Studia Geobotanica* 8: 89- 97. [1]
Roccardi, A. and Piervittori, R. (1998) The aerodiffused lichen-component: problems and methods. In: Book of Abstracts, 6th Congress of International Congress on *Aerobiology – IAA* (International Association for Aerobiology), (Perugia, 31 Agosto - 5 Settembre 1998): 268. Perugia, Italy. [6]
Roccardi, A. and Piervittori, R. (2000) Aerobiologia e Beni Culturali: la componente lichenica aerodiffusa. In: Abstracts del Convegno annuale della Società Lichenologica Italiana *Licheni e Ambiente* (D. Ottonello, ed.), (Palermo, 9-12 Dicembre 1995). *Notiziario della Società Lichenologica Italiana* 13: 67-68. [6]
Roccardi, A., Brunialti, G., Modenesi, P. and Senarega, C. (2002) Studi preliminari sull'effetto della presenza di un lichene endolitico sulla permeabilità del travertino. In: Abstracts del Convegno annuale della Società Lichenologica Italiana (G. Massari and S. Ravera, eds.), (Roma, 26-28 Ottobre 2001). *Notiziario della Società Lichenologica Italiana* 15: 36-37. [2]
Rodriguez-Hidalgo, J.M., Garcia-Rowe, J. and Saiz-Jimenez, C. (1994) Mosaicos de Italica: ejemplos de deterioro. In: Proceedings of the *Mosaicos 5, Conservacion* in situ, ICCROM, (Palencia, 1990): 293-303. Diputacion Provincial, Palencia. [1, 4]
Romão, P.M.S. and Del Monte, M. (1991) The biodeterioration of granitic monuments: preliminary studies on the contribution of lichens. In: Proceedings of the Simposio Internacional sobre *Biodeterioro*: 74. Madrid. [1]
Romão, P.M..S. and Rattazzi, A. (1996) Biodeterioration on Megalithic monuments. Study of lichens' colonization on Trapadão and Zambujeiro dolmens (Southern Portugal). *International Biodeterioration and Biodegradation* 37: 23-35. [1]
Romão, P.M.S., Prudêncio, M.I., Trindade, M.J., Nasraoui, M., Gouveia, M.A., Figueiredo, M.O. and Silva, T. (2000) The Sao Sebastiao Church of Terceira Island (Azores, Portugal) – Characterization of the stones and their biological colonization. In: Proceedings of the 9th Internatinal Congress *Deterioration and Conservation of Stone* (V. Fassina, ed.), (Venice, 19-24 June 2000): 493-497. Elsevier, Amsterdam. [1]
Rooney-Dawn, F. (1994) Cambodia: the condition of the temples at Angkor in 1993. *Newsletter (Oriental Ceramic Society)* 2: 8-9. [1]
Rosato, V.G. and Traversa, L.P. (2000) Lichen growth on a concrete Dam in a rural environment (Tandil, Buenos Aires Province, Argentina). In: Proceedings of the 1st International Workshop *Microbial Impact on Building Materials*: 1-6. Sao Paulo. [1, 2]

Rosato, V.G., Traversa, L. and Cabello, M.N. (2000) The action of *Caloplaca citrina* on concrete surfaces: a preliminary study. In: Proceedings of the 9th International Congress *Deterioration and Conservation of Stone* (V. Fassina, ed.): 507-511. Elsevier, Amsterdam. [1, 2]

Rossi Manaresi, R., Grillini, G.C., Pinna, D. and Tucci, A. (1989) La formazione di ossalati di calcio su superfici monumentali: genesi biologica o da trattamenti. In: Atti del Convegno *Scienza e Beni Culturali, Le pellicole ad ossalato: origine e significato nella conservazione delle opere d'arte* (V. Fassina, ed.): 113-125. Centro del C.N.R. "Gino Bozza", Milano. [3]

Rossi Manaresi, R., Grillini, G.C., Pinna, D. and Tucci, A. (1989) Presenza di ossalati di calcio su superfici lapidee esposte all'aperto. In: Atti del Convegno *Scienza e Beni Culturali, Le pellicole ad ossalato: origine e significato nella conservazione delle opere d'arte* (V. Fassina, ed.): 195-205. Centro del C.N.R. "Gino Bozza", Milano. [3]

Rossi Manaresi, R., Tucci, A., Grillini, G.C., Pinna, D. and Di Francesco, C. (1989) Indagini multidisciplinari per lo studio di un monumento esemplare: casa Romei a Ferrara. In: Atti del Convegno *Scienza e Beni Culturali, Il Cantiere della Conoscenza, il Cantiere del Restauro* (G. Biscontin, M. Dal Col and S. Volpin, eds.): 403-416. Libreria Progetto Editore, Padova. [1, 4]

Roux, C. (1991) Phytogéographie des lichens saxicoles-calcicoles d'Europe méditerranéenne. *Botanika Chronika* 10: 163-178. [1]

Russ, J., Kaluarachchi, W.D., Drummond, L. and Edwards, H.G.M. (1999) The nature of a whewellite-rich rockcrust associated with pictographs in southwestern Texas. *Studies in Conservation* 44: 91-103. [3]

Russell, N.C., Edwards, H.G.M. and Wynn-Williams, D.D. (1998) FT-Raman spectroscopic analysis of endolithic microbial communities from Beacon sandstone in Victoria Land, Antarctica. *Antarctic Science* 10: 63-74. [3, 2]

Sabbioni, C. and Zappia, G. (1991) Oxalate patinas on ancient monuments: the biological hypothesis. *Aerobiologia* 7: 31-37. [3]

Saiz-Jimenez, C. (1981) Weathering of building materials of the Giralda (Seville, Spain) by lichens. In: Proceedings of the 6th Triennal Meeting ICOM Committee for Conservation (Ottawa, 4 October 1981). Ottawa. [1]

Saiz-Jimenez, C. (1984) Weathering and colonization of limestones in an urban environment. *Soil Biology and Conservation of the Biosphere* 2: 65-71. Budapest. [1, 5]

Saiz-Jimenez, C. (1990) The mosaics of Italica. A natural laboratory of stone colonization. *European Cultural Heritage, Newsletter on Research* 4: 34-38. [1]

Saiz-Jimenez, C. (1991) Deterioro de materiales petreos por microorganismos. In : Jornadas Restauracion y Conservacion de Monumentos: 31-39. Instituto de Conservacion y Restauracion de Bienes Culturales, Madrid. [1, 2]

Saiz-Jimenez, C. (1994) Biodeterioration of stone in historic buildings and monuments. In: Biodeterioration Research 4, *Mycotoxin, Wood Decay, Plant Stress, Biocorrosion, and General Biodeterioration* (G.C. Llewellyn, W.V. Dashek and C.E. O'Rear, eds.): 587-604. Plenum Press, New York. [1, 2]

Saiz-Jimenez, C. (1999) Biogeochemistry of weathering processes in monuments. *Geomicrobiology Journal* 16: 27-37. [2]

Saiz-Jimenez, C. and Ariño, X. (1995) Colonizacion biologica y deterioro de morteros por organismos fototrofos. Biological colonization and deterioration of mortars by phototrophic organisms. *Materiales de Construccion* 45: 5-16. [1, 2]

Saiz-Jimenez, C. and Garcia-Rowe, J. (1991) Biodeterioration of marbles and limestones in Roman pavements. In: Proceedings of the 2nd International Symposium *The Conservation of Monuments in the Mediterranean Basin* (D. Decrouez, J. Chamhay and F. Zezza, eds.): 263-271. Genéve. [1]

Saiz-Jimenez, C. and Garcia-Rowe, J. (1991) Biodeterioro de mosaicos y pavimentos de Italica. In: Preprints of Simposio Internacional sobre *Biodeterioro*: 35. Madrid. [1]

Saiz-Jimenez, C., Grimalt, J., Garcia-Rowe, J. and Ortega-Calvo, J.J. (1991) Analytical pyrolysis of lichen thalli. *Symbiosis* 11: 313-326. [2]

Saiz-Jimenez, C., Garcia-Rowe, J. and Rodriguez-Hidalgo, J.M. (1991) Biodeterioration of polychrome Roman mosaics. *International Biodeterioration, Special Issue Biodeterioration of Cultural Property* 28: 65-79. [1]

Salter, J.W. (1856) On some reaction of oxalic acid. *Chemical Gazette* 14: 130-131. [3]

Salvadori, O. (2002) Colonizzazione dei manufatti lapidei da parte di organismi endolitici: un fenomeno sottostimato. In: Abstracts del Convegno annuale della Società Lichenologica Italiana (G. Massari and S. Ravera, eds.), (Roma, 26-28 Ottobre 2001). *Notiziario della Società Lichenologica Italiana* 15: 33. [1, 2]

Salvadori, O. and Lazzarini, L. (1991) Lichen deterioration on stones of Aquileian monuments (Italy). *Botanika Chronika* 10: 961-968. [1, 2, 3]

Salvadori, O. and Tretiach, M. (2002) Thallus-substratum relationships of silicicolous lichens occurring on carbonatic rocks of the Mediterranean region. In: *Progress and Problems in Lichenology at the Turn of the Millenium* (X. Llimona, H.T. Lumbsch and S.Ott, eds.). *Bibliotheca Lichenologica* 82: 57-64. Cramer, Berlin-Stuttgart. [3]

Salvadori, O. and Zitelli, A. (1981) Monohydrate and dihydrate calcium oxalate in living lichen incrustation biodeteriorating marble columns of the basilica of S. Maria Assunta on the island of Torecello (Venice). In: Proceedings of 2nd International Symposium *The Conservation of Stone II*, (R. Rossi Manaresi, ed): 759-767. Centro per la Conservazione delle Sculture all'Aperto, Bologna. [3]

Salvadori, O., Pinna, D. and Grillini, G.C. (1994) Lichen-induced deterioration on an ignimbrite of the Vulsini complex (Central Italy). In: Proceedings of the International Meeting *Lavas and Volcanic Tuffs, Easter Island* (A.E. Charola, R.J. Koestler and G. Lombardi, eds.), (Chile, 25-31 October 1990): 143-154. ICCROM, Roma. [1, 2, 3]

Salvadori, O., Sorlini, C. and Zanardini, E. (1994) Microbiological and biochemical investigations on stone of the Ca' d'Oro facade (Venice). In: Proceedings of 3rd International Symposium *Conservation of Monuments in the Mediterranean Basin* (V. Fassina, H. Ott and F. Zezza, eds.): 343-347. Soprintendenza ai beni artistici e storici, Venezia. [2]

Salvadori, O., Appolonia, L. and Tretiach, M. (2000) Thallus-substratum interface of silicicolous lichens occurring on carbonatic rocks of the Mediterranean regions. In: Book of Abstracts, Fourth IAL Symposium *Progress and Problems in Lichenology at the turn of the Millenium* (Barcelona, 3-8 September 2000): 34. Universitat de Barcelona, Barcelona. [2, 1]

Salvadori, O., Tretiach, M. and Appolonia, L. (2000) Relazione tallo-substrato in *Tephromela atra* (Norman) Hafellner v. *atra* e v. *calcarea* (Jatta) Clauz. In: Abstracts del Convegno annuale della Società Lichenologica Italiana *Licheni e Ambiente* (D. Ottonello, ed.), (Palermo, 9-12 Dicembre 1995). *Notiziario della Società Lichenologica Italiana* 13: 69. [2]

Samidi, S. (1981) How to control the organic growth on Borobodur stones after the restoration. In: Proceedings International Symposium *The Conservation of Stone* (R: Rossi

Manaresi, ed.): 759-767. Centro per la Conservazione delle Sculture all'Aperto, Bologna. [4]

Sampò, S. and Piervittori, R. (1990) Le malte come substrato elettivo per *Candelariella vitellina* (Ehrht.) Müll. Arg. In: Atti del Convegno Scienza e Beni Culturali, *Superfici dell'Architettura: Le Finiture* (G. Biscontin, S. Volpin, eds.), (Bressanone, 26-29 Giugno 1990): 313-316. Libreria Progetto Editore, Padova. [1, 2]

Sand, W. (1997) Microbial mechanisms of deterioration of inorganic substrates - a general mechanistic overview. *International Biodeterioration and Biodegradation* 40: 183-190. [2]

Sanders, W.B., Ascaso, C. and Wierzchos, J. (1994) Physical interactions of two rhizomorph-forming lichens with their rock substrate. *Botanica Acta* 107: 432-439. [2]

Savoye, D. and Lallemant, R. (1980) Evolution de la microflore d'un substrat avant et pendant sa colonisation per les lichens. I - Le case de toitures en amiante-cement en zone suburbaine. *Cryptogamie, Bryologie et Lichénologie* 1: 21-31. [1, 5]

Schabereiter-Gurtner, C., Piñar, G., Lubitz, W. and Rölleke, S. (2001) Analysis of fungal communities on historical church window glass by denaturing gradient gel electrophoresis and phylogenetic 18S rDNA sequence analysis. *Journal of Microbiological Methods* 47: 345-354. [1]

Schaffer, R.J. (1932) The weathering of natural building stones. *Department of Scientific and Industrial Research Building Research Special Report* 18: 1-149. [1]

Schatz, A. (1962) Pedogenic (soil forming) activity of lichen acids. *Naturwissenschaften* 49: 518-519. [2]

Schatz, A. (1963) Soil micro-organisms and soil chelation: the pedogenetic action of lichens and lichen acids. *Journal of Agricultural and Food Chemistry* 11: 112-118. [2]

Schatz, A. (1963) The importance of metal-binding phenomena in the chemistry and microbiology of the soil. Part. I: the chelating properties of lichens and lichen acids. *Advancing Frontiers of Plant Science* 6: 113-134. [2]

Schatz, A., Cheronis, N.D., Schatz, V. and Trelawny, G.S. (1954) Chelation (sequestration) as a biological weathering factor in pedogenesis. *Proceedings of the Pennsylvania Academy of Science* 28: 44-51. [2]

Schatz, V., Schatz, A., Trelawny, G.S. and Barth, K. (1956) Significance of lichens as pedogenic (soil-forming) agents. *Proceedings of the Pennsylvania Academy of Science* 30: 62-69. [2]

Seaward, M.R.D. (1985) Lichens and ancient monuments: conservation issues. In: Proceedings of the International Workshop *Biodeterioration of Ancient Stonework*. Aurangabad (India). [1, 4]

Seaward, M.R.D. (1988) Lichen damage to ancient monuments: a case study. *The Lichenologist* 20: 291-295. [2]

Seaward, M.R.D. (1989) Lichens and historic works of art. *British Lichen Society Bulletin* 64: 1-7. [1]

Seaward, M.R.D. (1997) Major impacts made by lichens in biodeterioration processes. *International Biodeterioration and Biodegradation* 40: 269-273. [2]

Seaward, M.R.D. (1998) Major environmental impacts of lichens. In: Book of Abstracts, Sixth International Mycological Congress (IMC6), (Jerusalem, 23-28 August 1998): 132. Jerusalem. [5]

Seaward, M.R.D. (2001) Lichen conservation: monuments and urban habitats. In: *Lichen Habitat Management* (A. Flechter, P. Wolseley and R. Woods, eds.): 15-1-15/5. British Lichen Society, London. [1]

Seaward, M.R.D. (2001) The role of lichens in the biodeterioration of ancient monuments with particular reference to central Italy. *International Biodeterioration and Biodegradation* 48: 1-4. [1]

Seaward, M.R.D., Edwards, H.G.M. and Farwell, D.W. (1998) Fourier-transform Raman microscopy of the apothecia of *Chroodiscus megalophthalmus* (Müll.Arg.) Vezda & Kantvilas. *Nova Hedwigia* 66: 463-472. [2]

Seaward, M.R.D. and Giacobini, C. (1988) Lichen-induced biodeterioration of Italian monuments, frescoes and other archeological materials. *Studia Geobotanica* 8: 3-11. [2]

Seaward, M.R.D. and Giacobini, C. (1989) Lichens as biodeteriorators of archaeological materials, with particular reference to Italy. In: Proceeding of the International Conference *Biodeterioration of Cultural Property* (O.P. Agrawal and S. Dhawan, ed.), (Lucknow, 20-25 February 1989): 195-206. Macmillan, Delhi. [1, 2]

Seaward, M.R.D. and Giacobini, C. (1989) Oxalate encrustation by the lichen *Dirina massiliensis* forma *sorediata* and its role in the deterioration of works of art. In: Atti del Convegno *Scienza e Beni Culturali, Le pellicole ad ossalato: origine e significato nella conservazione delle opere d'arte* (V. Fassina. ed.): 215-219. Centro del C.N.R. "Gino Bozza", Milano. [3]

Seaward, M.R.D. and Edwards, H.G.M. (1995) Lichen-substratum interface studies, with particular reference to Raman microscopic analysis. I. Deterioration of works of art by *Dirina massiliensis* forma *sorediata*. *Cryptogamic Botany* 5: 282-287. [2, 3]

Seaward, M.R.D. and Edwards, H.G.M. (1997) Biological origin of major chemical disturbances on ecclesiastical architecture studied by Fourier transform Raman spectroscopy. *Journal of Raman Spectroscopy* 28: 691-696. [1, 2]

Seaward, M.R.D., Capponi, G. and Giacobini, C. (1989) Biodeterioramento da licheni in Puglia. In: Proceedings of the 1st International Symposium *La conservazione dei monumenti nel bacino del Mediterrano* (F. Zezza ed.): 243-245. Grafo Edizioni, Bari. [1, 2]

Seaward, M.R.D., Giacobini, C., Giuliani, M.R. and Roccardi, A. (1989) The role of lichens in the biodeterioration of ancient monuments with particular reference to Central Italy. *International Biodeterioration*, 25, 49-55. [Reprinted in *International Biodeterioration and Biodegradation* 48: 202-208, 2001.] [1, 2]

Seaward, M.R.D., Giacobini, C. and Roccardi, A. (1990) I licheni a Villa Cordellina. In: Catalogo *Problemi di salvaguardia e restauro, Le sculture del Giardino. Vicende e problemi di conservazione*: 327-328. Electa, Milano. [2]

Seaward, M.R.D., Edwards, H.G.M. and Farwell, D.W. (1994) FT-Raman microscopic studies of *Haematomma ochroleucum* (Necker) Laundon var. *porphyrium* (Pers.) Laundon. In: *Contribution to Festschrift für Dr. Christian. Leuckert, Chemotaxonomy and Geography of Lichens* (J.G. Knoph, K. Schrüfer and H.J.M. Sipman, eds.). *Bibliotheca Lichenologica* 57: 395-407. Cramer, Berlin-Stuttgart. [3]

Sedelnikova, N.V. and Cheremisin, D.V. (2001) The use of lichens for dating of petroglyphs. *Siberian Journal of Ecology* 8: 479-481. [1]

Sengupta, R. (1979) Protecting our stone monuments. *Science Reporter* 16: 220-236. [4]

Seshadri,T.R. and Subramanian, S.S. (1949) A lichen (*Parmelia tinctorum*) on a Java monument. *Journal of Scientific and Industrial Research* B 8: 170-171. [1]

Sharma, B.R.N. (1978) Stone decay in tropical conditions - treatment of monuments at Khajuraho M.P. India. In: Proceedings of the International Symposium on *Deterioration and Protection of Stone Monuments*: 7-19. UNESCO/RILEM, Paris. [4]

Sharma, B.R.N., Chaturvedi, K., Samadhia, N.K. and Tailor, P.N. (1985) Biological growth removal and comparative effectiveness of fungicides from Central India temples for a decade *in situ*. In: Proceedings of the 5th International Congress *Deterioration and Conservation of Stone*: 675-683. Presses Polytechniques Romandes, Lausanne. [4]

Silva, B., Prieto, B., Rivas, T., Sanchez-Biezma, M.J., Paz, J. and Carballal, R. (1997) Rapid biological colonization of a granitic building by lichens. *International Biodeterioration and Biodegradation* 40: 263-267. [1]

Silva B., Rivas, T. and Prieto, B. (1999) Effects of lichens on the geochemical weathering of granitic rocks. *Chemosphere* 39: 379-388. [2]

Silverman, M.P. (1979) Biological and organic chemical decomposition of silicates. In: *Studies in Environmental Science/3, Biogeochemical Cycling of Mineral-forming Elements* (P.A. Trudinger and D.J. Swaine, eds.): 445-465. Elsevier, Oxford. [2]

Singh, A. (1987) Effects of lichens on material of cultural property - need for further study. In: *Conservation of Metals in Humid Climate* (O.P. Agarwal, ed.): 83-87. ICCROM, Natural Research Laboratory Conservation of Cultural Property, Lucknow. [1]

Singh, A. and Upreti, D.K. (1989) Lichen flora of Lucknow with special reference to its historical monuments. In: Proceedings of the International Conference *Biodeterioration of Cultural Property* (O.P. Agrawal and S. Dhawan, eds), (Lucknow, 20-25 February 1989): 219-231. Macmillan, Delhi. [1]

Singh, A. (1993) Biodeterioration of building materials. In: *Recent Advances in Biodeterioration and Biodegradation* (K.L. Garg, N. Garg and K.G. Mukerji, eds.), 1: 399-427. Naya Prokash, Calcutta. [1]

Singh, A. and Sinha, G.P. (1993) Corrosion of natural and monument stone with special reference to lichen activity. In: *Recent Advances in Biodeterioration and Biodegradation* (K.L. Garg, N. Garg and K.G. Mukerji, eds.), 1: 355-377. Naya Prokash, Calcutta. [2]

Sollas, W.J. (1880) On the action of a lichen on limestone. *Report of the British Association for the Advancement of Science*: 586. [2]

Souza-Egipsy, V., Wierzchos, J., Garcia-Ramos, J.V. and Ascaso, C. (2002) Chemical and ultrastructural features of the lichen-volcanic/sedimentary rock interface in a semiarid region (Almeria, Spain). *The Lichenologist* 54: 155-167. [2, 3]

Sparrius, L.B. (2000) Korstmossen op oude kerken in Nederland. *Buxbaumiella* 52: 32-36. [1]

Spry, A.H. (1981) *The Conservation of Masonry Materials in Historic Buildings*. The Australian Mineral Development Laboratories, Frewville, South Australia. [4]

Stambolov, T. and van Asperen De Boer, J.R.J. (1976) Deterioration by biological agents. In: *Deterioration and Conservation of Porous Building Materials in Monuments: a review of the literature*: 76-77. ICCROM, Roma. [1, 2]

Subbaraman, S. (1985) Conservation of Shore Temple, Mahabalipuram and Kailasanatha temple, Kancheepuram. In: Proceedings of the 5th International Congress *Deterioration and Conservation of Stone*: 1025-1033. Presses Polytechniques Romandes, Lausanne. [4]

Sundholm, E.G. and Huneck, S. (1980) ^{13}C NMR-spectra of lichen depsides, depsidones and depsones. *Chemica Scripta* 16: 197-200. [2]

Syers, J.K. (1960) Chelating ability of fumaroprotocetraric acid and *Parmelia conspersa*. *Plant and Soil* 31: 205-208. [2]

Syers, J.K. (1964) *A study of soil formation on carboniferous limestone with particular reference to lichens as pedogenetic agents*. Ph.D.thesis. University of Durham, England. [2]

Syers, J.K. (1969) Chelating ability of fumarprotocetraric acid and *Parmelia conspersa*. *Plant and Soil*, 31, 205-208. [2]

14. Lichens and Monuments

Syers, J. K. and Iskandar, I. K. (1973) Pedogenetic significance of lichens. In: *The Lichens* (V. Ahmadjian and M. E. Hale, eds.): 225-248. Academic Press, New York. [2]

Syers, J.K., Birnie, A.C. and Mitchell, B.D. (1967) The calcium oxalate content of some lichens growing on limestone. *The Lichenologist* 3: 409-414. [3]

Terrón, A.A., Barquín, P. and Marcos, R. (2000) Alteration analysis of the lichen-rock interface in San Isidoro of Leon (Spain). In: Book of Abstracts, Fourth IAL Symposium *Progress and Problems in Lichenology at the Turn of the Millennium*, (Barcelona, 3-8 September 2000): 112. Universitat de Barcelona, Barcelona. [2]

Tiano, P. (1978) Les traitements. In: Proceedings of the International Symposium on *Deterioration and Protection of Stone Manuments*, II: 4.8. UNESCO/RILEM, Paris. [4]

Tiano, P. (1986) Biological deterioration in stone on exposed works of art. In: *Biodeterioration of Constructional Materials* (L.G.H. Morton, ed.). *The Biodeterioration Society Occasional Publication* 3: 37-44. [2]

Tiano, P. (1986) Problemi biologici nella conservazione del materiale lapideo esposto. *La Prefabbricazione* 22: 261-272. [4]

Tiano, P. (1987) Problemi biologici nella conservazione delle opere in marmo esposte all'aperto. In: *Restauro del Marmo-Opere e Problemi* (A. Giusti, ed.): 47-53. Opus Libri, Firenze. [4]

Tiano, P. (1991) Problemi biologici nella conservazione del patrimonio culturale. *Kermes* 10: 56-73. [1]

Tiano, P. (1993) Biodeterioration of stone monuments: a critical review. In: *Recent Advances in Biodeterioration and Biodegradation* (K.L. Garg, N. Garg and K.G. Mukerji, eds.), 1: 301-321. Naya Prokash, Calcutta. [1, 2, 7]

Tiano, P. (1998) Biodegradation on cultural heritage: decay mechanisms and control methods. *Science and Technology for Cultural Heritage* 7: 19-38. [2, 4, 7]

Tolpisheva, T.Yu. (1991) On a succession of lichens on iron. *Moscow University Biological Sciences Bulletin* 46: 65-68. [1]

Tormo, R., Recio, D., Silva, I. and Muñoz, A.F. (2001) A quantitative investigation of airborne algae and lichen soredia obtained from pollen traps in south-west Spain. *European Journal of Phycology* 36: 385-390. [6]

Torres Montes, L., Alvarez Gasca, D., Reyes, M., Hernandez Rivero, J., Charola, A. E., Koestler, R. J. and Lombardi, G. (1994) The Cuauhcalli - a monolithic Aztec temple at Malinalco, Mexico: deterioration and conservation problems. In: Proceedings of the International Meeting *Lavas and Volcanic Tuffs, Easter Island* (A.E. Charola, R.J. Koestler and G. Lombardi, eds.), (Chile, 25-31 October 1990): 63-72. ICCROM, Rome. [1, 4]

Traversa, L.P. and Rosato, V. (1998) Algunas consideraciones sobre la colonizacion de liquenes en las superficies del Hormigon. *Ciencia y Tecnologia del Hormigon* 6: 9-18. [1]

Traversa, L.P., Rosato, V. and Vitalone, C. (1999) Colonizacion biologica en costrucciones de valor historico. Biological colonization on buildings of historical value. In: V Congreso Iberoamericano de Patologia de las Costrucciones VII Congreso de Control de Calidad (Montevideo, 18-21 October 1999): 1575-1580. [1]

Tretiach, M. (1995) Ecophysiology of calcicolous endolithic lichens: progress and problems. *Giornale Botanico Italiano* 129: 159-184. [2]

Tretiach, M. and Geletti, A. (1997) CO_2 exchange of the endolithic lichen *Verrucaria baldensis* from karst habitats in Northern Italy. *Oecologia* 111: 515-522. [2]

Tretiach, M. and Monte, M. (1991) Un nuovo indice di igrofitismo per i licheni epilitici sviluppato sui nuraghi di granito della Sardegna nord-occidentale. *Webbia* 46: 183-192. [1]

Tretiach, M. and Pecchiari, M. (1995) Gas exchange rates and chlorophyll content of epi- and endolithic lichens from the Trieste Karst (NE Italy). *New Phytologist* 130: 585-592. [2]

Tretiach, M., Monte, M. and Nimis, P.L. (1991) A new hygrophytism index for epilithic lichens developed on basaltic nuraghes in NW Sardinia (Italy). *Botanika Chronika* 10: 953-960. [1]

Uchida, E., Ogawa, Y., Maeda, N. and Nakagawa, T. (2000) Deterioration of stone materials in the Angkor monuments, Cambodia. *Engineering Geology* 55: 101-112. [2]

Upreti, D.K. (1994) Indian lichenology in 1993. *British Lichen Society Bulletin* 74: 16-18. [1]

Upreti, D.K. (1995) Loss of diversity in Indian lichen flora. *Environmental Conservation* 22: 361-363. [1]

Urquhart, D.C.M., Young, M.E. and Cameron, S. (1997) *Stone Cleaning of Granite Buildings*. Historic Scotland Technical Advice Note 9, The Stationery Office, Edinburgh. [4]

Vendrell-Saz, M., Garcia-Vallès, M., Alarcón, S. and Molera, J. (1996) Environmental impact on the Roman monuments of Tarragona, Spain. *Environmental Geology* 27 263-269. [5]

Videla, H.A., Guiamet, P.S. and de Saravaia, S.G. (2000) Biodeterioration of Mayan archaeological sites in the Yucatan Peninsula. Mexico. *International Biodeterioration and Biodegradation* 46: 335-341. [1]

Vidrich, V., Cecconi, C.A., Ristori, G.G. and Fusi, P. (1982) Verwitterung toskanischer Gesteine unter Mitwirkung von Flechten. *Zeitschrift für Pflanzenernahrung und Bodenkunde* 145: 384-389. [1]

Viles, H. (1995) Ecological perspectives on rock surface weathering: towards a conceptual model. *Geomorphology* 13: 21-35. [1, 2]

Wadsten, T. and Moberg, R. (1985) Calcium oxalate hydrates on the surface of lichens. *The Lichenologist* 17: 239-245. [3]

Wainwright, I. N. M. (1985) Rock art conservation research in Canada. *Bollettino del Centro Camuno di Studi Preistorici* 22: 15-46. [1, 4]

Wainwright, I.N.M. (1986) Lichen removal from an engraved memorial to Walt Whitman. *Association for Preservation Technology Bulletin* 8: 46-51. [4]

Wakefield, R.D. and Jones, M.S. (1998) An introduction to stone colonizing micro-organisms and biodeterioration of building stone. *Journal of Engineering Geology* 31: 301-313. [1, 2]

Walton, D.W.H. (1985) A preliminary study of the action of crustose lichens on rock surfaces in Antarctica. In: *Antarctic Nutrient Cycles and Food Webs* (W.R. Siegfried, P.R. Condy and R.M. Laws, eds.): 180-5. Springer-Verlag, Berlin. [1, 2]

Walton, D.W.H. (1993) The effects of cryptogams on mineral substrates. In: *Primary Succession on Land* (J. Miles and D.W.H. Walton, eds.): 33-53. Blackwell Scientific Publications, Oxford. [2]

Warscheid, T.H. and Braams, J. (2000) Biodeterioration of stone: a review. *International Biodeterioration and Biodegradation* 46: 343-368. [1, 2, 3, 4, 7]

Warscheid, Th., Barros, D., Becker, T.W., Braams, J., Eliasaro, S., Grote, G., Janssen, D., Jung, L., Mascarenhas, S.P.B., Mazzoni, M. L., Petersen, K., Simonoes, E.E.S., Moreira, Y.K. and Krumbein, W.E. (1992) Biodeterioration studies on soapstone, quartzite and sandstones of historical monuments in Brazil. In: Proceedings of the International Congress *Deterioration and Conservation of Stone* (J. Delgado Rodridues, F. Henriques and F. Telmo Jeremias, eds.): 491-500. LNEC, Lisbon. [2]

Watchman, A.L. (1990) A summary of the occurrences of oxalate-rich crusts in Australia. *Rock Art Research* 7: 44-50. [3]
Watchman, A.L. (1990) Rassegna delle croste ricche in ossalato rinvenute in Australia. *Arkos* 9/10: 11-18. [3]
Watchman, A.L. (1991) Age and composition of oxalate-rich crusts in the Northern Territory, Australia. *Studies in Conservation* 36: 24-32. [3]
Weber, B. and Büdel, B. (2001) Flechten an mittelalterlichen feldsteinkirchen in Mecklenburg; Vielfalt, Ökologie und mögliche Einwirkungen auf das Gestein. In: *Erhaltung und beispielhafte Instandsetzung von feldsteinkirchen in Mecklenburg* (R. Gesatzky, ed.): 69-76. Thomas Helms Verlag, Schwerin. [1]
Wendler, E. and Prasartset, C. (1999) Lichen growth on old Khmer-style sandstone monuments in Thailand: damage factor of shelter? In: Proceedings of the 12th Triennal Meeting of the ICOM Committee for Conservation, vol. 2: 750-754. Lyon. [2, 4]
Wessels, D.C.J. and Wessels, L.A. (1991) Erosion of biogenically weathered Clarens sandstone by lichenophagus bagworm larvae (*Lepidoptera; Psychida*e). *The Lichenologist* 23: 283-291. [2]
Wessels, D. and Wessels, L. (1995) Biogenic weathering and microclimate of Clarence sandstone in South Africa. *Cryptogamic Botany* 5: 288-298. [2]
Wiedemann, H.G. and Bayer, G. (1988) Formation of whewellite and weddellite by dispacement reactions. In: Atti del Convegno *Le pellicole ad ossalato: origine e significato nella conservazione delle opere d'arte*: 127-135. Centro del C.N.R. "Gino Bozza", Milano. [3]
Wierzchos, J. and Ascaso, C. (1994) Application of back-scattered electron imaging to the study of the lichen-rock interface. *Journal of Microscopy* 175: 54-59. [2]
Wierzchos, J. and Ascaso, C. (1996) Morphological and chemical features of bioweathered granitic biotite induced by lichen activity. *Clays and Clay Minerals* 44: 652-657. [2]
Wierzchos, J. and Ascaso, C. (1998) Mineralogical transformation of bioweathered granitic biotite, studied by HRTEM: evidence for a new pathway in lichen activity. *Clays and Clay Minerals* 46: 446-452. [2]
Wierzchos, J., Rios, A. and Ascaso, C. (1994) Backscattered electron imaging of the weathering processes of rock by lichen activity. *Icem* 13: 789-790. [2]
Williams, M. E. and Rudolph, E.D. (1974) The role of lichens and associated fungi in the chemical weathering of rock. *Mycologia* 66: 648-660. [2]
Wilson, M.J. (1995) Interactions between lichens and rocks: a review. *Cryptogamic Botany* 5: 299–305. [2, 7]
Wilson, M.J. and Jones, D. (1983) Lichen weathering of minerals and implication for pedogenesis. In: *Residual Deposits: surface related weathering processes and materials* (R.C.L.Wilson, ed.): 5-12. Blackwell, London. [2]
Wilson, M.J. and Jones, D. (1984) The occurrence and significance of manganese oxalate in *Pertusaria corallina*. *Pedobiologia* 26: 373-379. [3]
Wilson, M.J., Jones, D. and Russel, J.D. (1980) Glushinskite, a naturally occurring magnesium oxalate. *Mineralogical Magazine* 43: 837-840. [3]
Wilson, M.J., Jones, D. and McHardy, W.J. (1981) The weathering of serpentinite by *Lecanora atra*. *The Lichenologist* 13: 167-176. [2, 3]
Wilson, M.J. and Jones, D. (1985) Biological weathering of minerals. In: Incontro della Società Italiana della Scienza del Suolo-Association Internazionale pour l'etude des argilles (Gruppo Italiano) *Minerali argillosi ed ossidi di ferro nel suolo*: 57-65. Stresa e Valsesia. [2, 3]

Wirth, V. (1972) Die Silikatflechten-Gemeinschaften im außeralpinen Zentraleuropa. *Dissertationes Botanicae* 17: 1-325. [1]

Yarilova, E.A. (1950) Transformation of syenite minerals during the early stages of soil formation. *Trudy Institut Pochvenno Akademija Nauk SSSR* 34: 110-142. [2]

Zagari, M., Antonelli, F. and Urzì, C. (2000) Biological patinas on the limestones of the Loches Romanic tower (Touraine, France). In Proceedings of 9th Internatinal Congress on *Deterioration and Conservation of Stone* (V. Fassina, ed.), (Venice, 19-24 June 2000): 445-451. Elsevier, Amsterdam. [1, 2, 3]

Zelaya Rubi, V. and Hale, M.E. (1983) Osservazioni a Copan Honduras il 7-8 giugno 1979 e raccomandazioni per ulteriori trattamenti. In: Atti del Convegno Internazionale *La Pietra: Interventi, Conservazione, Restauro*: 159-170. Congedo Editore, Galatina (Lecce). [4]

Zezza, F., Macri, F., Decrouez, D., Barbin, V., Ramseyer, K., Garcia-Rowe, J. and Saiz-Jimenez, C. (1991) Origin weathering and biological colonization of the marble terminal column on the via Appia, Brindisi, Italy. In: 2nd International Symposium *The Conservation of Monuments in the Mediterranean Basin* (D.Decrouez, J. Chamay and F. Zezza, eds.): 99-112. Genève. [1]

Zezza, F., Urzì, C., Moropolou, T., Macri, F., Zagari, M., Biscontin, G. and Driussi, G. (1995) Indagini microanalitiche e microbiologiche di patine e croste presenti su pietre calcaree e marmi esposti all'aerosol marino e all'inquinamento atmosferico. In: Atti del Convegno *Scienza e Beni Culturali, La pulitura delle superfici dell'architettura*: 293-303. Libreria Progetto Editore, Padova. [2, 5]

Zitelli, A. and Salvadori, O. (1982) Studio dell'azione biodeteriogena e descrizione del lichene colonizzante le colonne della Basilica di Santa Maria Assunta dell'isola di Torcello (VE) - Considerazioni e proposte. *Rapporti e Studi, Istituto Veneto di Scienze, Lettere e Arti* 8: 153-169. [2, 3]

ACKNOWLEDGEMENT

We are most grateful to Dr Peter Crittenden, Senior Editor of *The Lichenologist*, for permission to use published material (albeit it greatly modified for use here).

ANALYTICAL INDEX

[1] - Biodeterioration of monuments: general studies; floristic, vegetational and ecological studies on lithic substrata; lichens as deteriorators of stonework; lichen-substratum relationships.

18-19-25-26-28-30-31-34-36-37-38-39-40-42-44-45-46-48-49-51-52-53-54-55-56-57-58-59-60-61-62-66-72-80-83-84-85-86-87-99-100-101-102-104-105-106-107-108-109-110-113-120-121-125-127-128-130-131-132-133-136-137-139-143-145-148-149-150-153-155-158-160-164-167-168-170-171-178-185-187-192-193-194-196-197-198-199-200-208-209-211-212-214-215-223-238-239-242-244-245-246-

14. Lichens and Monuments

247-248-249-252-253-254-257-260-262-264-266-267-268-269-271-273-274-275-276-277-282-283-286-287-289-290-292-293-294-296-301-304-308-321-322-324-337-338-339-343-344-345-346-347-349-360-362-365-367-368-369-373-374-384-388-390-391-393-395-397-398-399-400-402-403-404-405-406-407-409-411-412-413-414-417-418-419-421-424-425-426-427-428-429-431-432-433-434-435-436-437-438-440-441-442-444- 445-449-462-463-464-465-466-467-468-469-470-472-473-474-476-477-481-484-488-494-498-500-502-507-512-513-514-515-516-517-518-521-522-526-527-528529-530-532-533-534-536-538-539-542-544-547-550-551-552-554-558-560-563-564-567-570-571-572-575-577-580-583-584-585-589-591604-605-606-609-610-611-614-616-618-619-622-623-624-626-628-629-631-636-652-654-656-660.

[2] - Degradation of rocks and stonework by lichens; weathering processes due to lichens as biogeophysical and biochemical agents.

1-2-3-4-5-6-7-8-9-10-11-12-13-14-15-18-19-20-24-46-47-51-55-62-63-64-65-67-68-69-70-71-72-73-74-75-76-77-78-79-80-81-82-84-88-89-90-91-92-95-96-98-103-113-115-123-126-130-134-135-140-144-149-152-153-154-155-159-166-171-172-173-177-178-184-186-188-189-190-191-193-194-195-199-203-206-215-216-217-219-220-221-222-227-231-236-237-242-244-246-252-253-255-256-263-264-265-268-269-270-271-272-273-277-278-279-280-281-282-283-290-291-292-295-298-299-302-307-309-312-313-314-315-316-317-318-319-321-325-327-328-329-330-331-332-336-340-341-342-349-352-353-357-359-360-361-363-364-376-377-378-379-380-381-383-384-385-398-400-403-407-410-411-437-443-444-454-457-459-461-469-471-476-478-479-480-481-485-486-487-488-489-490-491-492-501-505-506-507-511-517-518-524-529-530-531-532-535-538-539-542-543-544-545-547-548-549-553-554-555-556-557-559-560-565-566-567-569-570-571-572-573-581-582-586-587-588-591-593-594-595-596-597-599-601-605-606-612-613-615-617-624-628-629-630-631-632-637-638-639-641-642-643-644-645-646-647-650-651-653-654-657-658.

[3] - Oxalate patinas.

22-23-32-41-114-124-129-142-153-159-173-176-179-180-199-201-202-203-204-205-206-213-217-218-219-220-221-222-223-224-225-226-227-228-229-230-232-233-234-235-236-241-242-243-248-249-251-258-272-289-300-311-325-327-334-351-372-386-388-407-413-455-478-479-480-481488-489-490-519-520-523-524-525-537-539-540-541-542-568-569-574-588-598-625-631-633-634-635-640-648-649-650-651-654-658.

[4] - Methods for prevention and control of lichen colonization on monuments.

16-17-27-28-29-30-36-37-38-40-49-82-86-87-93-94-111-116-117-118-119-122-125-136-138-139-144-146-150-153-156-161-165-169-170-174-175-181-182-195-210-240-242-247-250-259-284-285-288-289-293-297-302-303-305-306-320-324-339-341-343-344-348-350-354-355-356-366-367-370-371-382-388-394-395-396-401-407-409-413-415-416-420-422-423-424-429-430-431-436-440-441-456-458-460-463-475-482-493-495-496-497-499-501-503-504-512-521-546-558-576-578-579-590-592-600-602-603-606-609-620-626-627-631-637-655.

[5] - Biomonitoring and heritage buildings.

21-43-140-141-151-155-207-387-388-389-392-437-500-504-527-550-551-562-621-657.

[6] - Aerobiology and heritage buildings.

157-162-163-365-439-448-509-510-608.

[7] - Critical bibliographies and reviews on lichens and biodeterioration of rocks and stonework.

9-63-173-215-246-261-273-319-328-335-445-446-447-450-451-452-453-468-489-605-606-631-646.

Index

Aboriginal, 6, 248, 257, 260, 289
Acarospora, 135, 136, 154, 155, 156, 157, 171, 176, 180, 181, 182, 183, 184, 185, 275, 297
Accelerator carbon 14 method, 241
Acetate, 38, 94, 141, 142, 229, 239
Acetic acid, 142
Acid rain, 12
Acrylic emulsion, 30
Acrylic solution, 31
Acrylic/Epoxy, 30
Actinomycete, 35, 36
Aerobiology, 8, 73, 74, 75, 309, 321
Agency for the Supervision of Environmental and Architectural Heritage, 52
α-ketoglutarate, 141
Albite, 118
Algicides, 231
Algophase, 191, 207
Alicante, ix, 89, 90, 92, 100, 101
Alizarin red S staining, 93
Alkali silicate, 138
Alkaline, 85, 204, 221
Alkaline-earth cations, 85
Alkalinity, 238
Alkoxysilanes, 138
Altamira, 254, 258, 261
Aluminum (Al), 85, 95, 106, 141, 151
Aluminum silicates, 62, 96, 142
Aluminum soap mixtures, 241
Amandinea, 185
Ammonia, 229, 231
Ammonium chloride monohydrate, 230

Ammonium hydroxide, 229, 236, 237, 238
Anions, 114, 126, 141
Anziaic acid, 166
Aosta, 52, 58, 304, 305
Apothecia, 13, 95, 107, 154, 155, 156, 157, 263, 265, 267, 273, 275, 298, 313
Araldite, 62
Archaic rock art, 131
Areole, 154, 155, 156
Arezzo, 52
Argopsin, 166
Aromatic compounds, 273
Aromatic polyphenolic lichen acids, 264
Asci, 95, 96, 109
Ascocarp, 96, 106
Ascomycete, 2, 150
Ascospores, 96, 109
Aspergillus, 34, 37, 38
Aspicilia, 78, 80, 155, 156, 157, 165, 166, 167, 168, 169, 170, 172, 185, 195, 197, 205, 213, 243, 244, 264
Associazione Italiana di Aerobiologia, 73, 75, 305
Atrazine, 230
Aureobasidium, 34, 35, 36
Authigenic minerals, 93
Bacidia, 214
Backscattered electron imaging (BSE), 92, 240
Baelo Claudia, 99, 191, 192, 193, 195, 198, 199, 201, 203, 204, 205, 206, 208, 281, 308
Barbatate, 166

Bateig stone, 89, 93, 100
Bayhydrol, 30, 38, 40, 41, 42
Bemalite (Fe(OH)$_3$), 117
Benzoic acid, 166
Beryllium foil, 153
Besovy Sledki, 252
Beta-carotene, 272, 273
Bibliographic Database of the Conservation Information Network (BCIN), 135
Binders, 243, 262
Bioaerosol, 74
Biocalcarenite (Bateig stone), 64, 89, 90, 93, 96, 97, 297
Biocide, 5, 7, 17, 21, 26, 28, 29, 30, 32, 38, 39, 40, 41, 42, 43, 44, 45, 46, 47, 48, 49, 69, 72, 78, 85, 86, 91, 98, 99, 100, 131, 132, 133, 134, 135, 142, 143, 144, 147, 191, 207, 208, 226, 227, 228, 230, 231, 235, 236, 237, 238, 241, 242, 243, 245, 248, 249, 250, 251, 254, 260, 283, 284, 285, 293, 302, 304, 307, 308
Bioclasts, 89, 95
Biocomplexity, 89, 90, 98
Biodegradation, 21
Biofilm, xiii, 7, 8, 89, 90, 91, 96, 98
Biogeochemical, 51, 61, 62, 97, 264
Biogeophysical, 51, 61, 97, 320
Biomineralization, 96, 97, 98, 282, 285
Bioreceptivity, 64, 192
Biotite, 77, 83, 85, 101, 318
Bis(tri N-butyltin) oxide, 32
Blood antigen, 245
Borate, 17, 31
Borax, 231
Botryolepraria, 193, 194, 195, 197
Bragg Brentano geometry, 117
Bromide, 230
Bryozoan, 93
Buellia, 78, 214
Bureau of Land Management (BLM), xi, 130, 131, 132, 133, 144, 145, 146, 225, 258, 261
Burgos, xi, 91, 99, 274, 282
Caceres, 206
Calcareous litharenite, 97, 99, 282
Calcareous rocks, 235, 236, 282, 289
Calcite, 11, 14, 29, 93, 95, 96, 118, 195, 200, 203, 231, 234, 236

Calcite cement, 93, 200
Calcium (Ca), 5, 10, 11, 14, 15, 16, 18, 20, 21, 27, 51, 53, 65, 66, 67, 68, 69, 77, 83, 84, 89, 94, 95, 96, 97, 98, 106, 110, 113, 118, 151, 183, 201, 205, 215, 216, 222, 224, 233, 242, 243, 244, 257, 261, 263, 268, 269, 270, 273, 281, 283, 287, 290, 291, 294, 299, 306, 311, 312, 315, 317
Calcium biocarbonate, 27
Calcium carbonate, 5, 21, 27, 77, 84, 151, 201
Calcium caseinate, 66
Caloplaca, 62, 80, 135, 136, 137, 155, 156, 157, 158, 165, 166, 167, 168, 169, 170, 172, 185, 188, 193, 194, 195, 197, 199, 201, 202, 204, 205, 206, 213, 214, 233, 295, 310
Calploicin, 166
Calycin, 166, 275
Candelaria, 215, 219, 220
Candelariella, 13, 62, 65, 80, 155, 156, 157, 158, 165, 166, 167, 168, 169, 170, 171, 173, 174, 183, 185, 213, 312
Candida, 37
Capillary mounting, 117
Caprarola, 14
Carbonate, 5, 7, 21, 27, 61, 62, 77, 84, 92, 95, 97, 98, 99, 107, 151, 201, 202, 203, 283, 288, 295
Carbonate rocks, 61, 92, 98, 295
Carbonic acid, 5, 150
Carmona, 191, 192, 195, 198, 199, 202, 203, 206, 208, 281
Casein, 226
Catapyrenium, 89, 93, 185, 193, 195, 197, 199
Cation leaching rate, 243
Cation-ratio, 242
Cations, 5, 62, 85, 114, 126, 151, 243
Cayo, 22, 29
Cellulase, 250
Cellulose, 38, 143, 250
Cementing matrix, 128
Charcoal crayons, 243
Charged carboxylic sites, 142
Chelate, 5, 10, 151, 220
Chemical attack, 27, 120
Chemi-lithotropics, 140
China wood oil, 241

Index 285

Chininase, 143
Chitin, 250
Chitinase, 250
Chlorite, 117, 118
Chromatic changes, 57, 191, 204, 208, 303
Chromatic effects, 204
Chroococcodiopsis, 268
Chroococcoid, 89
Chroococcus, 89, 93, 95, 104
Chroococcus-type cyanobacteria, 95
Citric acid, 141
Cladonia, 185, 188, 215
Cladosporium, 37, 38
Clauzadea, 194, 197, 199
Clorox, 226, 228, 231, 238
Coccocarpia, 215, 220
Coenosis, 54
Collema, 89, 93, 154, 155, 156, 158, 165, 166, 167, 168, 169, 170, 185, 188, 193, 194, 197, 199, 204, 205, 302
COLORMOD, 113, 114, 123, 124, 126, 127, 129, 130, 137
Confluentic acid, 166
Conservare OH®, 31, 138
Conservation, 7, 8, 16, 17, 18, 26, 32, 43, 49, 51, 52, 53, 54, 55, 56, 57, 61, 63, 67, 72, 73, 74, 77, 78, 90, 100, 113, 114, 116, 117, 120, 131, 132, 133, 134, 135, 138, 142, 144, 145, 146, 148, 208, 211, 225, 232, 236, 237, 241, 247, 249, 251, 254, 255, 257, 258, 259, 261, 279, 280, 281, 283, 287, 289, 296, 297, 299, 302, 303, 304, 306, 313, 316, 317
Consolidant, 5, 21, 26, 28, 29, 30, 38, 40, 46, 48, 49, 50, 114, 126, 137, 138, 144, 208, 226, 231, 232, 239, 240, 241, 242, 245, 249, 281
Consolidation, 21, 28, 29, 30, 48, 62, 113, 114, 126, 131, 137, 138, 139, 206, 229, 231, 239, 240
Copan, 29, 297, 319
Copper (Cu), 10, 19, 32, 41, 44, 46, 48, 49, 50, 117, 230, 237, 289, 308
Copper (I) chloride, 32
Copper (I) thiocyanate, 32
Copper (II) fluoride, 32, 42

Copper (II) hydroxyphosphate, 32
Copper radiation, 117
Cosmogenic nuclides, 244, 246, 259
Cross-polarized lenses, 249
Crustose, 10, 61, 66, 81, 96, 150, 154, 155, 156, 157, 165, 184, 185, 186, 187, 193, 195, 204, 217, 219, 221, 238, 263, 265, 288, 295, 298, 317
Cryogenic, 264
Cryptoendolithic microorganisms, 94, 103
Cryptoendolithic systems, 274
Crystalline, 117, 263, 266, 267, 268, 269, 270, 294
Crystallization, 17, 126, 236
Crystallography, 8
Cuneo, 52
Cyanoacrylate, 62
Cyanobacteria, 1, 2, 7, 64, 89, 90, 93, 94, 95, 97, 98, 101, 104, 105, 140, 150, 192, 199, 208, 268, 280, 281, 289, 290
Cyanophilous lichens, 64, 194, 289
Dakota, 149, 263, 266, 269
De-cohesion, 113
Depsides, 264, 315
Dermatocarpon, 155, 156, 158, 185
Dibutyl phthalate, 229
Dichlorophen, 229, 230
Diffractometer, 117
Dihydrate, 10, 11, 18, 69, 83, 224, 263, 273, 291, 311
Dilute formol, 229
Dione, 166
Dioxin, 231, 237
Diploicia, 11
Diploicin, 166
Diploschistes, 62, 80, 185, 206, 215
Dirina, 8, 11, 12, 14, 15, 16, 18, 20, 64, 66, 67, 68, 69, 80, 197, 202, 268, 274, 275, 292, 306, 313
Dirinaria, 214, 215, 219, 220
Dispersion-phase, 30
Dissolution rate, 141, 142
Divaricatic acid, 166, 167, 218
Dixit, 212, 224
Dolomite, 61, 89, 93, 94, 95, 96, 101, 103, 108, 110, 195, 257, 264, 290
Dolomite rhombus, 94, 96, 103

Driwall, 240
Dye, 113, 124
Early Cretaceous Lakota, 115
Echinoderm, 93
El Castillo, 22, 23, 24, 26, 46
El Morro National Monument, viii, 147, 148, 149, 151, 184, 186, 189, 240, 260
Electrochemical impedance spectroscopy, 38
Elemental analysis, 147, 152, 183, 184
Encrustation, 9, 11, 14, 15, 16, 18, 19, 66, 67, 68, 69, 114, 263, 266, 267, 268, 269, 270, 271, 272, 273, 274, 291, 292, 313
Endocarpon, 197, 215, 217, 218, 221, 224
Endolithic, 4, 62, 64, 65, 66, 68, 89, 90, 94, 95, 99, 102, 177, 178, 179, 180, 181, 182, 193, 199, 201, 233, 237, 264, 283, 285, 290, 294, 295, 299, 303, 306, 310, 316
Energy dispersive spectroscopy (EDS), 89, 90, 92, 93, 94, 105, 233
Enzymatic cleaning, 143
Enzyme digestion plots, 189
Epilithic, 4, 60, 61, 66, 89, 90, 94, 95, 193, 194, 199, 203, 275, 295, 297, 309, 316
Epiphorellic acid, 167
Epiphylls, 264
Epiphytic lichens, 68, 73
Eponex 1510, 31, 38, 40, 41, 42, 46
Epoxy, 30, 31
Epoxy resin, 42, 48, 61, 94
Eriodermate, 167
Erosion, 3, 19, 21, 29, 38, 39, 40, 42, 46, 48, 84, 140, 234, 246, 252, 253, 264, 266, 302
Etching, 120, 128, 129, 236
Ethanol, 93, 230
Ethanol series, 93
Ethyl silicate, 113, 126, 127, 138, 144, 239, 240, 241, 242, 260
Euendolithic, 89, 90, 94, 97, 98
European caves, 254
Eurotia, 149
Eutrophication, 54, 204
Evernic acid, 167
Extracellular polymeric substances (EPS), 98

Fall River Formation, 115, 118
Feldspar, 83, 85, 93, 118, 195
Feldspar crystal, 83
Figure of Merit (FOM), 38
Fiskerton, 15
Flavopunctelia, 185
Fluoride, 32, 41, 46, 236
Fluorine (F), 17, 8, 46, 49, 56, 57, 58, 59, 63, 64, 65, 66, 67, 68, 70, 71, 99, 100, 145, 146, 188, 190, 195, 197, 198, 199, 202, 208, 222, 223, 257, 258, 259, 260, 261, 274, 275, 278, 279, 280, 282, 283, 284, 286, 287, 288, 289, 290, 291, 292, 293, 294, 295, 297, 299, 300, 301, 306, 308, 310, 311, 312, 314, 316, 317, 318, 319
Foliose, 150, 171, 204, 214, 219, 220, 222, 233
Foraminifera, 29, 93, 95
Formol, 229, 254
Forum, 193, 204, 205
Fresco, 11, 14, 15, 18, 49, 59, 67, 68, 291, 292, 293, 313
Fruticose, 150
FT-Raman spectroscopy, 7, 11, 14, 18, 19, 67, 86, 190, 263, 264, 266, 267, 268, 269, 270, 271, 272, 273, 274, 275, 289, 291, 292, 293, 310, 314
Fulgensia, 197
Fungicides, 238, 314
Fusarium, 35, 37
Getty Conservation Institute (GCI), ix, x, 26, 33, 49, 50, 120, 129, 130, 133, 144, 145, 259, 299
Glacial moraine, 246
Glauconite, 93, 96, 110
Globigerina, 95, 107
Gloeocapsa, 89, 93, 94
Glomellic acid, 167
Glue, 142, 153, 241
Glutaraldehyde, 61, 92, 93
Gneiss, 61, 294
Göblemirrors, 117, 119
Goethite, 117
Granite, 19, 77, 78, 80, 81, 82, 83, 84, 85, 86, 87, 100, 214, 223, 229, 230, 253, 264, 265, 275, 281, 294, 295, 299, 300, 307
Graphite, 118
Ground charcoal, 243

Index

Guadalupe Monastery, 206
Guinier camera, 117
Gypsum, 11, 14, 62, 77, 84, 87, 195, 275, 307
Haematite, 263, 269, 274
Heinrich, 243
Heliophilic species, 194
Heterodermia, 212, 215, 219, 220
Hiascic acid, 167
Hispalis, 191, 192
Holocene, 60, 245, 246
Holocene dating, 60
Homogeneous, 66, 113, 115, 117, 125, 128, 129
Homosekikiac acid, 167
Huesca, 97
Hydrofluoric acid, 139, 153
Hydrogen peroxide, 229, 293
Hydrophase, 206
Hydroxide, 62, 63, 278
Hydroxyaluminium-vermiculite, 77, 85
Hygrophytism index, 54, 60, 316
Hypertrophication, 12, 19
Hyphae, 3, 4, 9, 27, 29, 34, 35, 36, 37, 48, 62, 81, 82, 83, 96, 107, 113, 120, 126, 132, 140, 144, 147, 148, 150, 153, 172, 177, 178, 179, 180, 181, 182, 189, 200, 201, 202, 203, 206, 285, 298, 299
Hypogeal tombs, 195
Hyvar-X, 85
Iberolevantine-mesomediterranean, 89, 92
Imbibition coefficient, 123
Indian subcontinent, 211, 212
Infrared (FTIR), 61
Inorganic salts, 241
Inscription, 17, 148, 149, 151, 189, 240, 255, 258
Inscription Rock, viii, 147, 148, 149, 151, 152, 154, 155, 156, 184, 189
Insolation, 198, 199, 204, 230, 251, 254, 263, 274
Interglacial periods, 245
Ionic chromatography analysis, 137
Iron (Fe), 63, 85, 115, 116, 117, 118, 141, 151, 183, 213, 234, 236, 240, 243, 244, 245, 263, 269, 271, 274, 278, 291, 300, 316
Iron film, 234, 240, 244
Iron minerals, 117
Iron oxide, 115, 116, 117, 118, 236, 244
Isopropanol, 30, 31, 32
Istituto Centrale per il Restauro, 53, 70, 279
Jaca cathedral, 97
Jurassic, 149
Kaolinite, 84
Karanataka, 218
Katabatic winds, 264
Khondalite, 214
Kinomagewapkong, 248
Kokopele, 157
La Mola quarry, 89, 93, 94, 97
Lactate, 141
Lacustrine beaches, 268
Lakota, 115, 118
Lasallia, 80, 85, 186
Lascaux, 253, 259
Latium, 53, 54
Leaching curve, 242
Lead citrate, 94
Lecania, 194, 195, 197, 205, 206
Lecanora, 13, 20, 79, 80, 135, 136, 154, 155, 156, 158, 165, 166, 167, 168, 169, 170, 186, 194, 195, 197, 205, 206, 213, 216, 218, 219, 222, 282, 318
Lecanorate, 168
Lecidea, viii, 62, 63, 80, 89, 93, 154, 155, 156, 157, 158, 165, 172, 183, 186, 197, 263, 265, 266, 267, 278, 289, 301
Lecidella, 79, 156, 158, 172, 186
Lepraria, 63, 155, 158, 197, 216, 218, 219, 277
Leprolomin, 168
Leproplaca, 195, 197
Leptogium, 186, 216, 220
Lichen acids, 10, 61, 217, 220, 232, 264, 273, 282, 312
Lichen-mortar interactions, 191
Lichenometry, 53, 228, 245, 246
Lichexanthone, 168
Lichina, 216
Lime-containing masonry, 27

Limestone, 7, 20, 21, 23, 25, 26, 27, 28, 29, 30, 32, 38, 39, 41, 42, 43, 44, 45, 46, 48, 50, 61, 62, 64, 67, 93, 95, 100, 103, 105, 199, 214, 223, 229, 241, 243, 264, 268, 278, 285, 288, 289, 290, 294, 298, 300, 301, 302, 311, 315, 318
Limestone clasts, 95
Limestone consolidation, 21
Limestone porosity, 29
Limestone stabilization, 21
Limiting zone, 94
Liquid chromatography, 114, 126
Lithic, 51, 54, 61, 62, 89, 90, 91, 92, 96, 97, 98, 101, 105, 319
Lithium silicate, 30
Lithobionts, 90, 97, 98, 191, 192
Lithochemism, 61
Lobaridone, 168
Lyases, 140
Lyophilized powder, 143
Lysing enzyme, 143
Macro porosity, 124
Mafic rocks, 61, 63, 278
Magnesium (Mg), 5, 10, 94, 95, 106, 151, 229, 231, 236, 238, 281, 291, 318
Magnesium fluorosilicate, 229, 236, 238
Magnesium silicate, 5, 10, 151
Manganiferous rock varnish, 244
Marble, 57, 64, 69, 214, 223, 224, 229, 230, 231, 235, 262, 287, 289, 300, 301, 303, 311, 319
Maya, vii, 21, 22, 23, 26, 27, 29, 32, 49, 50, 223, 251, 258, 297, 317
Mean atomic number, 92
Mechanical action, 120, 132, 134, 135, 142, 143, 294
Medullary zone, 96
Megalithic works, 54
Megaspora, 186
Meghalaya, 222
Melanelia, 156, 158, 165, 166, 167, 168, 170, 171, 175, 177, 178, 179, 183, 186, 188
Mercury porosimetry, 114, 129, 130
Mesic conditions, 244
Mesoamerica, 231
Metal, 3, 14, 17, 19, 214, 264
Metal-oxygen bonds, 141, 269
Metaquartzites, 93

Methylated spirit, 229
Methylene blue, 62
Mica, 117, 118, 214
Mica schist, 214
Micacious minerals, 85, 282
Micarea, 79, 213
Micritic matrix, 93
Micro porosity, 124
Microanalysis (EDXRA), 61
Microbial biofilm, 90, 92, 96
Microcolonial fungi, 243
Microcrystalline quartz, 93
Micropores, 124
Microscopy
  electron, 9, 10, 92, 98, 138, 189, 240, 259, 283, 298
  environmental scanning electron (ESEM), 113, 114, 119, 122, 128, 130, 134, 138, 206
  light, 91, 93, 115, 137, 147, 148, 171, 172
  low temperature scanning electron (LTSEM), 98
  optical (OM), 61, 100, 124, 285
  polarized, 114, 116
  polarized light, 93
  reflected light, 115
  scanning electron (SEM), 33, 61, 89, 90, 91, 93, 97, 98, 101, 110, 114, 119, 124, 128, 138, 147, 153, 171, 172, 189, 259, 282, 295, 298, 300
  transmission electron (TEM), 10, 61
Mineral dissolution stoichiometry, 141
Mollusk, 93
Mono-functional mono-protic acid, 142
Monohydrate, 10, 11, 18, 83, 230, 263, 268, 269, 270, 273, 291
Monoliths, 222
Monomeric silica, 142
Monsanto Silbond, 31, 35, 36, 113, 127
Monument, 2, 5, 6, 7, 9, 10, 11, 12, 16, 17, 18, 19, 52, 53, 55, 56, 57, 58, 59, 64, 66, 67, 68, 69, 72, 73, 74, 77, 78, 79, 80, 85, 86, 87, 89, 90, 91, 92, 93, 95, 96, 100, 101, 151, 184, 191, 204, 206, 207, 208, 209, 211, 212, 214, 215, 217, 218, 220, 221, 222, 223, 224, 225, 231, 232, 233, 236, 237, 238, 239, 242, 251, 257, 275, 279, 280, 281, 283, 286, 287, 288, 289,

Index 289

290, 294, 295, 296, 297, 299, 300, 302, 303, 306, 307, 309, 310, 311, 313, 314, 315, 316, 317, 318, 319, 320
Monuron, 230
Mortar, 11, 26, 27, 28, 29, 43, 191, 192, 194, 195, 198, 199, 200, 201, 202, 203, 204, 206, 214, 217, 308
Multifunctional acids, 142
Multi-protic acids, 142
Muscovite, 117, 118
Mycobiont, 2, 4, 10, 66, 89, 92, 95, 96, 97, 104, 147, 148, 150, 201, 264, 268
Nanaimo Petroglyph Park, 252
National Park Service, 148, 151, 152, 189
National Register of Historic Places, 132
Native American, 226, 247, 248
Necropolis, 191, 192, 195, 199, 206, 208, 281
Necropolis of Carmona, 191, 192, 195, 206, 208, 281
Neodesogen, 85
Nimite, 117, 118
Nitrification, 207
Nitrogen (N), 2, 80, 153, 267
Nitrogenous, 52, 286
Nitrophilous, 12, 73
Nitrophytic taxa, 195
Nitrophytism index, 54
Non-aqueous systems, 30
Non-calcareous, 11, 268
Non-crystalline, 117
Non-ionic detergent, 229, 230
Nostoc, 268
Novelda, viii, 89, 92, 93, 95, 100, 105, 106
Nucleation sites, 98
Ochrolechia, 11, 19, 80, 86, 87, 275, 307
Ojibwa, 248
Opegrapha, 194, 197
Organic acids, 5, 27, 83, 84, 141, 142, 144, 145, 147, 148, 151, 166, 189
Organic ligands, 141
Organic salts, 4, 150
Organic solvents, 48, 127
Organomineral interface, 92
Orissa, 214, 215, 218, 220, 221
Ornithocoprophilous, 204
Orthoclase, 83

Orthophenylphenol, 226, 228
Orvieto Cathedral, 54
Osmium tetroxide, 92
Oxalate, 5, 9, 10, 11, 14, 15, 16, 18, 19, 20, 51, 53, 62, 65, 66, 67, 68, 69, 77, 83, 89, 96, 97, 111, 118, 142, 151, 190, 201, 205, 222, 224, 234, 240, 243, 244, 245, 252, 257, 263, 268, 269, 270, 273, 275, 280, 286, 287, 289, 290, 291, 292, 293, 294, 295, 297, 298, 299, 306, 308, 311, 315, 317, 318
  calcium, 5, 11, 14, 15, 16, 18, 20, 51, 53, 65, 66, 67, 68, 69, 77, 83, 89, 96, 110, 118, 151, 201, 205, 222, 224, 243, 244, 257, 263, 268, 269, 270, 273, 287, 290, 291, 294, 306, 311, 315
  copper, 10, 19, 289, 308
  ferric, 62
  magnesium, 10, 318
  manganese, 10, 20, 318
Oxalic acid, 5, 10, 66, 83, 85, 141, 151, 201, 263, 264, 268, 273, 285, 311
Oxide, 62, 63, 115, 116, 117, 118, 244, 278
Oxido-reductases, 140
Ozone, 188
Palazzo Farnese, 14
Paleoenvironmental data, 244
Paleoenvironments, 225
Paleo-Indian, 131
Paludosic acid, 168
Pannarin, 168
Paraffin, 240, 241
Paraphyses, 95
Parietin, 168
Parmelia, 63, 80, 224, 278, 281, 282, 297, 314, 315
Parmelinella, 216, 220
Parmotrema, 213
Particle induced X-ray emission (PIXE), 147, 149, 152, 153, 183, 190
Patina, 67, 68, 202, 223, 290, 300
Pedogenesis, 3, 4, 8, 19, 148, 150, 190, 224, 258, 263, 298, 313, 318
Pelican paint, 226, 228
Peltigera, 186

Peltula, 135, 136, 154, 156, 158, 213, 216, 217, 221, 224
Pencapsula, 241
Penicillium, 34, 37
Pentachlorophenol, 229
Periodic Acid Schiff (PAS), 62
Perithecia, 201
Pertusaria, 20, 79, 80, 208, 213, 223, 298, 318
Petroglyph, 7, 113, 114, 116, 120, 129, 131, 132, 146, 148, 149, 189, 225, 226, 228, 229, 230, 231, 232, 233, 236, 239, 242, 244, 246, 247, 249, 250, 251, 252, 253, 257, 258, 261, 262, 289, 314
Petroleum distillates, 241
pH, 4, 5, 54, 79, 80, 84, 141, 142, 151, 204, 217
Phaeophyscia, 156, 159, 186, 213, 214
Phenol, 226, 228
Phenolic biocides, 236
Phenotypic plasticity, 98
Phosphate analysis, 237
Photobiont, 2, 92, 104, 204, 214, 272, 273
Photogrammetry, 249
Photophytism index, 54
Phylliscum, 216, 217, 221, 224
Physcia, 155, 156, 158, 159, 165, 166, 168, 169, 170, 171, 172, 175, 183, 187, 188, 213, 214, 216, 219
Physciella, 135
Physconia, 187
Pigment, 14
Pixels, 124
Placidium, 187
Placodioid, 154, 202, 204
Plagioclase, 84
Planaic acid, 169
Polarized light, 93, 116
Poleophile, 13
Polyblastia, 197
Polybor, 31, 39, 40, 41, 46, 49
Polycarbonate film, 153
Polymer particles, 30
Polymerization, 30, 94, 127, 128
Polymethyl methacrylate, 239
Polyphenolic lichen acids, 264, 273
Polysilicate, 31
Polysporina, 135, 136, 187

Polystyrene, 153
Polyurethane Dispersion, 30
Polyvinyl, 229, 239
Porosimetry, 113, 114, 129, 130
Porosity, 4, 29, 81, 97, 113, 114, 115, 116, 123, 124, 126, 127, 128, 129, 137, 138, 192, 217, 239, 241
Porosity gradient, 113, 124, 129
Postfixing, 92
Potassium (K), 83, 85
Prehistoric paint formulas, 236, 241
Prehistoric rock art, 225, 237
Preservation, xiii, 1, 8, 144, 223, 225, 226, 232, 233, 235, 237, 239, 242, 243, 247, 250, 251, 255, 257, 259, 294, 299, 300
Programa Ramon y Cajal, 86
Propionate, 141
Protease, 143, 250
Protein, 250
Proteolithic enzymes, 70
Protolichens, 94, 103
Protolichestericic acid, 169
Pseudosagedia, 194, 197
Psoromic acid, 169, 308
Pyruvate, 141
Pyxine, 216, 219, 220, 221
Quartz, 85, 93, 95, 105, 107, 113, 114, 115, 116, 117, 118, 119, 120, 123, 124, 126, 127, 128, 129, 138, 139, 140, 141, 142, 144, 145, 146, 195, 203, 240, 245, 268, 269, 273
Quartz dissolution, 141
Quartz matrix, 114, 115
Quaternary ammonium, 17, 39, 44, 45, 48, 70, 86, 229, 230, 231, 298
Quaternary ammonium chloride, 230, 298
Radiation protectants, 264
Radiocarbon dating, 237, 262
Raman Spectroscopy, viii, 7, 8, 9, 11, 14, 18, 19, 20, 61, 67, 68, 86, 87, 145, 190, 263, 265, 266, 267, 268, 269, 270, 271, 272, 273, 274, 275, 276, 289, 291, 292, 293, 297, 307, 310, 313, 314
Renaissance, 15, 18, 67, 68, 291, 292
Resin, 31, 42, 48, 92, 94, 113, 124, 125, 126, 206, 241
Rhizine, 3, 4, 62, 150, 232, 233, 237

Rhizocarpon, 56, 80, 156, 157, 159, 187, 188, 280, 282, 294
Rhizoplaca, 187
Rinodina, 79, 80, 156, 159, 197
Roccella, 216, 281
Rock art, 2, 7, 130, 131, 132, 133, 134, 135, 136, 138, 139, 142, 143, 144, 146, 148, 225, 226, 228, 230, 231, 232, 233, 234, 236, 237, 238, 239, 240, 241, 242, 243, 244, 245, 246, 247, 248, 249, 250, 251, 252, 253, 254, 255, 256, 257, 258, 259, 260, 261, 293, 308, 317
Roman, 11, 13, 55, 57, 58, 64, 67, 99, 100, 191, 192, 204, 208, 212, 223, 280, 281, 287, 290, 295, 302, 303, 304, 308, 311, 317
Roman Hispania, 192
Roman settlements, 191, 192
Rome, 12, 13, 50, 53, 54, 55, 59, 64, 66, 71, 73, 223, 260, 279, 287, 288, 289, 309, 316
Rune-stones, 251
Rutile, 269, 273
Sahara, 245, 257
Salicylanilide, 229
Salicylic acid, 142
Salt crystallization, 236
Salt efflorescence, 235
San Ignacio, 22
San Jose Succotz, 22
Sandstone, viii, 7, 20, 63, 113, 114, 115, 116, 117, 118, 119, 120, 121, 123, 124, 127, 128, 129, 131, 132, 133, 134, 135, 136, 137, 138, 139, 140, 142, 143, 144, 145, 147, 148, 149, 151, 153, 204, 214, 226, 228, 229, 230, 239, 240, 241, 244, 253, 258, 263, 264, 265, 266, 268, 269, 274, 283, 285, 298, 309, 310, 318
Sandstone-lichen relationship, 120
Sanit-S, 85
Sarcogyne, 80, 135, 136, 187, 194, 197, 199, 206
Saxicolous, 4, 54, 61, 73, 99, 100, 148, 278, 279, 281, 282, 283, 298
Schist, 214, 228, 294
Scoliciosporum, 79

Scrobiculin, 169
Scurano Portico, 207
Scytonemin, 263, 268, 272, 273, 274
Secondary chemistry, 5, 147, 149, 151, 165, 264
Sepentinite, 10, 20, 318
Serpentine rocks, 61, 63, 278
Shishkino, 230
Siberia, 229
Silane, 31
Silica (Si), 84, 95, 106, 114, 128, 138, 139, 140, 141, 142, 144, 146, 151, 153, 234, 236, 240, 241, 244, 300
Silica aqueous systems, 142
Silica based consolidant, 114
Silica gel, 84, 128, 241
Silica glaze, 244
Silicate, 5, 10, 43, 62, 100, 113, 126, 127, 138, 141, 142, 144, 146, 151, 239, 240, 241, 242, 260, 284, 286, 314
Silicate dissolution, 141
Silicate mineral dissolution, 141
Silicicolous epilithic, 61
Silicon dioxide (SiO_2), 138, 269
Silicone, 206, 207, 239, 285
Silver nitrate, 42, 48, 230, 237
Silver oxide, 237
Simazine, 230
Slates, 93
Societá Lichenologica Italiana (SLI), 52
Sodium Chloride (NaCl), 124
Sodium hypochlorite, 226, 228, 238
Sodium pentachlorphenate, 231
Sodium salt, 236
Solar irradiation, 192
Solorinic acid, 169
Soluble salts, 17, 85, 114, 126, 130, 252
Soluble soaps, 241
Soredium, 69, 96
Spaeric acid, 169
Spectral footprint, 267
Spectral signature, 268, 273, 274
Sphaeric acid, 169
Spinifex, 252
Squamarina, 197
Squamatic acid, 169
Squamulose, 10, 204, 217

Staurothele, 155, 159, 165, 166, 168, 169, 170, 171, 172, 183, 187, 197
Stictic acid, 85, 170
Streptomyces, 35
Stucco, 11, 22, 26, 27, 29
Sulfur dioxide (SO_2), 84, 188
Superficial layer, 113, 120, 128, 202, 204
Synthetic polyurethane resin, 241
Tamil Nadu, 213
Tephromela, 65, 80, 86, 214, 307, 312
Terracotta, 3, 13
Terricolous, 54, 198
Terrigenous components, 93
Tesserae, 199
Thelidium, 198
Thin-layer chromatography (TLC), 147, 149, 152, 154, 165, 166, 167, 168, 169, 170
Thiomelin, 170
Tikal, 29
Tin compound, 45
Tinorgano-compound, 229
Tirunevelli, 213
Titanium (IV) oxide, 269, 272
Toluene, 42, 153, 229
Toninia, 156, 157, 159, 193, 198, 206
Torre de Belem, 98
Torrey, 226, 228, 234, 238
Toxic dioxin, 231
Trapelia, 80
Treatment, 7
Trichoderma, 143
Tuckermannopsis, 187
U. S. Army Corps of Engineers, 132
UCLA Institute of Archaeology, 43, 49
Ultraviolet-radiation (UV), 268, 274
Umbria, 54
Uranyl acetate, 94
Urea, 70
Usnea, 147, 183, 184, 187, 188, 189
Usnic acid, 85, 170, 218, 275
Vacuum, 119

Valcamonica, 226
Varnish microlaminations, 243, 259
Varnishes, 241, 244
Venetia, 53
Verrucaria, 62, 65, 79, 80, 193, 194, 195, 198, 199, 201, 202, 205, 206, 213, 316
Vicanicin, 170
Wacker OH, 113, 127, 128
Wandjina, 252
Water absorption coefficient, 129
Water imbibition, 114
Water retention, 10
Water-borne systems, 30
Water-miscible systems, 30
Weathering rind, 132, 234, 243
Weddellite, 10, 65, 68, 118, 273, 301, 318
Whewellite, 10, 65, 67, 68, 84, 260, 268, 272, 289, 301, 310, 318
Xanthone, 170
Xanthoparmelia, 154, 156, 159, 165, 167, 168, 169, 170, 171, 172, 183, 188, 289
Xanthoria, 80, 85, 187, 188, 190, 204, 205, 213, 274, 281, 293
Xeric climatic episodes, 244
Xeric conditions, 217
Xerophytic, 195, 221
Xerophytic species, 195
X-ray diffraction (XRD), 10, 61, 114, 117, 119, 120, 124, 138
Xunantunich, 21, 22, 25, 26, 27, 29, 33, 37, 43, 46, 49, 50
Xunantunich Archaeological Project (XAP), 25
Yellow ochre, 117
Younger Dryas, 243
Zeorin, 218
Zinc (Zn), 229, 231, 236, 238, 300
Zinc fluorosilicate, 231, 236
Zuni, 149, 248, 262